Charles Thomas Millis

Metal-plate work. Its patterns and their geometry, also notes on metals, and rules in mensuration

Third Edition

Charles Thomas Millis

Metal-plate work. Its patterns and their geometry, also notes on metals, and rules in mensuration
Third Edition

ISBN/EAN: 9783337156718

Printed in Europe, USA, Canada, Australia, Japan

Cover: Foto ©berggeist007 / pixelio.de

More available books at **www.hansebooks.com**

Finsbury Technical Manuals

METAL-PLATE WORK

ITS PATTERNS AND THEIR GEOMETRY

ALSO

NOTES ON METALS, AND RULES IN MENSURATION

FOR THE USE OF

TIN, IRON, AND ZINC PLATE WORKERS, COPPERSMITHS,
BOILER-MAKERS, PLUMBERS AND OTHERS

BY

C. T. MILLIS, M.I. Mech. E.

EXAMINER IN METAL PLATE WORK TO THE CITY AND GUILDS OF LONDON INSTITUTE;
PRINCIPAL, EDUCATIONAL DEPARTMENT, BOROUGH POLYTECHNIC INSTITUTE

THIRD EDITION, CONSIDERABLY ENLARGED

London
E. & F. N. SPON, Ltd., 125 STRAND

New York
SPON & CHAMBERLAIN, 12 CORTLANDT STREET

PREFACE

TO

THE THIRD EDITION.

———•———

THE *Third Edition* of this work contains, with 50 new diagrams and illustrations, more than 80 pages of additional matter. Books I. and II. remain as before. Book III. contains several new and important problems in "penetrations." Book IV. is almost new. Towards Chapter I., which describes Pipe-Bending, help was furnished to the author by his former student and assistant, Mr. W. H. Bennett, now Instructor in *Metal-Plate Work* at the Regent Street Polytechnic. Chapter II. treats of the making up of Pipe Bends out of flat sheet, and deals with that subject as it has never before been dealt with—systematically and comprehensively. To Book V. has been added an article on "Tinning and Retinning."

The system of geometric construction for the patterns of sheet-metal articles set forth in this book, came of investigation made by the author about twenty-two years ago. In 1885, after several years of thorough testing, the system was illustrated in model and diagram at the South Kensington Inventions Exhibition, and in 1887 further made public in the First Edition. A workman, studying the book, learns to recognise easily the nature of any surface he meets with, whether flat or curved, and how to deal with it for his purposes. Indirect evidence of the value of the book is afforded by the extent to which other writers—with no acknowledgement, however, of their indebtedness—have turned to it for "copy" and model; and direct evidence in the frequent communications received from actual sheet-metal workers, who, by the author's system, have readily solved problems, very difficult if not impossible of solution otherwise.

As with the Second Edition, so now, I have to acknow-ledge the very valuable assistance given me by my friend, Mr. H. W. Ley, in revising the new matter, and seeing the book through the press.

C. T. M.

Borough Polytechnic Institute, London :
April, 1899.

PREFACE

THE SECOND EDITION.

———

THIS *Second Edition*, long delayed through pressure of work, is the First Edition augmented by about 70 more pages of original matter, making the book still more useful to the sheet-metal worker who desires to know his trade, and become an all-round hand, instead of being cramped up in a narrow everyday groove. The specialisation in the shops brought Technical Education about ; the increasing tendency to it marks the need of a wide basis for such education.

The divisions of tin-plate, iron-plate, and zinc workers did not formerly exist; they do not exist now among the older workmen, and should not among the younger. The shapes treated in this volume come before all sheet-metal workers ; the main differences of their work are not those of shape, but of size and consequent manipulation.

Blinn's ' Workshop Companion for Tin, Sheet Iron, and Copper Plate Workers,' is a book published in America and circulating here. More than one-quarter of that portion which deals with patterns in its 1891 edition is neither more nor less than a word for word appropriation, without acknowledgment, of original problems, diagrams and all, from my pages. The *Preface* to the book can well afford to make much of " the addition of new matter."

viii   PREFACE TO THE SECOND EDITION.

My thanks are due to Mr. H. W. Ley, Secretary of the Westminster Technical Institute during my Directorship of it, for many useful suggestions, and for very careful revision of its proof sheets.

C. T. M.

BOROUGH POLYTECHNIC INSTITUTE, LONDON,
March 1893.

PREFACE.

In the pages that follow, the setting-out of patterns for metal-plate workers is *systematised* (for the first time, so far as I am aware), and by the system laid down, nearly all the patterns required by sheet-metal workers can be set out on one general geometrical principle. Thus treated, the subject becomes a comparatively easy one, and the workman learns how, on the given principle, to develop for himself the surface of any article he may have to make, to the saving, as against various methods practised hitherto, of both time and material.

Of the methods heretofore taught in the workshop or otherwise, some have no application beyond the particular article with respect to which they are described, some are absolutely impracticable. The methods that these pages explain are applicable always; and have been proved by abundant experience in my classes.

The commencement of the book was a series of articles that were written by me for *The Ironmonger*. These and my class and lecture notes have greatly aided me in my effort to make the book comprehensive, as well as at the same time a welcome workshop companion. Not only does it contain patterns which are essentially those of numerous articles of every-day manufacture, but a *special feature* is made of large work;—and how it may be dealt with under ordinary workshop conditions. Each of the problems is complete in itself; but although solved independently they follow each other in due order.

It is my hope that the book will be an aid to students in engineering and geometry, as well as to those for whom it is particularly written; that it will be a serviceable addition to the scanty and insufficient literature on the application of geometry to metal-plate work; and that it will in some degree assist the cause of technical education.

C. T. M.

CITY AND GUILDS OF LONDON TECHNICAL COLLEGE,
FINSBURY, LONDON,
June 1887.

CONTENTS.

BOOK I.

CHAPTER I.

PAGE

CLASSIFICATION 1

CHAPTER II.

INTRODUCTORY PROBLEMS 3

Definitions, 3–5; Problems on Angles, Lines, Circles, Polygons, Ovals, Ellipses, and Oblongs, 6–20; Measurement
of Angles, 20–3.

CHAPTER III.

ARTICLES OF EQUAL TAPER OR INCLINATION OF SLANT 24

Right cone defined and described, with problems, 24–7.

CHAPTER IV.

PATTERNS FOR ROUND ARTICLES OF EQUAL TAPER OR INCLINA-
TION OF SLANT 28

Development of right cone, and problems on same, 28–31;
allowance for lap, seam, and wiring, 32–3; frustum (round
equal-tapering bodies) defined and described, with problems, 33–37; patterns for round equal-tapering bodies
(frusta) in one, two, or more pieces, for both small and
large work, 37–45.

CHAPTER V.

PAGE

EQUAL-TAPERING BODIES AND THEIR PLANS   46

     Plans and elevations, 46–50; plans of equal-tapering bodies;
     their characteristic features, 50–6; problems on the plans
     of round, oblong, and oval equal-tapering bodies, 56–65

CHAPTER VI.

PATTERNS FOR FLAT-FACED EQUAL-TAPERING BODIES   66

     Definition of right pyramid, and development, 66–7; pattern
     for right pyramid, 68–9; frustum of right pyramid defined
     and described, 69–71; patterns for frusta of right pyramids
     (hoppers, hoods, &c.) for both small and large work, 71–7;
     baking-pan pattern in one or more pieces, 77-83.

CHAPTER VII.

PATTERNS FOR EQUAL-TAPERING BODIES OF FLAT AND CURVED
     SURFACES COMBINED   84

     Pattern for oblong body with flat sides and semicircular ends,
     84–90; for oblong body with round corners, 90–6; for
     oval body, 96–104.

         The patterns for each of these bodies are given in one, two,
     and four pieces, as well as for both small and large work.

BOOK II.
—

CHAPTER I.

PATTERNS FOR ROUND ARTICLES OF UNEQUAL TAPER OR INCLI-
     NATION OF SLANT   105

     Oblique cone, definition, description, and development,
     105–111; oblique cone frustum (round unequal-tapering
     body) and oblique cylinder, 111–3; patterns for round
     unequal-tapering bodies (oblique cone frusta), also
     oblique cylinder—work small or large, 113-23

CHAPTER II.

PAGE

UNEQUAL-TAPERING BODIES AND THEIR PLANS 124

Plans of unequal-tapering bodies and their characteristic
features, 124-9; problems on plans of oblong, oval, and
other unequal-tapering bodies, 130-3; plans of Oxford
hip-bath, Athenian hip-bath, sitz bath and oblong taper
bath, 134-42

CHAPTER III.

PATTERNS FOR FLAT-FACED UNEQUAL-TAPERING BODIES 143

Oblique pyramids and their frusta, 143-5; pattern for oblique
pyramid, 145-8; patterns for frusta of oblique pyramids
(flat-faced unequal-tapering bodies)—work small or large,
148-54; pattern for a hood which is not a frustum of
oblique pyramid, 154-6.

CHAPTER IV.

PATTERNS FOR UNEQUAL-TAPERING BODIES OF FLAT AND CURVED
SURFACES COMBINED 157

Pattern for equal-end bath (unequal-tapering body having
flat sides and semicircular ends), 157-68; for oval bath
(oval unequal-tapering body), 168-83; for tea-bottle top
(unequal-tapering body having round top, and oblong
bottom with semicircular ends), 183-91; for oval-canister
top (unequal-tapering body having round top and oval
bottom), 192-205; for unequal-tapering body having
round top and oblong bottom with round corners, 205-16.
(*The patterns for each of these bodies are given in one, two, or
four pieces, as well as for both small and large work.*)
Pattern for Oxford hip-bath, 216-30 (two methods); for
oblong taper bath, 230-6 (two methods).

Book III.

———

PAGE

PATTERNS FOR MISCELLANEOUS ARTICLES   239

Elbow patterns, 239–41; T patterns various, and at any angle, 242–54; patterns for pipe joining two unequal or equal circular pipes not in line with one another, 255; Y patterns, where the pipes joined are equal or unequal, 255–63; elbows in oblong pipes, patterns for, 264–7; T on oblique cylinder, pattern for, 267. Sections, patterns for sectioned cones, 271–86; definitions, 271; shape of section and pattern for elliptically cut cylinder, 274; shape of section and pattern for elliptically cut cone, 274–8; shape of section and pattern for cone cut in parabolic section, 278–81; shape of section and pattern for cone cut in hyperbolic section, 281–4; circular pipe to fit elliptic section of conical pipe, pattern for, 284; square and oblong pipes fitting on conical caps, patterns for, 286–92; patterns for tall-boy bases (bodies having oblong bases and circular tops), 293–301; to fit on flat surface, 293; for dripping pan with well, 295; tall-boy base to fit on slant of roof, 295; to fit on ridge of roof, 301; patterns for compound bent surfaces, as vases, aquarium bases, mouldings, 302–23; introduction, 302; sections of, and patterns for mouldings meeting at any angle, 303–10; aquarium bases formed of like mouldings, 310; of unlike mouldings, 314; patterns for inclined or raking mouldings, 318; problems of penetrations, 323–43; conical body penetrating flat surface, 323; penetrating cylinder, 328; penetrating cone, 333; to draw oval, 344; parabola, 345; hyperbola, 346.

———

Book IV.

———

CHAPTER I.

PIPE BENDING   347

CHAPTER II.

PAGE

PATTERNS FOR PIPE BENDS   359
  Prefatory.   Bend of equal circular section, 359–85; the bend
    in two pieces (one pattern-piece), 360; in two pieces (two
    pattern-pieces), 364; in four pieces (two pattern-pieces),
    369; in four pieces (three pattern-pieces), 375; in circular
    segments, 379; in circular segments, special case of lobster-
    back cowl, 382; round neck T to connect equal circular
    pipes, 385; bend of square section to connect unequal
    square pipes, 389; ventilator, square section, 397; tapering
    circular bend, 398–407; of circular segments, 398; in four
    pieces, 402; tapering circular ventilator, 407.

Book V.

CHAPTER I.

METALS; ALLOYS; SOLDERS; SOLDERING FLUXES; TINNING AND
    RE-TINNING   408
  Metals, their characteristics, properties, specific gravity, melting
    points, 408–13; iron and steel, 413–15; copper, 416; zinc,
    417; lead, 417; alloys, 418–22; of copper and tin, 419;
    of copper and zinc, 420–22; of tin and lead, 422; solders,
    422–25; soldering fluxes, 426; tinning and re-tinning,
    428–32; of copper, 429; wrought-iron, 430; cast-iron, 431

CHAPTER II.

SEAMS OR JOINTS   433

CHAPTER III.

USEFUL RULES IN MENSURATION; TABLES OF WEIGHTS OF
    METALS   436

INDEX   443

METAL-PLATE WORK.

BOOK I.

CHAPTER I.

CLASSIFICATION.

(1.) NOTWITHSTANDING the introduction of machinery and the division of labour in the various branches of metal-plate work, there is as great a demand for good metal-plate workers as ever, if not indeed a greater demand than formerly, while the opportunities for training such men are becoming fewer. An important part of the technical education of those connected with sheet-metal work is a knowledge of the setting-out of patterns. Such knowledge, requisite always by reason of the variety of shapes that are met with in articles made of sheet-metal, is nowadays especially needful; in that the number of articles made of sheet metal, through the revival of art metal-work, the general advance of science, and the introduction of new designs (which in many cases have been very successful), in articles of domestic use, has considerably increased. It is with the setting-out of patterns that this volume principally deals. To practical men, the advantages in saving of time and material, of having correct patterns to work from, are obvious. Whilst, however, the method of treatment here of the subject will be essentially practical, an

B

amount of theory sufficient for a thorough comprehension of the rules given will be introduced, a knowledge of rules without principles being mere 'rule of thumb,' and not true technical education.

(2.) Starting in the following pages with some introductory problems and other matter, we shall proceed from these to the articles for which patterns are required by sheet-metal workers and which may be thus conveniently classed and subdivided:

CLASS I.—*Patterns for Articles of equal taper or inclination* (pails, oval teapots, gravy strainers, &c.). Subdivisions.
  a. Of round surfaces.
  b. Of plane or flat surfaces.
  c. Of curved and plane surface combined.

CLASS II.—*Patterns for Articles of unequal taper or inclination* (baths, hoppers, canister-tops, &c.). Subdivisions.
  a. Of round surfaces.
  b. Of plane or flat surfaces.
  c. Of curved and plane surface combined.

CLASS III.—*Patterns for Miscellaneous Articles* (elbows, and articles of compound bent surface, as vases, aquarium stands, mouldings, &c.).

All these articles will be found dealt with in their several places.

We shall conclude with a few technical details in respect of the metals that metal-plate workers mostly make use of.

(3.) The setting out of patterns in sheet-metal work belongs to that department of solid geometry known as "Development of Surfaces," which may be said to be the spreading or laying out without rupture the surfaces of solids in the plane or flat, the plane now being sheet metal.

CHAPTER II.

INTRODUCTORY PROBLEMS; WITH APPLICATIONS.

DEFINITIONS.

Straight Line.—A straight line is the shortest distance between two points.

NOTE.—If not otherwise stated, lines are always supposed to be straight.

Angle.—An angle is the inclination of two lines, which meet, one to another. The lines A B, C B in Fig. 1 which are inclined to each other, and meet in B, are said to form an *angle* with one another. To express an angle, the letters which denote the two lines forming the angle are employed, the letter at the *angular point* being placed in the middle; thus, in Fig. 1, we speak of the angle A B C.

FIG. 1.	FIG. 2.	FIG. 3.

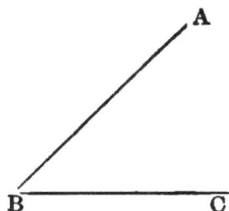 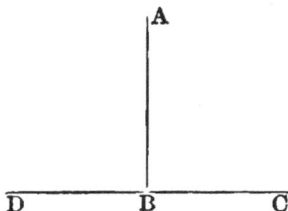 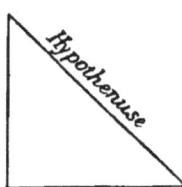

Perpendicular. Right Angles. — If a straight line, A B (Fig. 2), meets or stands on another straight line, C D, so that the adjacent angles (or angles on either side of A B) A B D, A B C, are equal, then the line A B is said to be *perpendicular* to, or *at right angles* with ('square with') D C, and each of the angles is a right angle.

Parallel Lines.—Parallel lines are lines which, if produced ever so far both ways, do not meet.

Triangle.—A figure bounded by three lines is called a triangle.

A triangle of which one of the angles is a right angle is called a *right-angled* triangle (Fig. 3); and the side which joins the two sides containing the right angle is called the *hypothenuse* (or hypotenuse). If all the sides of a triangle are equal, the triangle is *Equilateral*. If it has two sides equal, the triangle is *Isosceles*. If the sides are all unequal, the triangle is *Scalene*.

Polygon.—A figure having more than four sides is called a *polygon*. Polygons are of two classes, *regular* and *irregular*.

Irregular Polygons have their sides and angles unequal.

Regular Polygons have all their sides and angles equal, and possess the property (an important one for us) that they can always be inscribed in circles; in other words, a circle can always be drawn through the angular points of a regular polygon (Figs. 12 and 13).

Special names are given to regular polygons, according to the number of sides they possess; thus, a polygon of five sides is a *pentagon*; of six sides, a *hexagon*; of seven, a *heptagon*; of eight, an *octagon*; and so on.

Quadrilaterals.—All figures bounded by four lines are

FIG. 4a.   FIG. 4b.

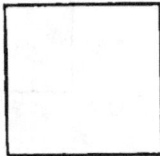   

called quadrilaterals. The most important of these are the *square* and *oblong* or *rectangle*. In a square (Fig. 4a) the sides are all equal and the angles all right angles, and consequently equal. An oblong or rectangle has all its angles right angles, but only its opposite sides are equal. (Fig. 4b.)

Circle.—A circle is a figure bounded by a curved line such that all points in the line are at an equal distance from a certain point within the figure, which point is called the *centre.*

The bounding line of a circle is called its *circumference.* A part only of the circumference, no matter how large or small, is called an *arc.* An arc containing a quarter of the circumference is a *quadrant.* An arc containing half the circumference is a *semicircle.* A line drawn from the centre to any point in the circumference is a *radius* (plural, *radii*). The line joining the extremities of any arc is a *chord.* A chord that passes through the centre is a *diameter.*

A line drawn from the centre of, and perpendicular to, any chord that is not a diameter of a circle, will pass through its centre.

In practice a circle, or arc, is 'described' from a chosen, or given, centre, and with a chosen, or given, radius.

If two circles have a common centre, their circumferences are always the same distance apart.

FIG. 5.

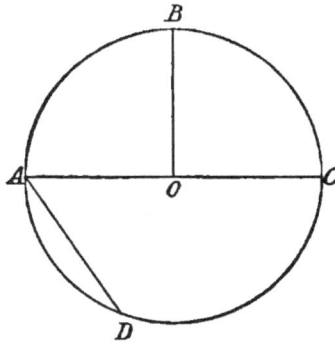

In Fig. 5.
O is the centre.
A D (the curve) is an arc.
A B (or B C) is a quadrant.
A C B is a semicircle.
O A (or O B, or B C) is a radius.
A D (the straight line) is a chord.
A C is a diameter.

PROBLEM I.

To draw an angle equal to a given angle.

CASE I.—Where the 'given' angle is given by a drawing.

This problem, though simple, is often very useful in practice, especially for elbows, where the angle (technically called 'rake' or 'bevil') is marked on paper, and has to be copied.

FIG. 6.

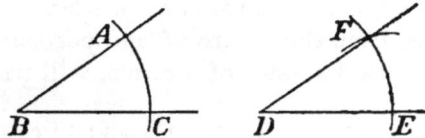

Let A B C (Fig. 6) be the given angle. With B as centre and radius of any convenient length, describe an arc cutting B A, B C (which may be of any length, *see* Def.) in points A and C. Draw any line D E, and with D as centre and same radius as before, describe an arc cutting D E in E. With E as centre and the straight line distance from A to C as radius, describe an arc intersecting in F the arc just drawn. From D draw a line through F; then the angle F D E will be equal to the given angle A B C.

CASE II.—Where the given angle is an angle in already existing fixed work.

The angle to which an equal angle has to be drawn, may be an angle existing in already fixed work, fixed piping for instance; or in brickwork, when, suppose, a cistern may have to be made to fit in an angle between two walls. In such cases a method often used in practice is to open a two-fold rule in the angle which is to be copied. The rule is then laid down on the working surface, whatever it may be (paper, board, &c.), on which the work of drawing an angle equal to the existing angle has to be carried out, and lines are drawn on that surface, along either the outer or inner edges of the rule. The rule being then removed, the lines are produced; meeting, they give the angle required.

CASE III.—Where the given angle is that of fixed work, and the method of CASE II. is inapplicable.

With existing fixed work, the method of CASE II. is not always practicable. A corner may be so filled that a rule cannot be applied. The method to be now employed is as follows. Draw lines on the fixed work, say piping, each way from the angle; and on each line, from the angle, set off any the same distance, say 6 in., and measure the distance between the free ends of the 6-in. lengths. That is, if A C, A B (Fig. 7)

FIG. 7.

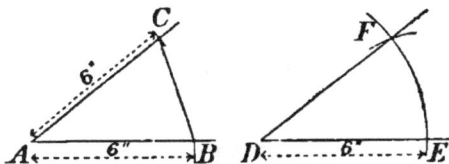

represent the lines drawn on the piping, measure the distance between B and C. Now on the working surface on which the drawing is to be made, draw any line D E, 6 in. long; and with D as centre and radius D E, describe an arc. Next, with E as centre, and the distance just measured between B and C as radius, describe an arc cutting the former arc in F. Join F D; then the angle F D E will be equal to the angle of the piping.

NOTE.—When points are 'joined,' it is always by *straight* lines.

PROBLEM II.

To divide a line into any number of equal parts.

Let A B (Fig. 8) be the given line. From one of its extremities, say A, draw a line A 3 at any angle to A B, and on it, from the angular point, mark off as many parts,—of any convenient length, but all equal to each other,—as A B is to be divided into. Say that A B is to be divided into three equal parts, and that the equal lengths marked off on A 3 are A to

FIG. 8.

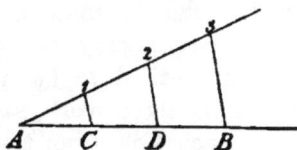

1, 1 to 2, and 2 to 3. Then join point 3 to the B extremity of A B, and through the other points of division, here 1 and 2, draw lines parallel to 3 B, cutting A B in C and D. Then A B is divided as required.

PROBLEM III.

To bisect (divide a line into two equal parts) a given line.

Let A B (Fig. 9) be the given line. With A as centre, and any radius greater than half its length, describe an indefinite arc; and with B as centre and same radius, describe an arc intersecting the former arc in points P and Q. Draw a line through P and Q; this will bisect A B.

FIG. 9.

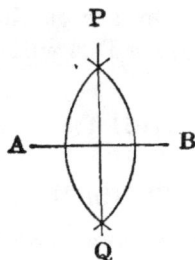

NOTE.—It is quite as easy to bisect A B by Problem II.; but the method shown gives, in P Q, not only a line bisecting A B, but a line perpendicular to A B. This must be particularly remembered.

PROBLEM IV.

To find the centre of a given circle.

Let A B C (Fig. 10) be the given circle. Take any three points A, B, C, in its circumference. Join A B, B C; then A B, B C, are chords (*see* Def.) of the circle A B C. Bisect A B, B C; the point of intersection, O, of the bisecting lines is the centre required.

Fig. 10.

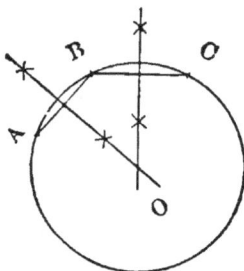

PROBLEM V.

To describe a circle which shall pass through any three given points that are not in the same straight line.

Let A, B, C (Fig. 10) be the three given points. Join A B, B C. Now the circle to be described will not be a circle through A, B, C, unless A B, B C, are chords of it. Let us therefore assume them such, and so treating them, find (by Problem IV.) O the centre of that circle. With O as centre, and the distance from O to A as radius, describe a circle; it will pass also through B and C, as required.

PROBLEM VI.

Given an arc of a circle, to complete the circle of which it is a portion.

Let A C (Fig. 10) be the given arc; take any three points in it as A, B, C; join A B, B C. Bisect A B, B C by lines intersecting in O. With O as centre, and O to A or to any point in the arc, as radius the circle can be completed.

PROBLEM VII.

To find whether a given curve is an arc of a circle.

Choose any three points on the given curve, and by Problem V describe a circle passing through them. If the circle coincides with the given curve, the curve is an arc.

PROBLEM VIII.

To bisect a given angle.

Let A B C (Fig. 11) be the given angle. With B as centre and any convenient radius describe an arc cutting A B, B C in D and E. With D and E as centres and any convenient distance, greater than half the length of the arc

Fig. 11.

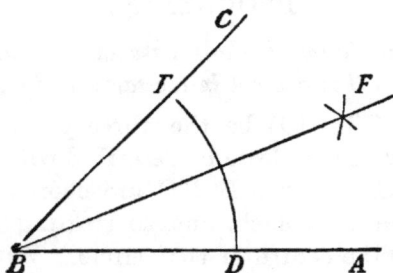

D E as radius describe arcs intersecting in F. Join F to B; then F B bisects the given angle.

PROBLEM IX.

In a given circle, to inscribe a regular polygon of any given number of sides.

Divide (PROBLEM II.) the diameter A C of the given circle (Fig. 12) into as many equal parts as the figure is to have sides, here say five. With A and C as centres, and C A as radius, describe arcs intersecting in P. Through P and the second point of division of the diameter draw a line P B

Fig. 12.

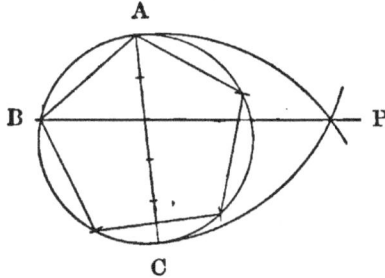

cutting the circumference in B; join B A, then B A will be one side of the required figure. Mark off the length B A from A round the circumference until a marking off reaches B. Then, beginning at point A, join each point in the circumference to the next following; this will complete the polygon.

Note.—By this problem a circumference, and therefore also one-half of it (semicircle), one-third of it, one-fourth of it (quadrant), and so on, can be divided into any number of equal parts.

PROBLEM X.

To describe any regular polygon, the length of one side being given.

Let A B (Fig. 13) be the given side of, say, a hexagon. With either end, here B, as centre and the length of the given side as radius, describe an arc. Produce A B to cut the arc in X. Divide the semicircle thus formed into as many equal parts (PROBLEM IX., NOTE) as the figure is to have sides (six), and join B to the second division point of the semicircle counting from X. This line will be another side of the required polygon. Having now three points, A, B, and the second division point from X, draw a circle through them (Problem V.), and, as a regular polygon can always be inscribed in a circle (*see* Def.), mark off the length B A round

FIG. 13.

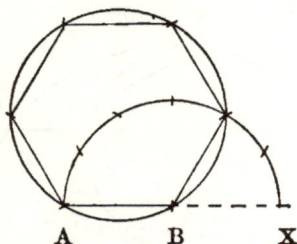

A   B   X

the circumference from A until at the last marking-off, the free extremity of the second side (the side found) of the polygon is reached, then, beginning at A, join each point in the circumference to the next following; this will complete the polygon (hexagon).

PROBLEM XI.

To find the length of the circumference of a circle, the diameter being given.

Divide the given diameter A B (Fig. 14) into seven equal parts (Problem II.). Then three times A B, with C B, one of the seven parts of A B, added, that is with one-seventh of

FIG. 14.

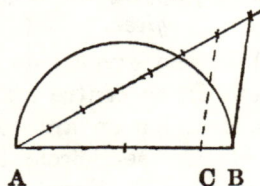

A   C B

A B added, will be the required length of the circumference. The semicircle of the figure is superfluous, but may help to make the problem more clearly understood.

PROBLEM XII.

To draw an oval, its length and width being given.

Draw two lines A B, C D (the axes of the oval), perpendicular to one another (Fig. 15), and intersecting in O. Make

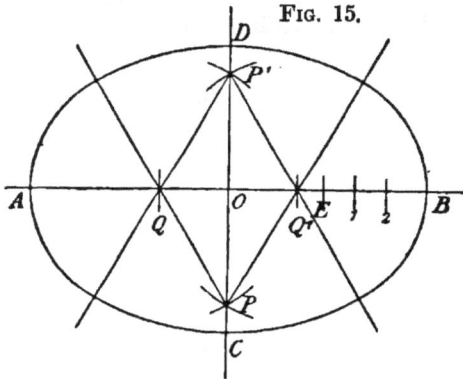

FIG. 15.

O A and O B each equal to half the length, and O C and O D each equal to half the width of the oval. From A mark off A E equal to C D the width of the oval, and divide E B into three equal parts. With O as centre and radius equal to two of the parts, as E 2, describe arcs cutting A B in points Q and Q'. With Q and Q' as centres and Q Q' as radius describe arcs intersecting C D in points P and P'. Join P Q, P Q', P' Q and P' Q'; in these lines produced the end and side curves must meet. With Q and Q' as centres and Q A as radius, describe the end curves, and with P and P' as centres and radius P D, describe the side curves; this will complete the oval.

NOTE.—Unless care is taken, it may be found that the end and side curves will not meet accurately, and even with care this may sometimes occur. It is best if great accuracy be required in the length, to draw the end curves first, and then draw side curves to meet them; or, if the width is most important, to draw the side curves first. The centres (P and P') for the side curves come inside or outside the curves, according as the oval is broad or narrow. This figure is sometimes erroneously called an ellipse. It is, however, a good approximation to one, and for most purposes where an elliptical article has to be made, is very convenient.

PROBLEM XIII.

To draw an egg-shaped oval, having the length and width given.

Make A B (Fig. 16) equal to the length of the oval, and
from A set off A O equal to half its width. Through O draw

FIG. 16.

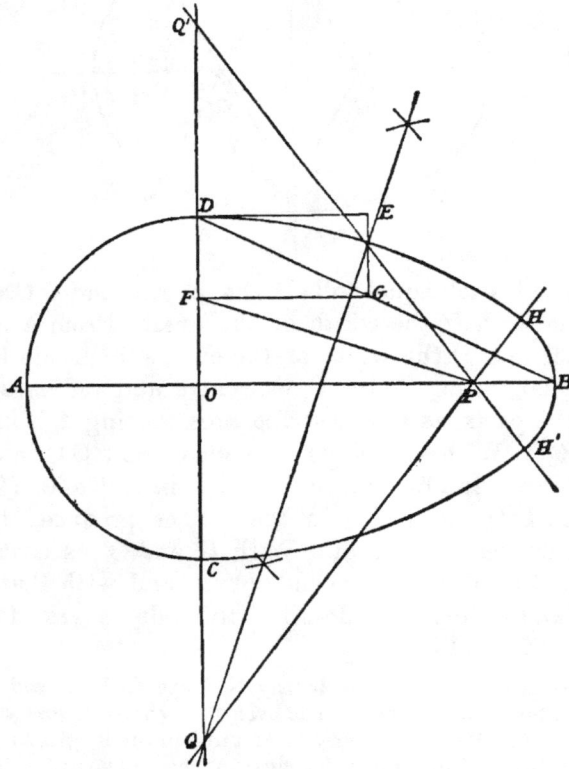

an indefinite line Q Q' perpendicular to A B, and with O as
centre and O A as radius describe the semicircle C A D.
Join D B; and from D draw D E perpendicular to Q Q' and
equal to O D. Also from E draw E G parallel to Q Q' and

intersecting D B in G, and from G draw G F parallel to D E
and intersecting Q Q' in F. From B set off B P equal to D F,
and join P F. Bisect F P and through the point of bisection
draw a line cutting Q Q' in Q. Join Q P and produce it
indefinitely, and with Q as centre and Q D as radius
describe an arc meeting Q P produced in H. Make O Q'
equal to O Q, and join Q' P and produce it indefinitely.
With Q' as centre, and Q' C (equal to Q D) as radius,
describe an arc meeting Q' P produced in H'. And with P
as centre and P B as radius describe an arc to meet the arcs
D H and C H' in H and H'; and to complete the egg-shaped
oval.

PROBLEM XIV.

To describe an ellipse.

Before working this as a problem in geometry, let us draw
an ellipse non-geometrically and get at some sort of a definition. This done, we will solve the problem geometrically,
and follow that with a second mechanical method of describing the curve.

METHOD I.—MECHANICAL.

A. *Irrespective of dimensions.*—On a piece of cardboard or
smooth-faced wood, mark off any two points F, F' (Fig. 17)
and fix pins securely in those points. Then take a piece of
thin string or silk, and tie the ends together so as to form a
loop; of such size as will pass quite easily over the pins.
Next, place the point of a pencil in the string, and take up
the slack so that the string, pushed close against the wood,
shall form a triangle, as say, F D F', the pencil point being
at D. Then, keeping the pencil upright, and always in the
string, and the string taut, move the pencil along from left
to right say, so that it shall make a continuous mark. Let
us trace the course of the mark. Starting from D, the
pencil, constrained always by the string, moves from D to P,

then on to B, P', C, P², P³, A, P⁴, and D again, describing a
curve which returns into itself; this curve is an ellipse.

Having drawn the ellipse, let us remove the string and
pins, draw a line from F to F', and produce it both ways to
terminate in the curve, as at B and A. Then A B is the
major axis of the ellipse, and F, F' are its *foci*. The mid-
point of A B is the *centre* of the ellipse. Any line through
the centre and terminating both ways in the ellipse is a
diameter. The major axis is the longest diameter, and is
commonly called the length of the ellipse. The diameter
through the centre at right angles to the major axis is the
shortest diameter, or *minor axis*, or width of the ellipse.

Referring to the Fig. :—

A D P B C is an ellipse.

F, F' are its foci (singular, *focus*).

A B is the major axis.

C D is the minor axis.

    O is the centre.

Fig. 17.

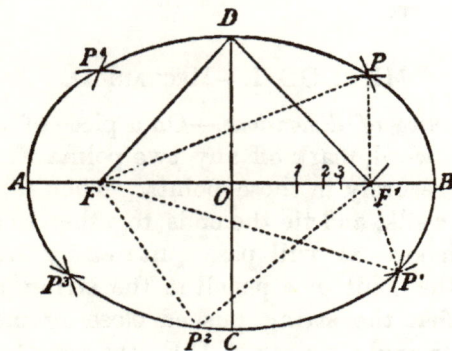

We notice with the string and pencil that when the
pencil point reaches P, the triangle formed by the string is
F P F'; when it reaches P', the triangle is F P' F'; when it
reaches P² the triangle is F P² F'; and when P³ is reached,
it is F P³ F'. Looking at these triangles, it is obvious that

F F′ is one side of each of them; from which it follows, seeing that the loop of string is always of one length, that the sum of the other two sides of any of the triangles is equal to the sum of the other two sides of any other of them; that is to say, F D added to D F′ is equal to F P added to P F′, is equal to F P′ added to P′ F′, and so on.

Which leads us to the following definition.

DEFINITION.

Ellipse.—The ellipse is a closed curve (that is, a curve returning into itself), such that the sum of the distances of any point in the curve from certain two points (foci), inside the curve is always the same.

B. *Length and width given.* — Knowing now what an ellipse is, we can work to dimensions. Those usually given are the length (major axis), and width (minor axis). Draw A B, C D (Fig. 17), the given axes, and with either extremity, C or D, of the minor axis as centre, and half A B, the major axis as radius, describe an arc cutting A B in F and F′. Fix pins securely in F, F′ and D (or C). Then, having tied a piece of thin string or silk firmly round the three pins, remove the pin at D (or C); put, in place of it, a pencil point in the string; and proceed to mark out the ellipse as above explained.

METHOD II.—GEOMETRICAL.—THE SOLUTION OF THE PROBLEM. LENGTH AND WIDTH GIVEN.

Draw A B, C D (Fig. 17), the major and minor axes. With C or D as centre, and half the major axis, O B say, as radius, describe arcs cutting A B in F and F′. On A B, and between O and F′, mark points—any number and anywhere, except that it is advisable to mark the points closer to each other as they approach F′. Let the points here be 1, 2, and 3. With F and F′ as centres and A 2, B 2 as radii respectively, describe arcs intersecting in P; with same centres and A 3, B 3 as radii respectively, describe arcs intersecting in P′. With F′ and F as centres and A 3, B 3 as radii respectively,

c

describe arcs intersecting in P³. With same centres and
A 2, B 2 as radii respectively, describe arcs intersecting in
P⁴. Similarly obtain P². We have thus nine points, D, P,
B, P', C, P², P³, A and P⁴, through wh' *h* an even curve may
be drawn which will be the ellipse required. A greater
number of points through which to draw the ellipse may of
course be obtained by taking more points between O and F',
and proceeding as explained.

METHOD III.—Mechanical.—Length and Width given.

As it is not always possible to proceed as described
at end of *Method I.*, for pins cannot always be fixed in the
material to be drawn upon, we now give a second mechanical method. Having drawn (Fig. 18*b*) A B, C D, the

Fig. 18*a*.   Fig. 18*b*.

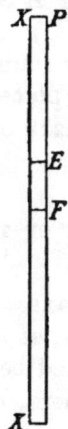

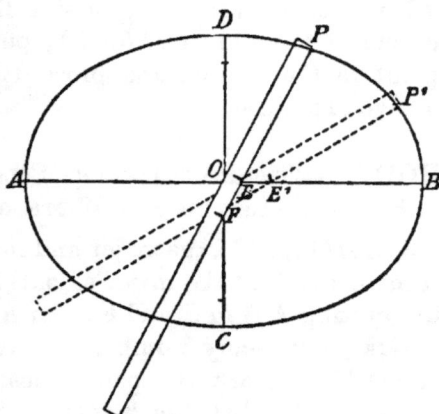

given axes, then, on a strip of card or stiff paper X X (Fig.
18*a*), mark off from one end P, a distance P F equal to half
the major axis (length), and a distance P E equal to half the
minor axis (width). Place the strip on the axes in such
a position that the point E is on the major axis, and the

point F on the minor, and mark a point against the point P. Now shift X X to a position in which E is closer to B, and F closer to C, and again mark a point against P. Proceed similarly to mark other points, and finally draw an even curve through all the points that have been obtained.

The following problems deal with shapes often required by the metal-plate worker, and will give him an idea of how to adapt to his requirements the problems that precede. The explanation of the measurement of angles that concludes the chapter will further assist him in his work.

PROBLEM XV.

To draw an oblong with round corners.

Draw two indefinite lines A B, C D (Fig. 19) perpendicular to one another and intersecting in O. Make O A

Fig. 19.

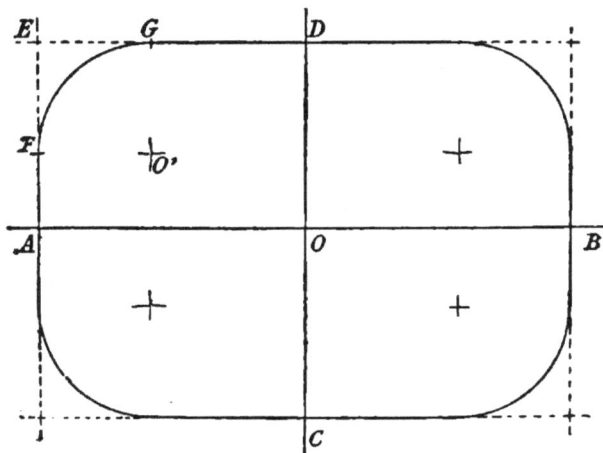

and O B each equal to half the given length; and O C and O D each equal to half the given width. Through C and D

draw lines parallel to A B, and through A and B draw lines
parallel to C D. We now have a rectangle or oblong, and
require to round the corners, which are quadrants. Mark
off from E along E D and E A equal distances E G and E F
according to the size of corner required. With F and G as
centres and E F or E G as radius, describe arcs intersecting
in O'. With O' as centre and same radius describe the
corner F G. The remaining corners can be drawn in similar
manner.

PROBLEM XVI.

To draw a figure having straight sides and semicircular ends
(oblong with semicircular ends).

Draw a line A B (Fig. 20) equal to the given length;
make A O and B O' each equal to half the given width.

FIG. 20.

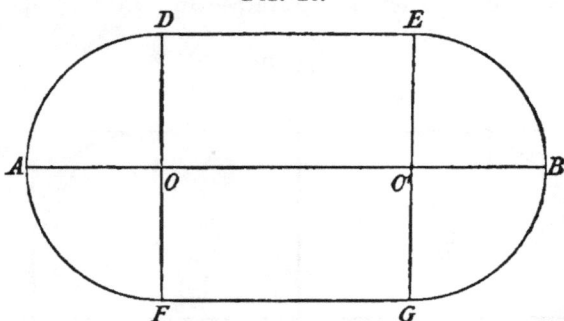

Through O and O' draw indefinite lines perpendicular to
A B; with O and O' as centres and O A as radius describe
arcs cutting the perpendiculars through O and O' in D F
and G E. Join D E, G F; this will complete the figure
required.

ANGLES AND THEIR MEASUREMENT.

The right angle B O C (Fig. 5) *subtends* the quadrant B C.
If we divide that quadrant into 90 parts and call the parts

degrees, then a right angle subtends or contains 90 degrees (written 90°), or as usually expressed, *is an angle of* 90 degrees, the degree being the unit of measurement. If each division point of the quadrant is joined to O, the right angle is divided into 90 angles, each of which subtends or is an angle of 1 degree. That is to say, an angle is measured by the number of degrees that it contains. Suppose the quadrant B A is divided as was B C, then B O A also is an angle of 90 degrees. If the division is continued round the semicircle A D C, this will contain 180 degrees, and the whole circumference has been divided into 360 degrees. As an angle of 90, which is a fourth part of 360 degrees, subtends a quadrant or fourth part of the circumference of the circle, so an angle of 60, which is a sixth part of 360 degrees, subtends a sixth part of the circumference, and similarly an angle of 30 degrees subtends a twelfth part, an angle of 45 an eighth part, and so on. And this angular measurement is quite independent of the dimensions of the circle; the quadrant always subtends a right angle; the 60 degrees angle always subtends an arc of one sixth of the circumference; and the like with other angles. From our definition p. 5 we have it that a chord is the line joining the extremities of any arc. The chord of a sixth part of the circumference of any circle, we have now to add, is equal to the radius of that circle. This being the case, and as an angle of 60 degrees subtends the sixth part of the circumference of a circle, it follows that an angle of 60° subtends a chord equal to the radius.

SCALE OF CHORDS.

Construction.—We have now the knowledge requisite for setting out a scale of chords, by which angles may be drawn and measured.

On any line O B (Fig. 21) describe a semicircle O A B, and from its centre C draw C A perpendicular to O B. Divide O A into nine equal parts. Then, as O A, being a quadrant, contains 90°, each of the nine divisions will contain 10°. The points of division, from O, of the quadrant, are marked 10, 20, 30, &c., up to 90 at A. With O as centre, describe arcs from each of these division points, cutting the line O B. Note that the arc from point 60 cuts O B in C, the centre of the semi-circle; the chord from O to 60 (not drawn in the Fig.), that

FIG. 21.
FIG. 22.

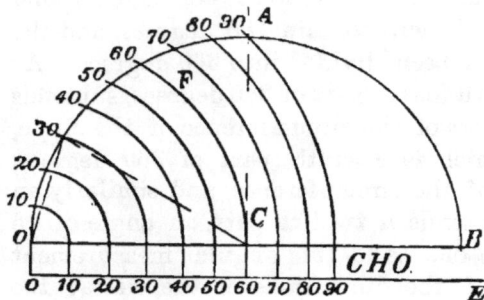
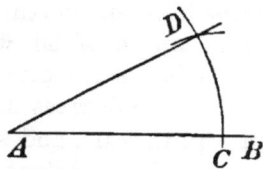

is, the chord of one-sixth of the circumference of the circle whose centre is C, being equal to the radius of that circle. Draw a line O E parallel to O B, and from O let fall O O perpendicular to O B. Also from each of the points where the arcs cut O B let fall perpendiculars to O B and number these consecutively to correspond with the numbers on the quadrant O A. The scale is now complete.

How to use. It is used in this way. Suppose from a point A in any line A B (Fig. 22) we have to draw a line at an angle of 30° with it. Then with A as centre and the distance from O to 60 on O E (Fig. 21) as radius, describe an arc C D cutting A B in C. And with C as centre and the distance from O to 30 on O E (Fig. 21) (the angle to be drawn is to be of 30°) as radius describe an arc intersecting arc C D in D. Join D A, then D A C will be the required angle of 30°. Similarly with angles of other dimensions.

In taking the lengths of arcs, we really take the length of their chords, and it is these lengths that (Fig. 21) we have set off along O E. The angle (Fig. 21) O C F (the point F is the point 60) being an angle of 60° subtends a chord equal to the radius; therefore in O to 60 we have the radius C O. In the example (Fig. 22), the distance C D (O to 30) is the chord of 30°; and it is clear that we must set this off on an arc C D of a circle of the same size as that employed in the construction of the scale, and this we do by making A C equal O to 60 on the same scale.

When a scale of chords has been constructed as explained, the semicircle may be cut away, and we thus get a scale convenient for shop use in the form of a rule.

CHAPTER III.

Patterns for Articles of Equal Taper or Inclination.

(CLASS I.)

(4.) It is necessary here at once to remark that ordinary workshop parlance speaks of ' slant,'—not as meaning an *angle*, but a *length ;* not as referring to the angle of inclination of a tapering body, but to the length of its slanting portion. It is in this sense that we shall use the word, and shall employ the word 'taper' or the term 'inclination of slant' when meaning an angle.

(5.) In order that the rules for the setting out of patterns for articles of equal taper or inclination may be better understood and remembered, it is advisable to consider the principles on which the rules are based, as a knowledge of principles will often enable a workman himself to find rules for the setting out of patterns for odd work. The basis of the whole of the articles in this Class is the right cone. It is necessary, therefore, to define the right cone and explain some of its properties.

Definition.

(6.) *Right Cone.*—A right cone is a solid figure generated or formed by the revolution of a right-angled triangle about one of the sides containing the right angle. The side about which the triangle revolves is the *axis* of the cone ; the other side containing the right angle being its *radius.* The point of the cone is its *apex ;* the circular end its *base.* The hypotenuse of the triangle is the slant of the cone. From the method of formation of the right cone, it follows that the axis is perpendicular to the base. The height of the cone is the length of its axis.

(7.) Referring to Fig. 1*a,* O B E represents a cone gene-

rated or formed by the revolution of the right-angled triangle O A B (Fig. 1b) about one of its sides containing the
right angle, here the side O A. Similarly the cone O D F,

FIG. 1a.　　　　　FIG. 1b.

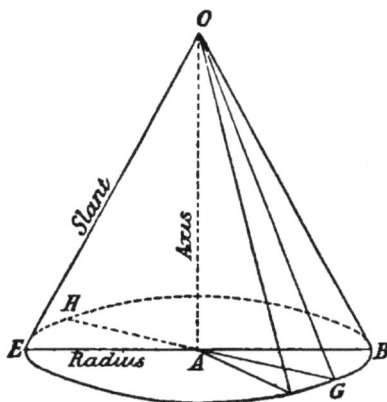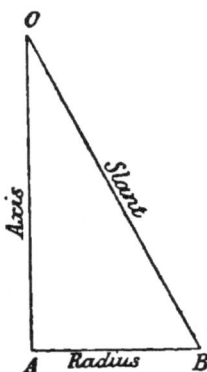

Fig. 2a, is formed by the revolution of O C D (Fig. 2b) about
its side O C. As will be seen from the figs., O A, O C are
respectively the axes of the cones O B E, O D F, as also their
heights. Their bases are respectively B G E H, D K F L,

FIG. 2a.　　　　　FIG. 2b.

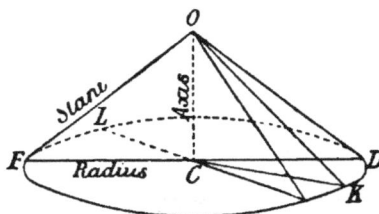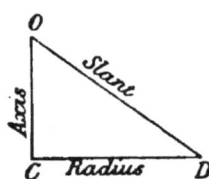

and the radii of the bases are A B and C D. The slants of
the cones are O B and O D, the apex in either being the point
O. Other lines will be seen in figs., namely, those representing the revolving triangle in its motion of generating

the cone.   The sides of these triangles that start from the
apex and terminate in the base are all equal, it must be borne
in mind ; and each of them is the slant of the cone.   Likewise
their sides that terminate in A are all equal, and each shows
a radius of the base of the cone.   How these particulars of
the relations to one another of the several parts of the *right
cone* apply in the setting-out of patterns will be seen in the
problems that follow.

PROBLEM I.

*To find the height of a cone, the slant and diameter of the base
being given.*

Draw any two lines O A, B A (Figs. 3 and 4) perpendicular to each other and intersecting in A.   On either line

Fig. 3.   Fig. 4.

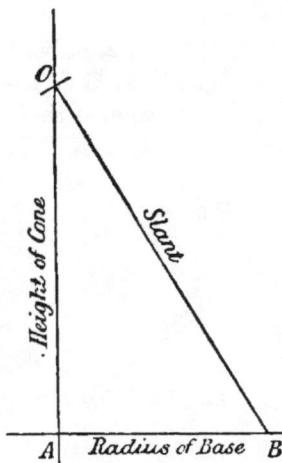
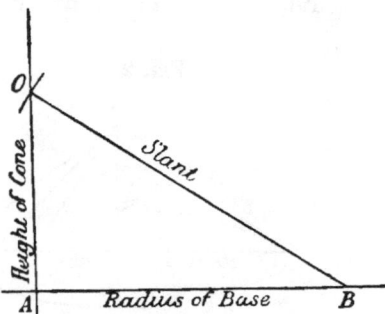

mark off from A half the diameter of the base, in other
words, the radius of the base, as A B.   With B as centre, and
radius equal to the slant, describe an arc cutting A O in O.
Then O A is the height of the cone.

PROBLEM II.

To find the slant of a cone, the height and diameter of the base being given.

Draw any two lines O A, B A (Figs. 3 and 4) perpendicular to each other and intersecting in A. On either line mark off from A half the diameter of the base (radius of the base), as A B, and make A O on the other line equal to the height of the cone; join O B. Then O B is the required slant

CHAPTER IV.

PATTERNS FOR ROUND ARTICLES OF EQUAL TAPER OR INCLINATION OF SLANT.

(CLASS I.   *Subdivision a.*)

(8.) If a cone has its inclined or slanting surface painted' say, white, and be rolled while wet on a plane so that every portion of the surface in succession touches the plane, then the figure formed on the plane by the wet paint (see Fig. 5)

FIG. 5.

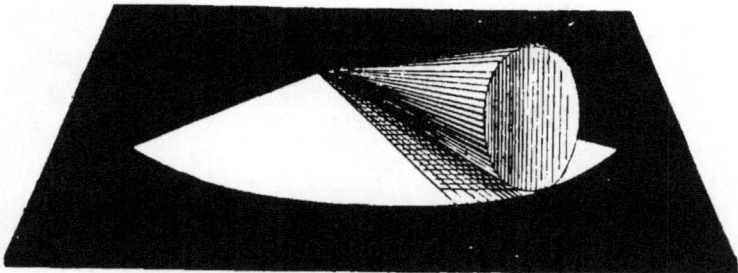

will be the pattern for the cone.   As the cone rolls (the figure represents the cone as rolling), the portion of it touching the plane at any instant is a slant of the cone (see § 7).

(9.) Examining the figure formed by the wet paint, we find it to be a *sector* of a circle, that is, the figure contained between two radii of a circle and the arc they cut off.   The length of the arc here is clearly equal to the length of the circumference of the base of the cone, and the radius of the arc evidently equal to the slant of the cone. From this it is obvious that to draw the pattern for a cone, we require to know the slant of the cone (which will be the radius for the pattern), and the circumference of the base of the cone.

PROBLEM III.

To draw the pattern for a cone, in one piece or in several pieces, the slant and diameter of the base being given.

PATTERN IN ONE PIECE.—With O A (Fig. 6*b*) equal to the slant as radius, describe a long arc A C E. What has now

FIG. 6*b*.

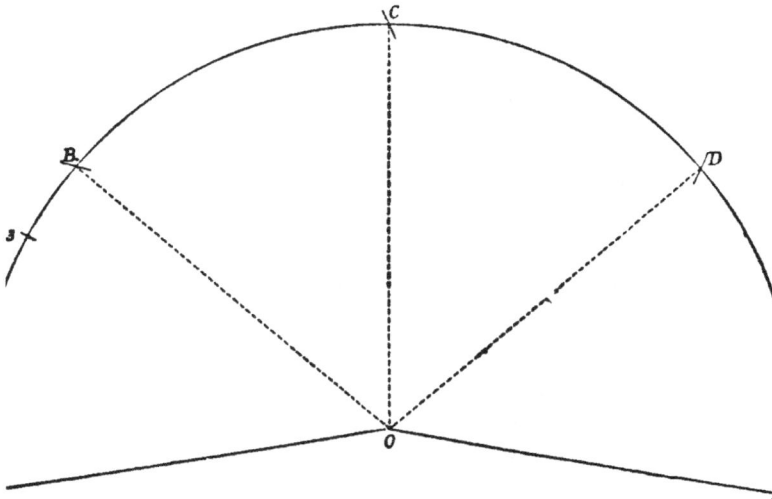

to be done is to mark off a length of this arc equal to the circumference of the base of the cone. The best and quickest way for this is as follows. Draw a line F B (Fig. 6*a*) equal to the given diameter of the base, and bisect it in G; then G B is a radius of the base. From G draw G E perpendicular

to F B ; and with G as centre and radius G B describe from
B an arc meeting G E in E.   The arc B E is a quadrant
(quarter) of the circumference of the base of the cone.
Divide this quadrant into a number of equal parts, not too

FIG. 6a.

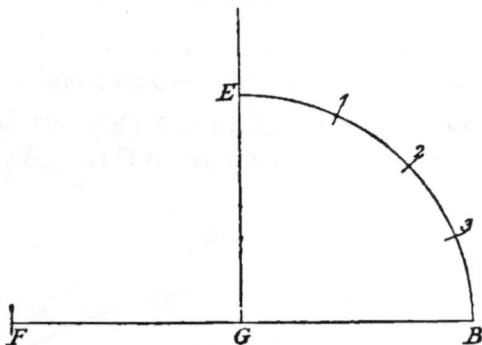

many, say four, by points 1, 2, 3.   From A (Fig. 6b) mark
off along arc A C E four parts, each equal to one of the
divisions of the quadrant, as from A to B.   Take this length
A B equal to the four parts, that is, equal to the quadrant,
and from B set it off three times along the arc towards E as
from B to C, C to D, D to E.   Join E to O; then O A C E O
will be the pattern required.

NOTE.—It must be noted that when this pattern is bent round to form the
cone, the edges O A and O E will simply butt up against each other,
for no allowance has been made for *lap* or *seam*.   Let us call the junction
of O A and O E the line of butting.   Nor, further, has any allowance been
made for *wiring* of the edge A C E.   These most essential matters will
be referred to immediately.

PATTERN IN MORE THAN ONE PIECE.—If B be joined to O,
then the sector O A B will be the pattern for *one-quarter*
of the cone.   If C be joined to O, then the sector O A C is
the pattern for *one-half* of it.   Similarly O A D will give

three-quarters of the cone. A cone pattern can thus be made in one, two, three, or four pieces. If the cone is required to be made in three pieces, then instead of dividing, as above, a quadrant of the circumference of the base, divide one-third of it into parts, say five; set off five of the parts along A C E from A, and join the last division point to the centre; the sector so obtained will be the pattern for *one-third* of the cone. If required to be made in five pieces, divide a fifth of the circumference of the base into equal parts, and proceed as before. Similarly for any number of pieces that the pattern may be required in.

PROBLEM IV.

To draw the pattern for a cone, the height and the diameter of the base being given.

First find the slant O B (Fig. 7a) by PROBLEM II. Then with A as centre and radius A B, describe B C a quadrant of

FIG. 7a.   FIG. 7b.

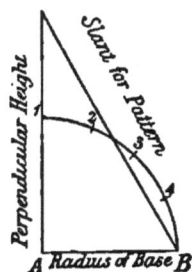

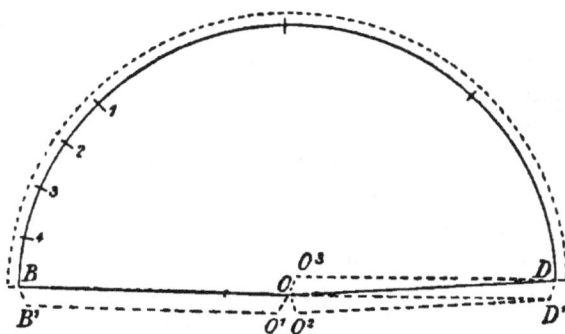

the circumference of the base, and proceed, as in PROBLEM III., to draw the pattern Fig. 7b (the plain lines).

ALLOWANCE FOR LAP, SEAM, WIRING, &c.

(10.) It has already been stated that the geometric pattern Fig. 6b has no allowance for seam, wiring, or edging. (For the present it is assumed that these terms are understood; we shall come back to them later on.) In the pattern Fig. 7b the dotted line O' B' parallel to the edge O B shows 'lap' for soldered seam. For a 'grooved' seam not only must there be this allowance, but there must be a similar allowance along the edge O D. These allowances, it must be distinctly remembered, are always *extras* to the geometric pattern; that is to say, the junction line of O D and O B, or line of butting (*see* NOTE, Problem III.) is not interfered with. And here a word of warning is necessary. Suppose instead of marking off a parallel slip or lap for soldered seam, a slip D O D' going off to nothing at the centre O, is marked off, and that then, for soldering up, there is actually used not this triangular slip, but a parallel one as D D' O O^3, the result brought about will be that the work will solder up untrue; there will be, in fact, a 'rise' at the base of the work. We can understand the result in this way. If the parallel slip D D' O O^3 used for soldering were cut off, there would remain a pattern which is not the geometric pattern, but a nondescript approximation, having a line of butting other than the true line. And it being thus to an untrue pattern that the parallel slip for seam is added, the article made up from the untrue pattern must of course itself necessarily be untrue. In the fig. the dotted line parallel to the curve of the pattern shows an allowance for wiring. For a grooved seam there must be on the edge O D an addition O D D' O^2 similar to the addition on the edge O B, as above stated.

(11.) In working from shop patterns for funnels, oil-bottle tops, and similar articles, workmen often find that if they take a good lap at the bottom, and almost nothing at the top of the seam, the pattern is true. And so it is, for these patterns have the triangular slip D O D' added. Whereas, if a parallel piece ·D O^3 O D' is used for lap, the pattern is untrue. Which again is the case, because, now, in addition to D O D', an extra triangular piece D O³ O is used, and this extra *is taken off the geometric pattern.* Consequently, the line of butting is interfered with; that is to say, the two lines O B and O D, instead of meeting, overlap; O B forming a junction with O³ D instead of with O D; with which O B must always form a junction, for the pattern to be true. In setting out patterns, to prevent error, the best rule to follow and adopt is, to first mark them out independent of any allowance for seams, or wiring, or edging, and to afterwards add on whatever allowances are intended or requisite. In future diagrams, allowances, where shown, will be mostly shown by dotted lines.

DEFINITION.

(12.) FRUSTUM.—If a right cone is cut by a plane parallel to

FIG. 8a.   FIG. 8b.

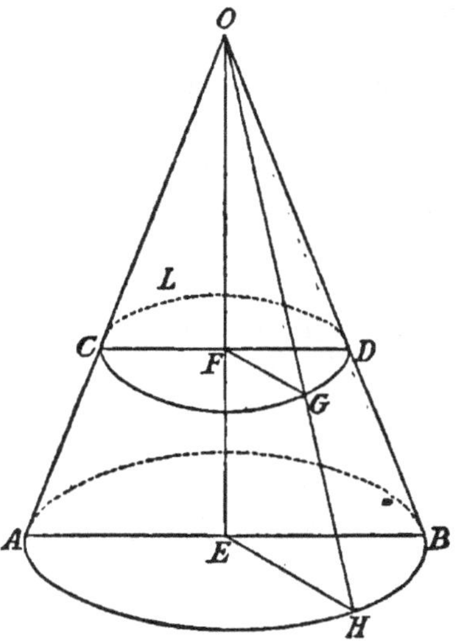

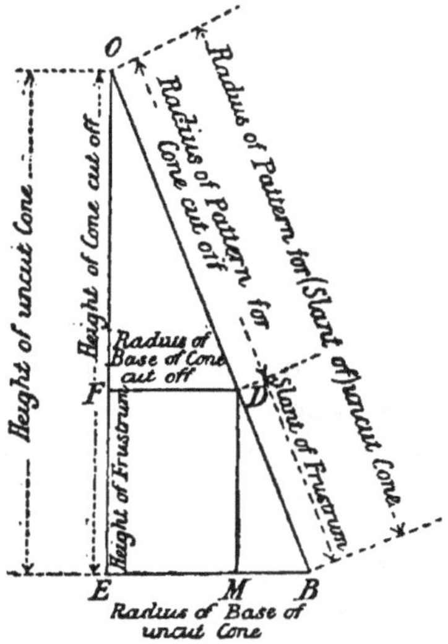

its base, the part containing the apex is a complete cone, as O C G D L (Fig. 8a), and the part C A B D containing the base A H B K is a *frustum* of the cone. In other words a *frustum* of a right cone is a solid having circular ends, and of equal taper or inclination of slant everywhere between the ends. Conversely a round equally tapering body having top and base parallel is a frustum of a right cone.

(13.) Comparing such a solid with round articles of equal taper or inclination of slant, as pails, coffee-pots, gravy

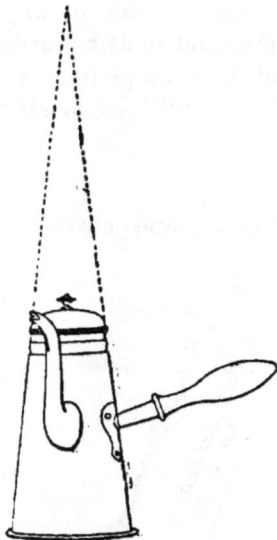

FIGS. 9.

strainers, and so on (Fig. 9), it will be seen that they are portions (frusta) of right cones.

(14.) In speaking here of metal-plate articles as portions of cones, it must be remembered that all our patterns are of *surfaces*, seeing that we are dealing with metals in *sheet*; and that these patterns when formed up are not solids, but merely simulate solids. It is, however, a convenience, and leads to no confusion to entirely disregard the distinction; the method of expression referred to is therefore adopted throughout these pages.

(15.) By Fig. 8*b* is shown the relations of the cone O A B of Fig. 8*a* with its portions O C D (complete cone cut off), and C A B D (frustum). The portion O C D is a complete cone, as it is the solid that would be formed by the revolution of the right-angled triangle O F D (both figs.) around O F. The triangles O F G and O F C (Fig. 8*a*) represent the triangle O F D in progress of revolution. The triangle O E B (both figs.) is the triangle of revolution of the uncut cone O A B (Fig. 8*a*) and O E H, O E A represent O E B in progress of revolution. The height of the cone O A B being O E (both figs.), the height of the cone O C D is O F (both figs.). The radius for the construction of pattern of the uncut cone O A B will be O B (both figs.) , for the pattern of O C D, the cone cut off, the radius will be O D (both figs.). In F E, or D M, we have the height of the frustum. Just as (§ 8) the portion of the rolling cone touching the plane at any instant is a slant of the cone, so the slant of a frustum is that portion of it, which, if it were set rolling on a plane, would at any instant touch the plane. D B is a slant of the frustum C A B D. The extremities of a slant of a frustum are 'corresponding points.' Other details of cone and frustum are shown in Fig. 8*b*.

FIG. 10.

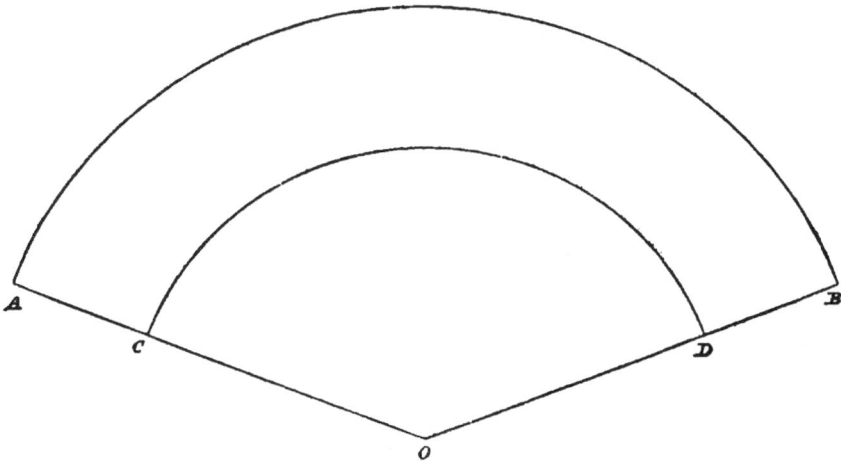

(16.) It is obvious that, if the patterns for the cones O A B, O C D (Fig. 8*a*) be drawn (Fig. 10) from a common centre O, the figure A C D B will be the pattern for the frustum.

ACDB (Fig. 8a). From which we see that in order to draw the pattern for the frustum of a cone, we must know the slant of the cone of which the frustum is a portion, that is, we must know the radius for the construction of the pattern of that cone, and also the slant (radius for pattern) of the cone cut off.

PROBLEM V.

Given the dimensions of the ends of a round equal-tapering body (frustum of right cone), and its upright height. To find the slant, or the height, of the cone of which it is a portion.

Draw any two lines O A, A B (Fig. 11) at right angles to each other and intersecting in A. From A on either line,

Fig. 11.

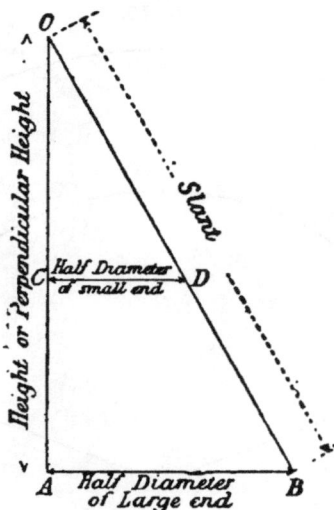

say on B A, mark off A B equal to half the diameter of the larger of the given ends, and from A on the other line make A C equal to the given upright height. Draw a line C D parallel to A B, or, which is the same thing, at right angles to A O, and make C D equal to half the diameter of the smaller end. Join B D, and produce it, meeting A O in O. Then O A is the height of the cone of which the tapering body is a portion, and O B the slant.

PROBLEM VI.

To draw the pattern for a frustum of a cone, the diameters of the ends of the frustum and its upright height being given.

The Frustum.—Draw any two lines O A, B A (Fig. 12a) perpendicular to each other and meeting in A; on one of the

Fig. 12a.

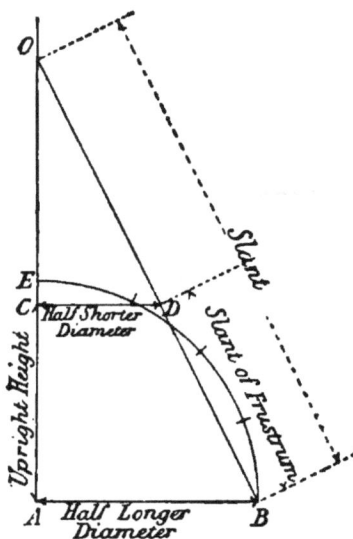

perpendiculars, say B A, make A B equal to half the longer diameter (radius), and on the other make A C equal to the given upright height. Draw a line C D perpendicular to A O and make C D equal to half the shorter diameter. Join B D, and produce it, meeting A O produced in O. With A as centre, and radius A B, describe quadrant B E, which divide into any convenient number of equal parts, here four.

To draw the pattern (Fig. 12b) take any point O' as

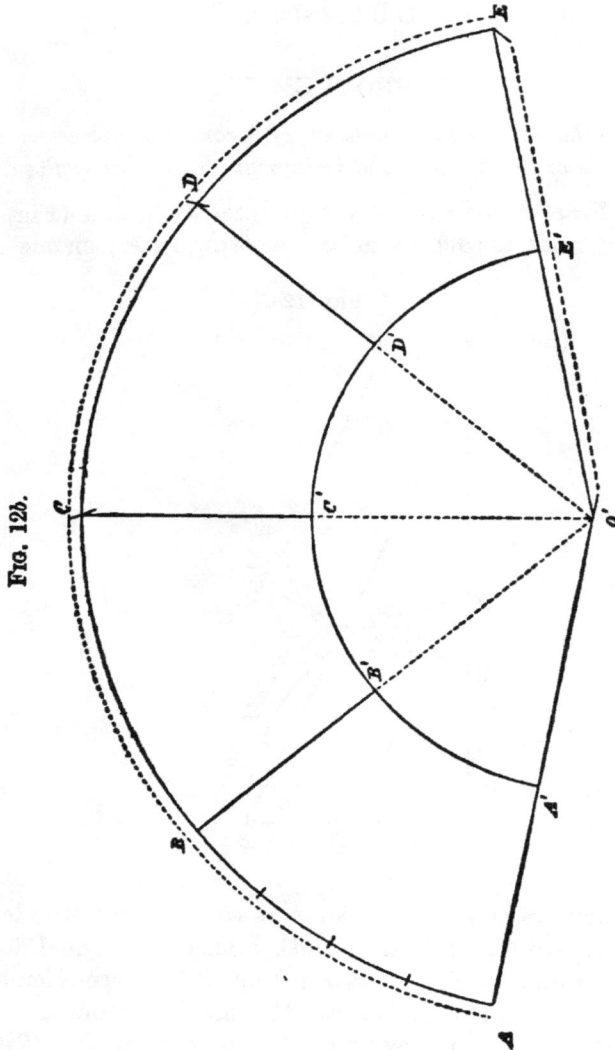

FIG. 12b.

centre, and with radius O B (Fig. 12a) describe an arc
A C E; also with same centre and radius, O D (Fig. 12a),
describe an arc A′ C′ E′. From any point in the outside
curve, as A, draw a line through O′, and cutting the
inner arc in A′. From A mark off successively parts equal
to those into which the quadrant B E (Fig. 12a) is divided,
and the same number of them, four, to B. And from B,
along the outer curve, set off B C, C D, D E, each equal to
A B. Join E O′, cutting the inner curve in E′. Then
A A′ E′ E is the pattern required.

Just as O B (Fig. 12a) is the slant of the cone that would
be generated by the revolution of right-angled triangle
O A B around O A, so D B is the slant of the frustum of
which A A′ E′ E (Fig. 12b) is the pattern. In the pattern
the slant D B appears as A A′, B B′, C C′, &c.

Parts of the Frustum —If B be joined to O′, the figure A A′
B′ B will be one-quarter of the pattern of the frustum; and
if C be joined to O′, the figure A A′ C′ C will be pattern for
one-half of it, and so on. The paragraph " Pattern in
more than one Piece " in PROBLEM III. should be re-read in
connection with this " Parts of a Frustum."

(17.) The problem next following is important, in that,
in actual practice, the slant of a round equal-tapering body
is very often given instead of its height, especially in cases
where the taper or inclination of the slant is great ; as for
instance in ceiling-shades. The only difference in the working out of the problem from that of PROBLEM VI. is that the
radii required for the pattern of the body are found from
other data. Let us take the problem.

PROBLEM VII.

*To draw the pattern for a round equal-tapering body (frustum of
right cone), the diameter of the ends and the slant being given.*

To find the required radii, draw any two lines O A, B A
(Fig. 13) perpendicular to one another, and meeting in A.

On either line, as A B, make A B equal to half the longer of
the given diameters and A C equal to half the shorter. From
C draw C D perpendicular to A B. With B as centre and

FIG. 13.

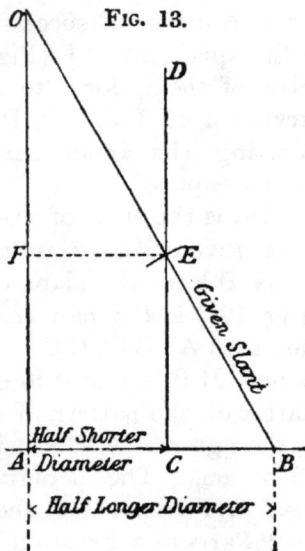

the given slant as radius, describe an arc cutting C D in E.
Join B E and produce it to meet A O in O. Then O B and
O E are the required radii. By E F being drawn parallel to
A B, comparison may be made between this Fig. and Fig.
12a, and the difference between PROBLEMS VI. and VII.
clearly apprehended. To draw the pattern, proceed as in
PROBLEM VI.

(18.) For large work and for round equal-tapering bodies
which approximate to round bodies without any taper at all,
the method of PROBLEM VI. is often not available, for want
of space to use the long radii that are necessary for the
curves of the patterns. The next problem shows how to
deal with such cases ; by it a working-centre and long radii
can be dispensed with. The method gives fairly good
results.

PROBLEM VIII.

To draw, **without long radii,** *the pattern for a round equal-tapering body (frustum of right cone), the diameters of the ends and the upright height being given.*

First draw one-quarter of the plan. (To do this, we fore-stall for convenience what is taught in the following chapter.)

FIG. 14.

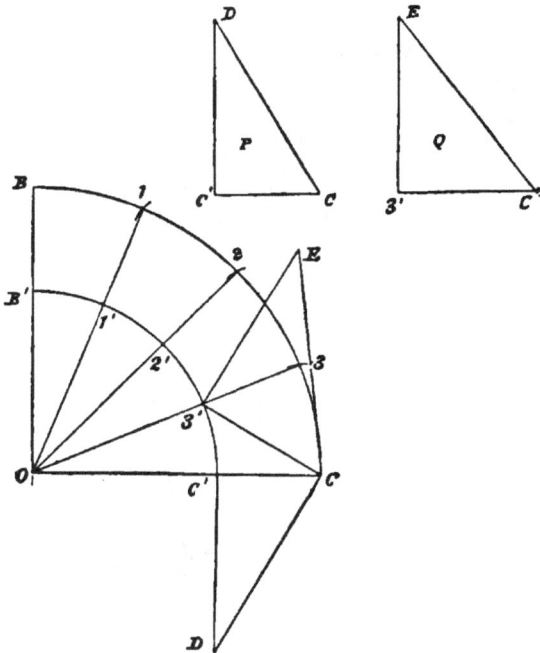

Draw any two lines B O, C O (Fig. 14) perpendicular to each other and meeting in O. With O as centre and radius equal to half the longer diameter, describe an arc meeting the lines B O, C O in B and C. With O as centre and radius equal to half the shorter diameter describe an arc B' C'. This completes the one-quarter plan.

Now divide B C, the largest arc, into any number of equal parts, say four; and join the points of division to O by lines cutting B' C' in 1', 2', 3'. Join 3' C, and through 3' draw 3' E perpendicular to 3' C, and equal to the given upright height. Join C E; then C E may be taken as the true length of C 3'. Through C' draw C' D perpendicular to C O and equal to the upright height. Join C D; then C D is the true length of C C'. If it is inconvenient to find these true lengths on the plan, it may be done apart from it, as by the triangles P and Q.

To set out the pattern. Draw (Fig. 15) any line C C' equal to C D (Fig. 14). With C' and C as centres and radii respectively C E and C 3 (Fig. 14) describe arcs intersecting in 3 (Fig. 15). With C and C' as centres and radii respectively C E and C' 3' (Fig. 14) describe arcs intersecting in 3' (Fig. 15). Then C and 3 are two points in the outer

FIG. 15.

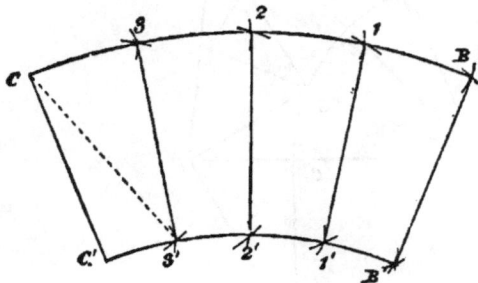

curve of the pattern, and C' 3' two points in the inner curve. To find points 2 and 2', proceed as just explained, and with the same radii, but 3' and 3 as centres instead of C' and C. Similarly, to find points 1' and 1, and B' and B. A curved line drawn from C through 3, 2, and 1 to B will be the outer curve of one-quarter of the required pattern, and a curved line from C' through 3', 2', and 1' to B' its inner curve; that is C C' B' B is one-quarter of the pattern. Four times the quarter is of course the required pattern complete.

NOTE.—In cases where this method will be most useful, the pattern is generally required so that the article can be made in two, three, four, or more pieces. If the pattern is required in three pieces, one-third of the plan must be drawn (see end of Problem III., p. 31) instead of a quarter, as in Fig. 14; the remainder of the construction will then be as described above.

(19.) It is often desirable in the case of large work to know what the slant or height, whichever is not given, of a round equal-tapering body (frustum of right cone) will be, before starting or making the article. Here the following problems will be of service.

PROBLEM IX.

To find the slant of a round equal-tapering body (frustum of right cone), the diameters of the ends and the height being given.

Mark off (Fig. 16) from a point O in any line O B the lengths of half the shorter and longer diameters, as O C, O B.

FIG. 16.

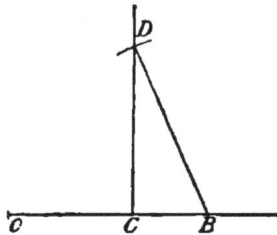

From C draw C D perpendicular to O B. Make C D equal to the given height, and join B D. Then B D is the slant required.

PROBLEM X.

To find the height of a round equal-tapering body (frustum of right cone), the diameters of the ends and the slant being given.

Mark off (Fig. 17) from a point O in any line O B the lengths of half the shorter and longer diameters, as in PROBLEM IX., and from C draw C D perpendicular to O B. With B as centre and radius equal to the given slant, describe an arc cutting C D in E. Then C E is the height required.

Essentially this problem has already been given, in the working of PROBLEM VII.

FIG. 17.   FIG. 18.

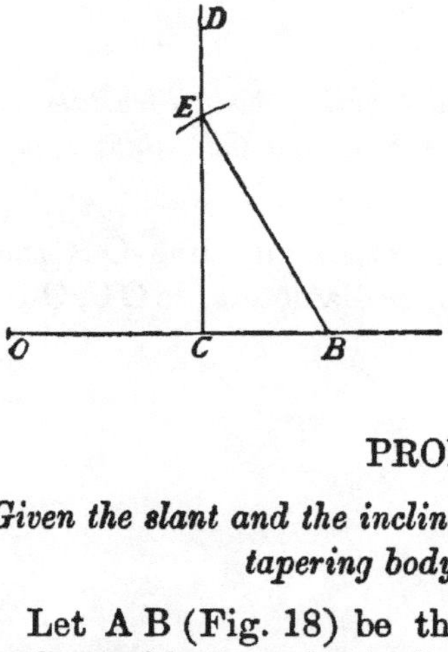

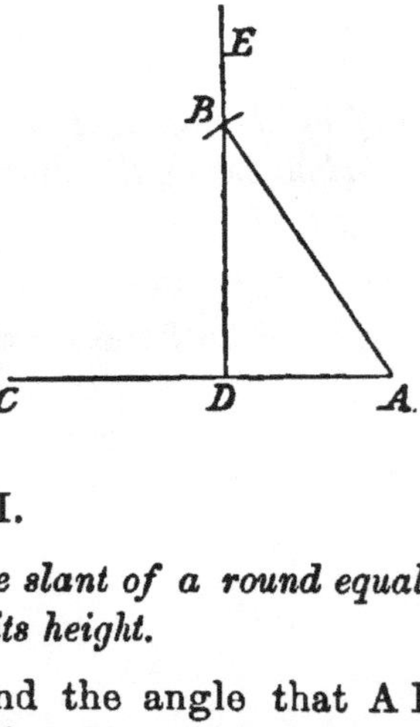

PROBLEM XI.

Given the slant and the inclination of the slant of a round equal-tapering body; to find its height.

Let A B (Fig. 18) be the slant, and the angle that A B makes with C A the inclination of the slant. From B let fall B D perpendicular to A C. Then B D is the height required.

(20.) In the workshop, the inclination of the slant of a tapering body is sometimes spoken of as the body being so many inches "out of flue." This will be explained in the following chapter. If the inclination of the slant is given in these terms the problem is worked thus. From any point D in any line C A (Fig. 18) make D A equal in length to the number of inches the body is " out of flue," and draw D E perpendicular to C A. With A as centre and radius equal to the given slant, describe an arc intersecting D E in B. Then B D will be the height required.

CHAPTER V.

EQUAL-TAPERING BODIES OF WHICH TOP AND BASE ARE PARALLEL, AND THEIR PLANS.

(21.) First let us understand what a plan is. Fig. 19 represents an object Z, made of tin, say, having six faces,

FIG. 19.

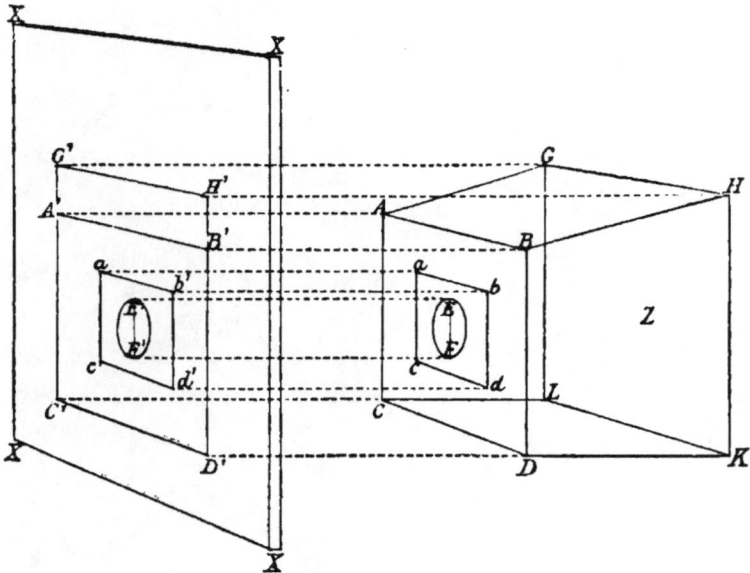

of which the A B C D and G H K L faces are parallel, as also the B D K H and C A L G. The A B C D and C D K L faces are square. The A B C D face has, soldered flat on it centrally, a smaller square of tin *abcd* with a central

circular hole in it. Now suppose wires (represented in the
fig. by dotted lines), soldered perpendicularly to the
A B C D face, at A, B, C, D, *a*, *b*, *c*, *d*, E, and F (the points
E and F are points at the extremities of a diameter of the
circular hole). Also suppose wires soldered at G and H
parallel to the other wires, and that the free ends of all the
wires are cut to such length that they will, each of them,
butt up against a flat surface (plane), of glass say, X X X X,
parallel to the A B C D face. Lastly suppose that all the
points where the wires touch the glass are joined by lines
corresponding to edges of Z (see the straight lines in the

FIG. 20.

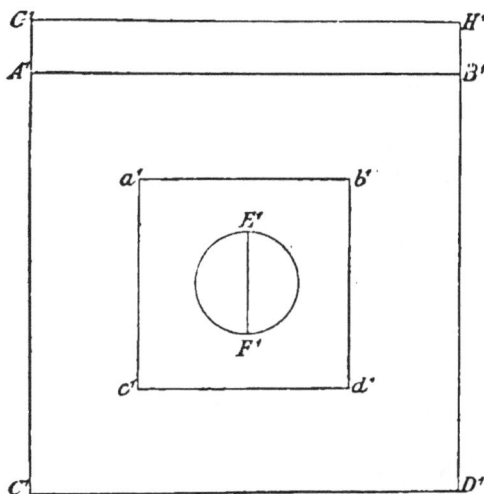

figure on the plane); also that F′ and F″ are joined, that
the line joining them is bisected, and a circle described
passing through E′ and F′. Then the complete representation obtained is a *projection* of Z. Instead of actually *projecting* the points by wires, we may make the doing of it
another supposition may, find, as if by wires, the required
points, and draw the projection. The A B C D face being,

say 2 inches square, the flat piece, say 1 inch square, and the hole $\frac{1}{2}$ inch diameter, and the back face G H K L, say $2\frac{1}{4}$ in. by 2 in., then the projection that is upon the glass would be as shown in Fig. 20. The plane X X X X is here supposed vertical, and the projection G' C' D' H' is therefore an *elevation*; if the plane were horizontal, the projection would be a *plan*, and we might regard A B C D as the top of the body, and G H K L as its base, or *vice versâ*. We may define a plan then as the representation of a body obtained by projecting it on to a horizontal plane, by lines perpendicular to the plane.

(22.) The plane X X X X was supposed parallel to the A B C D face of Z; the plan A' B' C' D' of it is therefore of the same shape as A B C D, and in fact A B C D may be said to be its own plan. Similarly the G' H' D' C' is the plan of the back-face G H K L and is of the same shape as that face. But the plan of the face A G H B to which the plane is not parallel is by no means the same shape as that face, for the long edges B H and A G of the face A G H B are, in plan, the short lines B' H' and A' G'. We need not, however, go farther into this, because in the case of the bodies that now concern us, the horizontal plane on which any plan is drawn is always supposed to be parallel to the principal faces of the body, so that the plans of those faces are always of the same shape as the faces. In this paragraph the plane X X X X is supposed to be horizontal.

(22a.) We are now in a position to explain the getting at the true length of C C' in the fig. of PROBLEM VIII., p. 42; or, putting the matter generally, to explain the finding the true lengths of lines from their apparent lengths in their plans and elevations. Horizontal lines being excepted, there is, manifestly, for any line, however positioned in space, a vertical plane in which its elevation will appear as (if not a point) a vertical line. Let B E (Fig. 17, p. 44) be any line in the plane of the paper, and let C D be the vertical plane seen edgeways on which the elevation E C of B E is a vertical line. Then if O B be a horizontal plane seen edgeways passing through C, the line joining the B extremity of B E to the C extremity of its elevation will be the plan of B E. We get thus the figure E C B, a figure in one plane, the plane of the paper, a right-angled triangle in fact, of which the E C side is the elevation of B E, the C B side its plan, and the hypotenuse the line itself ; a figure, which, as combining a line, its plan, and its elevation, we have under no other conditions than when the elevation in question is a vertical line. In the plane passing through E C and B E, that is, in the plane in which these lines wholly lie, we have in the line that we get by joining C with B the plan, full length, of B E. In respect of this plan of B E, we are concerned with no other measurement, because, in a right-angled triangle representing a line and its plan and elevation, no other measurement of the plan line can come in. Not so, however, with the elevation line of B E. Here other measurement of it than its length can and does come in, because that length varies according to the position of the vertical plane with regard to it ; the plan length is always the same. But to have in the three sides of E C B, the representation of B E, and its plan and elevation, it is evident that the plane which contains B E and its plan C B must also wholly contain the elevation E C ; that is, the plane must be perpendicular to the plane of the triangle. Now, no matter on what vertical plane the line B E is projected, although the length of the projection will vary, the vertical distance between its extremities, that is, its height, never varies. Hence, if, in any right-angled triangle, we have in the hypotenuse the representation of a line, in one of its sides the plan of the line, and in the other side, not necessarily the elevation that comes out vertical, but the *height* of *any* elevation of the line, it comes to the same thing as if in the latter side we had the actual elevation that is vertical. And hence, further, if we have given the plan-length of an unknown line, and the vertical distance between its extremities, we can, by drawing a line, say C B, equal to the given plan-length, then drawing from one of its extremities and at

E

right angles to it, a line, say C E, equal to the given vertical
distance, and finally joining the free extremities, as by B E,
of these two lines, construct a right-angled triangle, the
hypotenuse, B E, of which must be the true length of the
unknown line ; for there is no other line than B E of which
C B and C E can be, at one and the same time, plan and
elevation. We have explained this true-length matter fully,
because we have to make use of it abundantly in problems to
come.

(23.) Proceeding to the bodies we have to consider, we

FIG. 21.

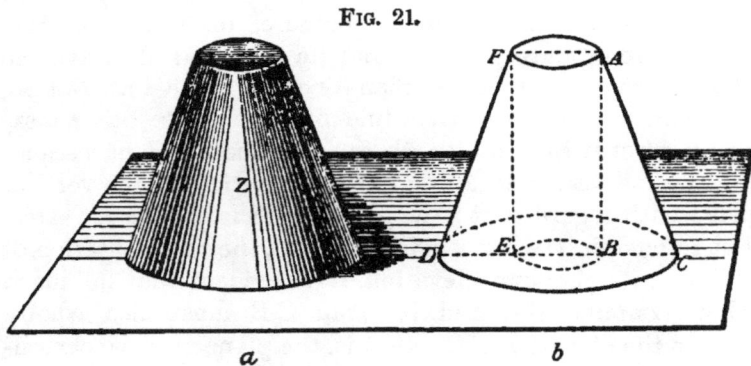

take first a frustum of a cone, Fig. 21a. To draw its plan, let
us suppose the extremities of a diameter of its smaller face
top (namely points A and F of the skeleton drawing Fig. 21b)
(neither drawing is to dimensions), to be projected, in the
way just explained, on to a plane parallel to the face, then,
also as there explained, we can draw the circle which is a
projection of that face. Suppose the smaller circle of Fig. 22
to be that circle, and to be to dimensions. Projecting now,
similarly, the extremities of a diameter of the larger face
(base), namely the points C and D of the skeleton drawing,
on to the same plane, we can get the projection of the larger
face. Let the larger circle of Fig. 22 be that projection.
The two circular projections will be *concentric* (having the

same centre) because the body Z is of equal taper, and they will, together, be the plan of Z, that is Fig. 22 is that plan.

A C and F D each show the slant, and B A and E F the height. B C and D E each show the distances between the plans of corresponding points.

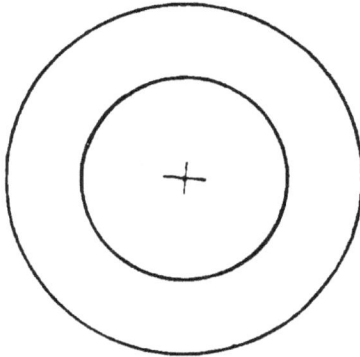

(24.) Turn to the skeleton drawing of Z. Here A C shows a slant of the frustum (§ 15), A B its height (see D M, Fig. 8b), and A and C are 'corresponding points' (§ 15). Looking at C D E B as at the plan of the frustum, we have, in the point B, the plan of the point A. Joining B C, we get a right-angled triangle A B C; the slant A C is its hypotenuse, the height A B is one of the sides containing the right angle, and the other side containing the right angle, B C, is the distance between the plans of the corresponding points A and C, as also between plans of corresponding points of Z anywhere. This distance is that of how much the body is 'out of flue' (a workshop expression that was referred to at the end of the previous chapter), in other words, how much A C is out of parallel with A B. What points, in the plan of a frustum, are the plans of corresponding points is shown by the fig., as the line joining the plans of corresponding points (the line joining B and C or that joining D and E, for instance) will always, if produced, pass through the centre of the circles that constitute the plan of the frustum; the centre of the circles being the plan of the apex of the cone of which the frustum is a part. Which leads us to this; that the distance, actually, between the plans of corresponding points in the plan of a frustum is equal to half the difference of the diameters of its two circles; for, the difference between E B and D C is the sum of D E and B C, and D E and B C are equal; in other words, either D E or B C is half the difference between E B and D C.

FIG. 23a.

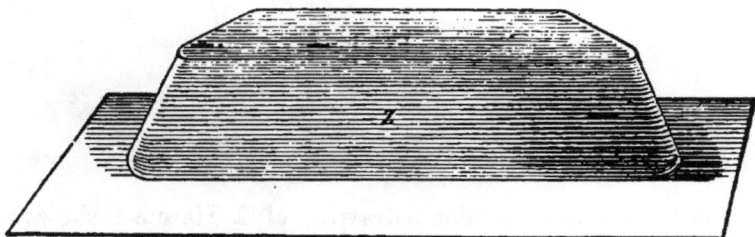

(25.) Let us now consider another equal-tapering body which has top and base parallel, and we will suppose it to have flat parallel sides, flat ends, and round (quadrant) corners. Such a body is represented, except as to dimensions, in Z, Figs. 23a and b; Fig. 23b being a skeleton drawing of the body represented in Fig. 23a. Extending our definition of 'slant' to apply to such a body, a 'slant' becomes the shortest line that can be drawn anywhere on the slanting surface; and 'corresponding points' become, in accordance, the extreme points of such line. Either of the lines F A, G B, E C, H L, or M O represent a slant of the body, and F and A are corresponding points; as also are G and B, E and C, H and L, and M and O. The height of the body is represented by either of the lines F A', G B', E C', H L', or M O'. The plane for the plan being parallel to the M P Q R face (here the top) of Z, the plan of that face is of the same shape as the face. The round-cornered rect-angle A' F' G' D' B' C' of Fig. 29 is the plan to dimensions. For the same reason the plan of the O A T S face (here the base) is of the same shape as that face. The round-cornered rectangle A F G D B C of Fig. 29 is the plan to dimensions. How actually to draw these plans we shall deal with presently as a problem. The two circles constituting the plan of the frustum were concentric, that is, symmetrically disposed with respect to one another, because the frustum

FIG. 23b.

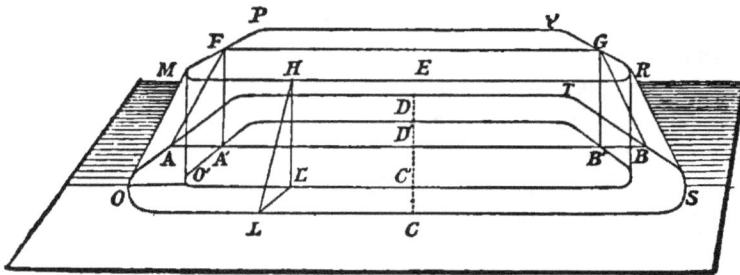

was an equal-tapering body; and the plans of top and base of the body we are now dealing with are symmetrical to each other for the same reason. The two plans (Fig. 29) together are the plan of the body Z.

(26.) Looking at A B C D A' B' C' D' of the skeleton draw-ing (Fig. 23b) as at the plan of Z, we have, just as with the cone frustum, in the point A' the plan of F, in the point B' the plan of G, in the point C' the plan of E, in the point L' the plan of H, and in the point O' the plan of M. Further as in the case of the frustum, if we join any point in the plan of the base, as A, to the plan of its corresponding point A', then we have a right-angled triangle, F A A', of which the

hypotenuse F A represents the slant of the body, F A', one
of the sides containing the right angle, its height, and A' A,
the other side containing the right angle, the distance
between the plans of the corresponding points F and A, which
is also the distance between B and B', C and C', L and L', O
and O', and between plans of corresponding points of the body
anywhere, the body being of equal taper. As with Fig. 21b
what points, in the plan, are the plans of corresponding points
is clear from the fig. Where the plan of the body consists
of straight lines, the plans of corresponding points are always
the extremities of lines joining these straight lines perpendicularly; the extremities of A A', B B', C C', and L L', for
instance. Where the plan of the body consists of arcs, the
plans of corresponding points (compare with cone frustum)
are the extremities of lines joining the arcs, and which, produced, will pass through the centre from which the arcs are
described; the line O O' for instance. To make all this quite
plain, reference should again be made to Fig. 29; also to
Fig. 28, which is the plan of an equal-tapering body with
top and base parallel, and having flat sides, and semicircular
ends. In Fig. 29, A A', B B', C C', D D', are lines joining
the plan lines of the flat sides and ends perpendicularly, and
the extremities of each of these lines are plans of corresponding points, that is to say, A and A' are plans of corresponding
points, as are also B and B', C and C', and D and D'. Also
F and F' are plans of corresponding points, being the
extremities of the line F F' which is a line joining the ends
of the arcs which are the plans of one of the quadrant
corners of the body. Similarly G and G' are plans of corresponding points. In Fig. 28, F F', G G', D D', E E', are lines
joining perpendicularly the plan lines of the flat sides of the
body at their extremities where the semicircular ends begin ;
and F and F', G and G', D and D', E and E' are plans of
corresponding points. Also A and A' are plans of corresponding points, and B and B', seeing that the lines joining
these points, produced, pass respectively through O and O',
the centres from which the semicircular ends are described.

In the cone frustum, the actual distance between the plans of corresponding points was, we saw, equal to half the difference of the diameters of the two circles constituting its plan. Similarly with the body Z of Fig. 23a, and indeed with any equal-tapering body of which the top and base are parallel, if we have the lengths of the top and base given, or their widths, the distance between the plans of corresponding points (number of inches 'out of flue') is always equal to half the difference between the given lengths or widths. Thus, the distance between the plans of corresponding points of Z is equal to half the difference between A B and A' B' (Fig. 29) or between C D and C' D'; and the distance between the plans of corresponding points of the body of which Fig. 28 is the plan, is equal to half the difference between A B and A' B' of that fig., or between F D and F' D'.

Summarising we have

a. In the plans of equal-tapering bodies which have their tops and bases parallel, there is, all round, an equal distance between the plans of corresponding points of the tops and bases.

b. *Conversely.*—If, in the plan of a tapering body with top and base parallel, there is an equal distance all round between the plans of corresponding points of the top and base, then the tapering body is an equal-tapering body, that is, has an equal inclination of slant all round.

c. The plan of a round equal-tapering body having top and base parallel, consists of two concentric circles. The plan of a portion of a round equal-tapering body having top and base parallel, consists of two arcs having the same centre.

The corners of the body Z (Fig. 23a) are portions (quarters) of a round equal-tapering body; their plans are arcs (quadrants) of circles having the same centre.

d. *Conversely.*—If the plan of a tapering body having top and base parallel, consists of two concentric circles, then the body is a frustum of a right cone. Also if the plan of a tapering body having top and base parallel, consists of two arcs having the same centre, then the body is a portion of a frustum of a right cone.

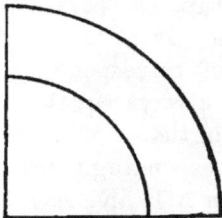

The plan of each end of the tapering body represented in plan in Fig. 28 consists of two arcs (semicircles) having the same centre; the ends are portions (halves) of a frustum of a right cone. The plan of each corner of the tapering body Z (Fig. 23a) consists of two arcs (quadrants, Fig. 29) having the same centre; the corners are portions (quarters) of a frustum of a right cone. The fig. annexed represents a quadrant corner in plan separately.

We conclude the chapter with some problems.

PROBLEM XII.

Given the height and slant of an equal-tapering body with top and base parallel; to find the distance between the plans of corresponding points of the top and base (number of inches ' out of flue').

Let C A' (Fig. 25) be the given height. Draw A'B perpendicular to A' C; with C as centre and the given slant as

FIG. 25.

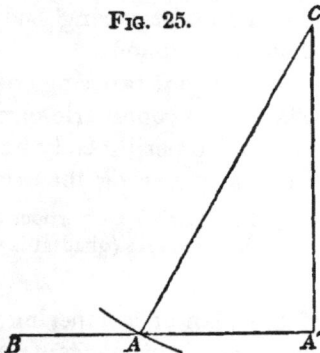

radius, describe an arc cutting B A' in A. Then A A' is the distance required.

PROBLEM XIII

*Given the height of an equal-tapering body with top and base
parallel, and the inclination of slant (number of inches ' out of
flue'); to find the distance between the plans of corresponding
points of the top and base.*

Let C A' (Fig. 26) be the given height. Through A' draw
a line A' B perpendicular to C A'; from any point, D, in A' B
draw a line D E making with A' B an angle equal to that of
the given inclination. From C draw C A parallel to E D
and cutting A' B in A; then A A' is the distance required.

Fig. 26.

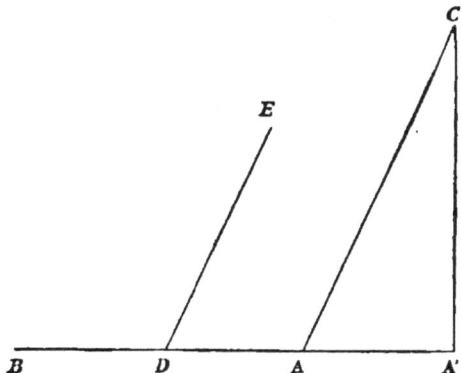

PROBLEM XIV.

*To draw the plan of a round equal-tapering body with top and base
?arallel (frustum of right cone), the diameter of either end
being given and the height and slant.*

Case I.—Given the height and slant and the diameter of the
smaller end.

On any line O B (Fig. 17) set off O C equal to half the
given diameter, and from C draw C D perpendicular to O B.

Mark off C E equal to the given height, and with E as centre and radius equal to the given slant, describe an arc intersecting O B in B; then C B will be the distance in plan between corresponding points anywhere in the frustum; that is to say (by c, p. 55) O C will be the radius for the plan of the smaller end of the frustum, and O B the radius for the plan of the larger end.

CASE II.—Given the height and slant and the diameter of the larger end.

On any line O B (Fig. 27), set off O B equal to half the given diameter, and now work from B towards O instead of from O towards B; thus.   From B draw B C perpendicular to O B.   Mark off B D equal to the given height, and with D as centre and radius equal to the given slant, describe an arc intersecting O B in E; then B E will be the distance in

FIG. 27.

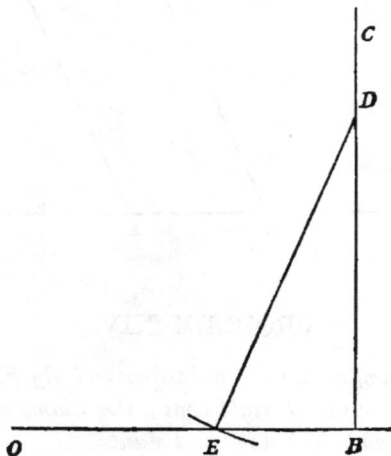

plan of corresponding points anywhere in the frustum; that is to say (by c, p. 55) O B will be the radius for the plan of the larger end of the frustum, and O E the radius for the plan of the smaller end.

PROBLEM XV.

To draw the plan of a round equal-tapering body with top and base parallel (frustum of right cone), the diameter of either end being given, and the number of inches ' out of flue' (distance between plans of corresponding points).

CASE I.—Given the number of inches ' out of flue,' and the diameter of the smaller end.

The radius for the smaller circle of the plan will be half the given diameter; the radius for the larger circle of the plan will be this half diameter with the addition of the number of inches ' out of flue.'

CASE II.—Given the number of inches ' out of flue,' and the diameter of the larger end.

The radius for the larger circle of the plan will be half the given diameter; the radius for the smaller circle of the plan will be the half diameter less the number of inches ' out of flue.'

(27.) It should be noted that with the dimensions given in this problem, we can draw plan only, we could not draw a pattern. To do that we must also have height given, for a plan of small height and considerable inclination of slant is also the plan of an infinite number of other frusta (*plural* of frustum) of all sorts of heights and inclinations of slant.

PROBLEM XVI.

To draw the plan of an oblong equal-tapering body with top and base parallel, and having flat (plane) sides and semicircular ends.

CASE I.—Where the length and width of the top are given, and the length of the bottom.

Commencing with the plan of the top, we know from § 25 that it will be of the same shape as the top; we have there-

fore to draw that shape. On any line A B (Fig. 28) mark off
A B equal to the given length of the top. From A set off
A O, and from B set off B O' each equal to half the given
width of the top. Through O and O' draw lines perpendicular to A B; and with O and O' as centres and O A or
O' B as radius describe arcs meeting the perpendiculars in
D F and E G. As D F and E G pass through the centres
O and O' respectively they are diameters, and the arcs are
semicircles; these diameters, moreover, are each equal to
the given width. Join D E, F G, and the plan of the top
is complete.

FIG. 28.

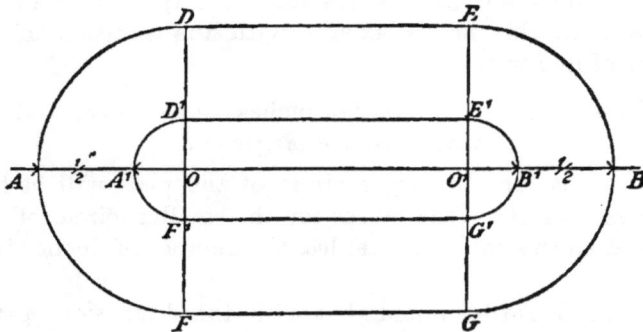

The plan of the base will be of the same shape as the
base, and we will suppose it smaller than the top. What we
have then to do is to draw a figure of the same shape as the
base, and to so place it in position with the plan of the top
that we shall have a complete plan of the body we are
dealing with. By a, p. 55, we know that the distances between
the plans of corresponding points of the top and base all
round the full plan will be equal. We have therefore first
to ascertain the distance between the plans of any two
corresponding points. This by § 26 will in the present
instance be equal to half the difference between the given

lengths of the top and base. Set off this half-difference, as the base is smaller than the top, from A to A'. Then with O and O' as centres, and O A' as radius, describe the semi-circles D' A' F', E' B' G'. Join D' E', F' G', and we have the required plan of the body.

CASE II.—Where the length and width of the top are given, and the height and slant, or the height and the inclination of the slant (number of inches 'out of flue ').

First draw the plan of the top as in Case I. Then if the height and slant are given, find by Problem XII. the distance between the plans of corresponding points. If the height and inclination of slant are given, find the distance by Problem XIII. If the inclination of the slant is given in the form of 'out of flue,' the number of inches 'out of flue ' is the required distance. Set off this distance from A to A' in the fig. of Case I., and complete the plan as in Case I.

CASE III.—Where the length and width of the base (bottom) are given, and the height and slant, or the height and the inclination of the slant.

On any line A B (Fig. 28) mark off A' B' equal to the given length of the bottom. From A' set off A' O and from B' set off B' O' each equal to half the given width of the bottom. Through O and O' draw indefinite lines D F, E G perpendicular to A B; and with O and O' as centres, and O A' as radius describe the semicircles F' A' D', G' B' E', join D' E', F' G', and we have the plan of the bottom. Now by Problem XII. or Problem XIII., as may be required, find the distance between the plans of corresponding points, or take the number of inches 'out of flue,' if this is what is given. Set off this distance from A' to A. With O and O' as centres and O A as radius describe semicircles meeting the perpendiculars through O and O' in D and F and in E and G. Join D E, F G, and the plan of the body is completed.

PROBLEM XVII.

To draw the plan of an oblong equal-tapering body with top and base parallel, and having flat sides, flat ends, and round (quadrant) corners.

CASE I.—Where the length and width of top and bottom (base) are given.

Draw any two lines A B, C D (Fig. 29) perpendicular to each other and intersecting in O. Make O A and O B each equal to half the length of the top, which we will suppose

FIG. 29.

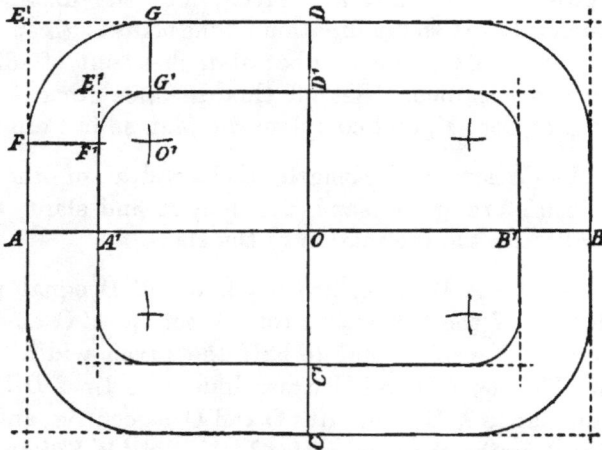

FIG. 29.

larger than the bottom, and O A′ and O B′ each equal to half the length of the bottom. Also make O C and O D each equal to half the width of the top, and O C′ and O D′ each equal to half the width of the bottom. Through C, D, C′, and D′ draw lines parallel to A B, and through A, B, A′, and B′ draw lines parallel to C D and intersecting the lines parallel to A B. We have now two rectangles or oblongs, and we require to draw th round corners, which are quarters of circles.

From the intersecting point E along the sides of the rectangle mark equal distances E F and E G, according to the size of quadrant corners required. With F and G as centres and E F or E G as radius, describe arcs intersecting in O'; and with O' as centre and same radius, describe the arc F G, which will be a quadrant because if the points F and G be joined to O' the angle F O' G will be a right angle (p. 21). Draw F F' parallel to A B and G G' parallel to C D, and with O' as centre and radius O' F' describe the arc F' G', which also will be a quadrant. We have now the plan of one of the quadrant corners; the other corners can be drawn in like manner.

(27a.) It is important to notice that the larger corner determines the smaller one. In practice it is therefore often best to draw the smaller corner first, otherwise it may sometimes be found, after having drawn the larger corner, that it is not possible to draw the smaller curve sufficiently large, if at all. To draw the smaller corner first, mark off from the intersecting point E' equal lengths E' F', E' G', according to the size determined on for the corner. With F' and G' as centres and E' F' or E' G' as radius describe arcs intersecting in O'. Then O' will be the centre for the smaller corner. It will also be the centre for the larger corner, which may be described in similar manner to the smaller corner in the preceding paragraph.

CASE II.—Where the dimensions of the top are given and the height and slant, or the height and the inclination of the slant.

Draw the plan of the top, A D B C. Find the distance between the plans of corresponding points of the top and base by Problem XII., or Problem XIII., according to what is given; and set off this distance, as the base is smaller than the top, from A and B inwards towards O on the line A B, and from D and C inwards towards O on the line D C. Complete the plan by the aid of what has already been explained.

CASE III.—Where the dimensions of the bottom are given, and the length and slant, or the height and the inclination of the slant.

Draw the plan of the bottom, A' D' B' C', find the distance between the plans of corresponding points of top and bottom, set this off outwards from A', D', B', and C' and complete the plan by aid of what has already been stated.

PROBLEM XVIII.

To draw the plan of an oval equal-tapering body with top and base parallel, the length and width of the top and bottom being given.

Draw (Fig. 30) any two lines A B, C D intersecting each other at right angles in O. Make O A and O B each equal

FIG. 30.

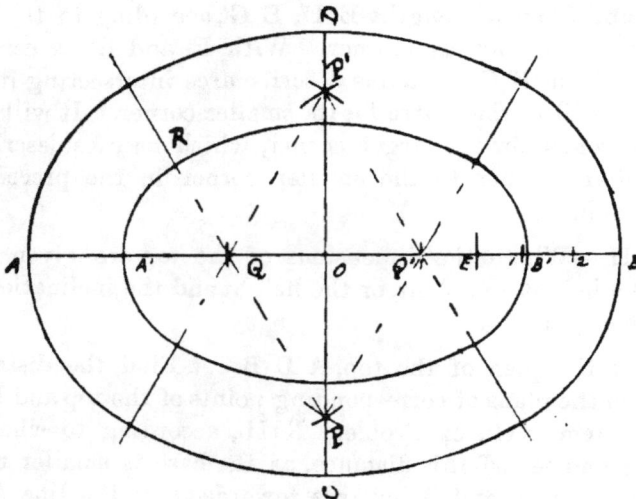

to half the given length of the larger oval (top or bottom, as may be), and O C and O D each equal to half its given width. A B and C D will be the *axes* of the oval. From A, on A B, mark off A E equal to C D the width of the oval, and divide E R into three equal parts. With O as centre and radius equal to two of these parts, as from E to 2, describe arcs cutting A B in Q and Q'. With Q and Q' as centres and Q Q' as radius describe arcs intersecting in P and P'; and from P and P' draw lines of indefinite length through Q and Q'. With P and P' as centres and radius P D describe arcs (the side arcs), their extremities terminating in the lines drawn through Q and Q'; and with Q and Q' as centres, and radius Q A, describe arcs (the end arcs) to meet the extremities of the side arcs. This completes the plan of the larger oval.

To draw the plan of the smaller oval. Make O A' and O B' each equal to half the length of the smaller oval, and with Q and Q' as centres and Q A' as radius describe the end curves, their extremities terminating, as do the outer end-curves, in the lines drawn through Q and Q'; the point R is an extremity of one of the smaller curves. With P and P' as centres and radius P R, describe the side curves. The plan of the oval equal-tapering body is then complete; of which either the larger or smaller ovals are plan of top and bottom according to the purpose the article may be required for.

(28.) The plans of corresponding points in the plan of an oval equal-tapering body will be the extremities of any line joining the inner and outer curves anywhere, and that, produced, will pass through the centre from which the curves where joined by the line are described.

To draw the plan of an oval equal-tapering body with top and base parallel, other dimensions than the above may be given. For instance the top or bottom may be given, and either the height and slant, or the height and the inclination of the slant (number of inches out of flue). It will be a useful practice for the student to work out these cases for himself by the aid of the instruction that has been given.

F

CHAPTER VI.

PATTERNS FOR ARTICLES OF EQUAL TAPER OR INCLINATION OF SLANT, AND HAVING FLAT (PLANE) SURFACES.

(CLASS I. *Subdivision b.*)

DEFINITION.

(29.) Pyramid.—A pyramid is a solid having a base of three or more sides and triangular faces meeting in a point above that base, each side of the figure forming the base being the base of one of the triangular faces, and the point in which they all meet being the *apex*. The shape of the base of a pyramid determines its name; thus a pyramid with a triangular base is called a *triangular* pyramid; with a square base, a *square* pyramid; with a hexagon base, an *hexagonal* pyramid (Fig. 31); and so on. The centre of the base of a pyramid is the point in which perpendicular lines bisecting all its sides will intersect. If the apex of a pyramid is perpendicularly above the centre of its base, the pyramid is a *right* pyramid (Fig. 31 represents a right pyramid), in which case the base is a regular polygon and the triangular faces are all equal and all equally inclined. In a pyramid, the line joining the apex to the centre of the base is called the *axis* (the line V V', Fig. 31) of the pyramid.

(30.) An important property that a right pyramid possesses is that it can be *inscribed in a right cone.*

(31.) A pyramid is said to be inscribed in a cone when both the pyramid and the cone have a common apex, and the base of the pyramid is inscribed in the base of a cone; in other words, when the angular points of the base of the pyramid are on the circumference of the base of the cone and the apex of cone and pyramid coincide.

(32.) Fig. 31 shows a right pyramid inscribed in a right cone. The apex V is common to both pyramid and cone, and the A, B, C, &c., of the base of the pyramid are on the circumference of the base of the cone. Also the axis V V' is common to both cone and pyramid. Further, the edges V A, V B, V C, &c., of the pyramid are lines on the surface of the cone, such lines or edges being each a slant of the cone, or in other words a radius of the pattern of the cone in

FIG. 31.

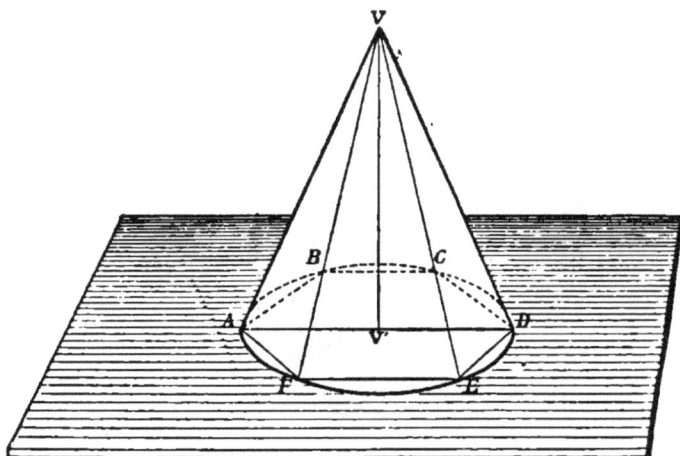

which the pyramid is inscribed. It hence follows, that if the pattern of the cone in which a right pyramid is inscribed be set out with the lines of contact of cone and pyramid, as V A, V B, &c., on it, and the extremities of these lines be joined, we shall have the pattern for the pyramid. Thus, the drawing a pattern for a right pyramid resolves itself into first determining the cone which circumscribes the pyramid, and next drawing the pattern of that cone with the lines of contact of pyramid and cone upon it.

F. 2.

PROBLEM XIX.

To draw the pattern for an hexagonal right pyramid, its height and base being given.

Draw (Fig. 32a) the plan A B C D E F of the base of the pyramid, which will be of the same shape as the base (*see* Chap. V.) ; the base in fact will be its own plan. Next draw any two lines O A, B A (Fig. 32b), perpendicular to each other and meeting in A ; make A B equal to the radius of the circumscribed circle (Fig. 32a), and A O equal to the given height of the pyramid. Join B O ; then B O is a slant of the

Fig. 32a.   Fig. 32b.

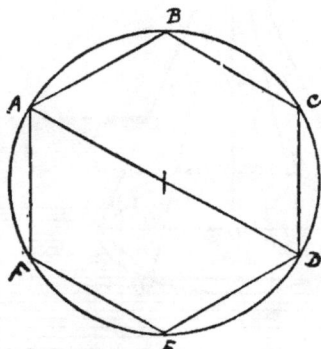 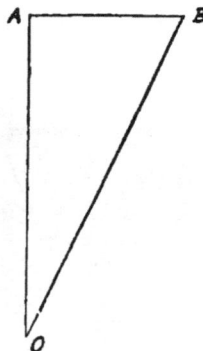

cone in which the pyramid can be inscribed, that is to say, is a radius of the pattern of that cone. The line B O is also a line of contact of the cone, in which the pyramid can be inscribed, that is, is one of the edges of the pyramid.

To draw the pattern (Fig. 32c). With any point O' as centre and B O (Fig. 32b) as radius, describe an arc A D A, and in it take any point A. Join A O', and from A mark off A B, B C, C D, D E, E F, and F A, corresponding to A B, B C, C D, D E, E F, and F A of the hexagon of Fig. 32a, and join the points B, C, D, E, F and A to O'. Join A B, B C, C D, D E, E F, and F A, by straight lines ; and the figure bounded by O' A, the straight lines from A to A, and A O', will be the pattern required. The lines B O', C O', &c., correspond to the edges of the pyramid, and show the lines on which to ' bend up ' to get the faces of the pyramid, the lines O' A and O' A then butting together to form one edge.

Similarly the pattern for a right pyramid of any number of faces can be drawn, the first step always being to draw the plan of the base of the pyramid ; the circle passing through the angular points of which will be the plan of the base of the cone in which the pyramid can be inscribed.

FIG. 32c.

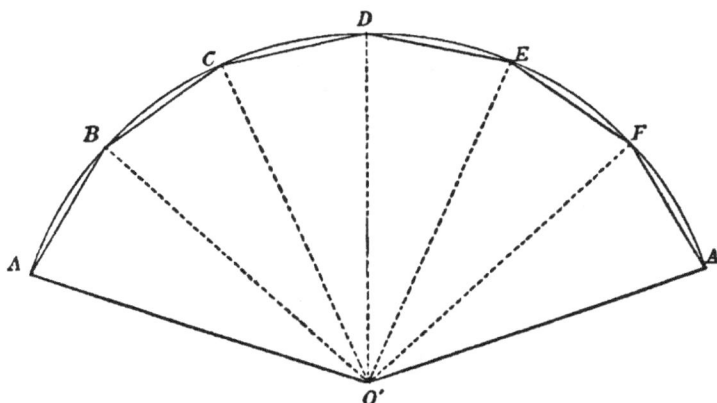

Suppose, instead of the dimensions from which to draw the plan of the base of the pyramid, the actual plan be given. The centre from which to strike the circumscribing circle can then be found by the Definition § 29.

DEFINITION.

(33.) Truncated pyramid. Frustum of pyramid.—If a pyramid be cut by a plane parallel to its base, the part containing the apex will be a complete pyramid, and the other part will be a tapering body, the top and base of which are of the same shape but unequal. This tapering body is called a *truncated pyramid*, or a *frustum of a pyramid*. The faces of a truncated pyramid which is a frustum of a right pyramid are all equally inclined. In Fig. 33 is shown such a frustum standing on a horizontal plane.

(34.) Comparison should here be made between this defini-

FIG. 33.

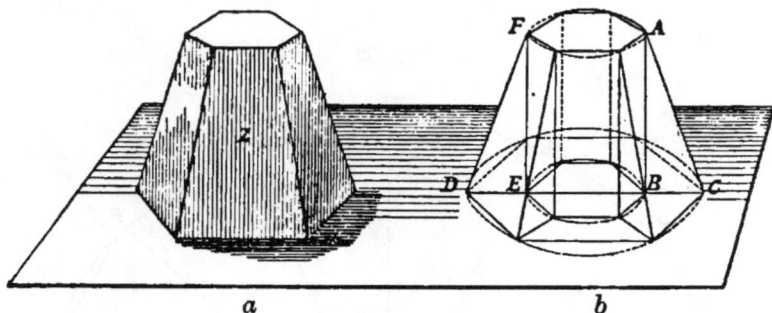

tion and that of a frustum of a cone (*see* § 12), which it closely follows; also between Fig. 21 and Fig. 33.

(35.) Articles of equal taper or inclination of slant and having flat (plane) surfaces and top and base parallel (hexagonal coffee-pots; hoods; &c.), are portions of right pyramids (truncated pyramids), or portions of truncated pyramids.

(36.) Exactly as a pyramid can be inscribed in a cone, so a truncated pyramid can be inscribed in a frustum of a cone, and the edges of the truncated pyramid are lines on the surface of that frustum. The skeleton drawing, Fig. 33*b*, shows a right truncated pyramid inscribed in a cone frustum. It also represents the plan of the cone frustum, and that of the pyramid frustum, with the lines of project'on (*see* Chapter V.), of the smaller end of the latter on to th. 'orizontal plane. This inscribing in a cone gives an easy construction for setting out the pattern of a truncated pyramid; which construction is, to first draw the pattern for the pyramid of

which the truncated pyramid is a portion; and then mark
off on this pattern the pattern for the pyramid that is cut off.
Here again comparison should be made with what has been
stated about the pattern of a frustum of a cone (*see* § 16), and
the resemblance noted.

PROBLEM XX.

*To draw the pattern for an equal-tapering body made up of flat
surfaces (truncated right pyramid), the height, and top and
bottom being given.*

Suppose the equal-tapering body to be hexagonal.

To draw the required plan of the frustum. The plans of
the top and bottom are respectively of the same shape as
the top and bottom (§ 25). Draw (Fig. 34a) A B C D E F the
larger hexagon (Problem X., Chap. II.) and its diagonals

Fig. 34a.   Fig. 34b.

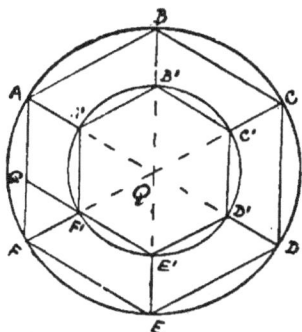
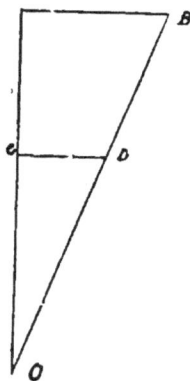

A D, B E, C F, intersecting in Q. On any one of the sides of
this hexagon mark off the length of a side of the smaller
hexagon, as A G on A F, and through G draw G F′ parallel
to the diagonal A D, and cutting the diagonal F C in F′.
With Q as centre and Q F′ as radius describe a circle. The

points in which this cuts the diagonals of the larger hexagon
will be the angular points of the smaller hexagon. Join each
of these angular points, beginning at F′, to the one next
following, as F′ E′, E′ D′, &c. Then F′ E′ D′ C′ B′ A′ is the
plan of the smaller hexagon, and, so far as needed for our
pattern, the plan of the 'equal-tapering body made up of
flat surfaces' is complete. The lines A A′, B B′, C C′, will be
the plans (see and compare lines D E and B C of Fig. 33b) of
the slanting edges of the frustum.

Next draw (Fig. 34b) two lines O A, B A, perpendicular to
each other, and meeting in A ; make A B equal to the radius
of the circle circumscribing the larger hexagon of plan, and

Fig. 34c.

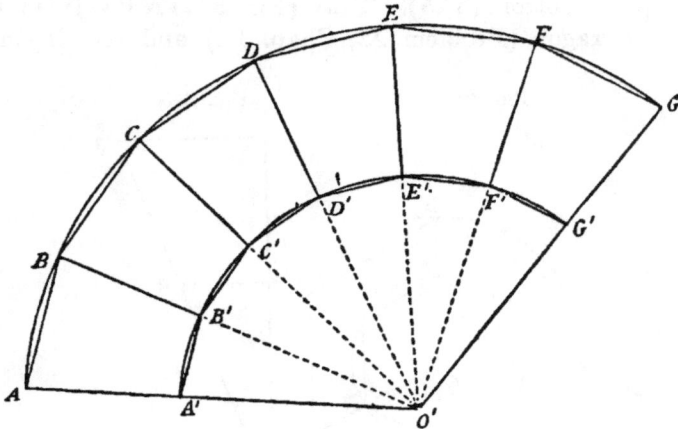

A C equal to the given height of the body. Through C draw
C D perpendicular to A O, and make C D equal to the radius
of the circle which passes through the angular points of the
smaller hexagon (Fig. 34a). Join B D, and produce it to
meet A O in O.

To draw the required pattern (Fig. 34c). Draw any line
O′ A, and with O′ as centre and O B, O D (Fig. 34b) as radii,
describe arcs A D G, A′ D′ G′ (Fig. 34c). Then take the

straight line length A B (Fig. 34a), and set it off as a chord from A (Fig. 34c) on the arc A D G. Do the same successively, from point B, with the straight line lengths (Fig. 34a) B C, C D, D E, E F, F G, the terminating point of each chord as set off, being the starting point for the next, the chord F G (Fig. 34c) corresponding to the straight line length F A (Fig. 34a). Join (Fig. 34c) the points B, C, D, E, F and G to O' by lines cutting the arc A' D' G' in B', C', D', E', F', and G'. Join A B, B C, C D, &c., and A' B', B' C', C' D', &c., by straight lines; then A D G G' D' A' is the pattern required.

The frustum of pyramid is here hexagonal, but by this method the pattern for any regular pyramid cut parallel to its base can be drawn. The next problem will show methods for larger work.

(37.) If O' A D G (Fig. 34c), the pattern for the cone in which the frustum of pyramid is inscribed, be cut out of zinc or other metal, and small holes be punched at the points A B. C, &c., and A', B', C', &c.; and if the cone be then made up with the lines O' A, O' B', &c., marked on it inside, and wires be soldered from hole to hole successively to form the top and bottom of the truncated pyramid, then (1) the whole pyramid of which the truncated pyramid is a portion, (2) the pyramid that is cut off, as well as (3) the truncated pyramid, will be clearly seen inscribed in the cone. The making such a model will amply repay any one who desires to be thoroughly conversant with the construction of articles of the kind now under consideration.

PROBLEM XXI.

To draw, **without using long radii,** *the pattern for an equal-tapering body made up of flat surfaces, the height and top and bottom being given.*

Again suppose the body to be hexagonal.

Case I.—For ordinarily large work.

Draw (Fig 35a) the plan as by last problem. Next join A B', and draw B' G perpendicular to A B' and equal to the given height of the body. Also draw B' H perpendicular to

B B'; and equal to the given height. Join A G and B H; then A G and B H are the true lengths of A B' and B B' respectively.

FIG. 35a.

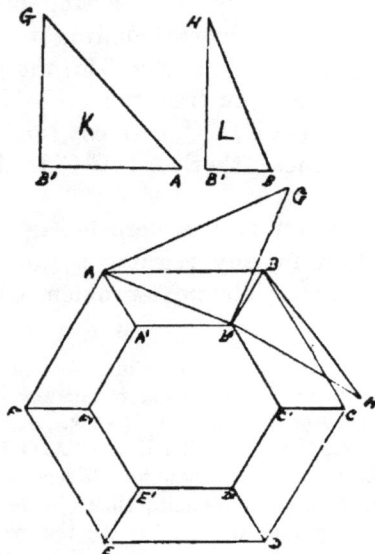

FIG. 35b.

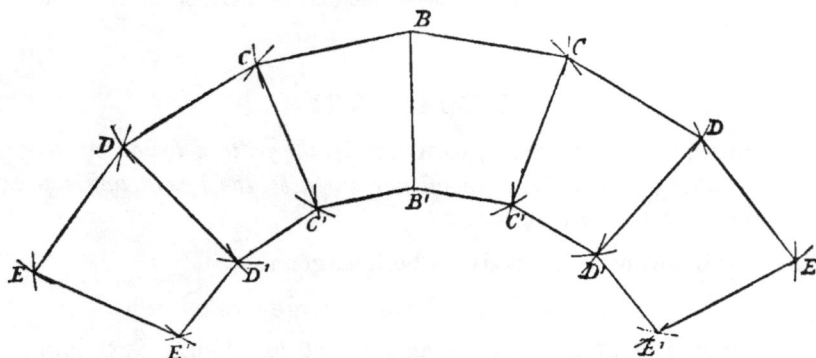

To draw the pattern. Draw (Fig. 35b) B B' equal to B H (Fig. 35a). With B as centre and radius equal to B A

(Fig. 35a), and with B' as centre and radius equal to G A (Fig. 35a) describe arcs intersecting in C, right and left of B B'. With B as centre and the same G A as radius, and with B' as centre and radius B' A', describe arcs intersecting in C', right and left of B' B. Join B C, C C', C' B' (Fig. 35b); then B B' C' C is the pattern of the face B B' C' C in plan (Fig. 35a). The other faces C D C' D', B C B' C', &c. (Fig. 35b) of the frustum are described in exactly similar manner, G A being the distance between diagonally opposite points of any face as well as of the face B B' C' C. The triangles A B' G, B B' H, can be drawn apart from the plan, as shown at K and L.

(38.) It should be observed that if the pattern is truly drawn, the top and bottom lines of each face will be parallel, as B C, B' C', of face B C B' C' (Fig. 35 b) ; and that this gives an easy method of testing whether the pattern has been accurately drawn.

CASE II.—For very large work; where it is inconvenient to draw the whole of the plan.

Draw A B (Fig. 36a) equal in length to the end-line of one of the faces of the frustum at its larger end, and produce it both ways. With B as centre and radius B A describe a semicircle, which divide into as many equal parts as the frustum has faces (Problem IX.). Here it is hexagonal, and the points of division working from point 1, are 2, C, 4, &c. Through the second division point, here C, draw a line to B, then A B C is the angle made in plan by two faces of the frustum one with another, and A B, B C are two adjacent end-lines of the plan of its larger ends. Bisect the angle A B C (Problem VIII.) by B E ; and draw a line C C' from C making the angle C' C B equal to the angle C B E (Problem I.). On B C set off B D equal to the end-line of one of the faces of the frustum at its smaller end ; and draw D C' parallel to B E, cutting C C' in C'. Through C' draw C' B' parallel to C B and meeting B E in B'; and draw B' A' parallel to B A and equal to B D or C' B'. Join A' A, and we

have in A' A B C C' the plan of two adjacent faces of the
tapering body or truncated pyramid. Next from B' let fall
B' G perpendicular to A B, and make G F equal to the given
height of the frustum. Join F B', then F B' is the true
length of B' G. Through B' draw B' H perpendicular to
B' B and equal to the given height. Join H B, then H B is
the true length of B B' one of the edges of the frustum.

To draw the pattern. Draw (Fig. 36b) B B' equal to H B
(Fig. 36a). With B and B' as centres and radii respectively
B G and F B' (Fig. 36a) describe arcs intersecting in G
(Fig. 36b). Join B G and produce it making B A equal to

Fig. 36a.   Fig. 36b.

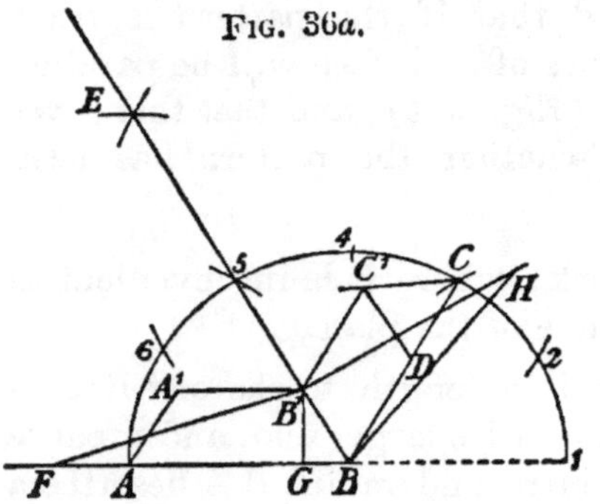

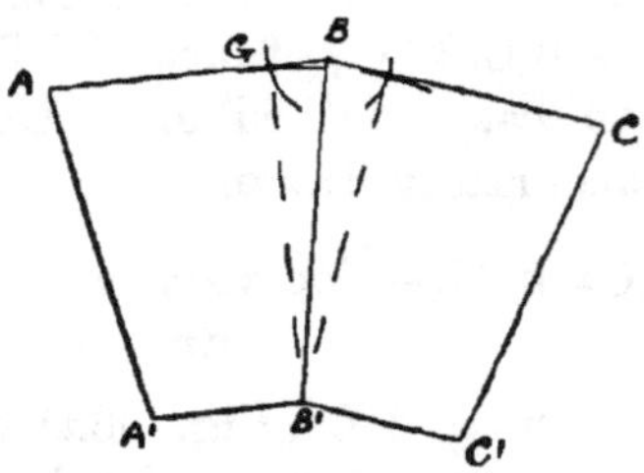

B A (Fig. 36a). Through B' draw B' A' parallel to B A and
equal to B' A' (Fig. 36a), and join A A'; then B A A' B' will
be the pattern of one face of the frustum. The adjacent face
B C C' B' is drawn in similar manner. The fig. shows the
pattern for two faces only of the equal-tapering body,
because in cases where this method would have to be
employed, two faces are probably the utmost that could be
cut out in one piece. Sometimes each face would have to be
cut out separately, or perhaps even one face would have to
be in portions. *Any point* in B' A' (Fig. 36a) instead of B' can
be chosen from which to let fall a perpendicular to B A, and
the true length of B' G found as explained. The choice of
position depends upon the means at hand for drawing large

arcs, the radius of the arc B' G (Fig. 36b) increasing in length as the point G approaches nearer to A. If the pattern be truly drawn, B A will be perpendicular to B' G.

It must not be forgotten that these methods are, both of them, quite independent of the number of the sides of the pyramid. Also it should be noted that B E does not of necessity pass through a division point, nor of necessity is C C' parallel to A B. These are coincidences arising from the frustum being here hexagonal.

PROBLEM XXII.

To draw the pattern for an oblong or square equal-tapering body with top and base parallel, and having flat sides and ends. (The bottom is here taken as part of the body, and the whole pattern is in one piece.)

NOTE.—This problem will be solved in the problem next following. We adopt this course because the article there treated of is so important an example of the oblong equal-tapering bodies in question, that it is desirable to make that, the special problem, the primary one, and this, the general problem, secondary to it. Its solution will be found at the end of Case I.

PROBLEM XXIII.

To draw the pattern for a baking-pan.

A baking-pan has not only to be water-tight, but also to stand heat; hence when made in one piece the corners are seamless.

CASE I.—Where the length and width of the bottom, the width of the top, and the slant are given.

Draw two lines X X, Y Y, intersecting at right angles in O (Fig. 37); make O A' and O B' each equal to half the length of the bottom, and O C' and O D' each equal to half its width. Through C' and D' draw lines parallel to Y Y; also through A' and B' draw lines parallel to X X and inter-

Fig. 37.

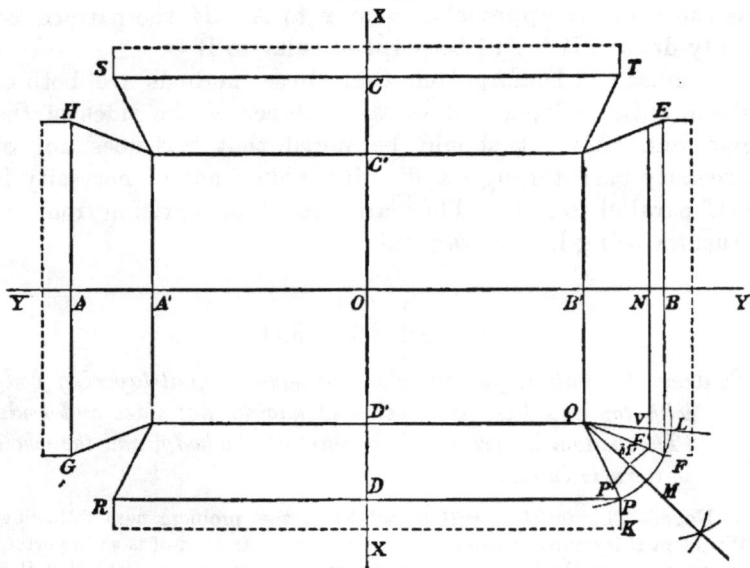

Fig. 38.

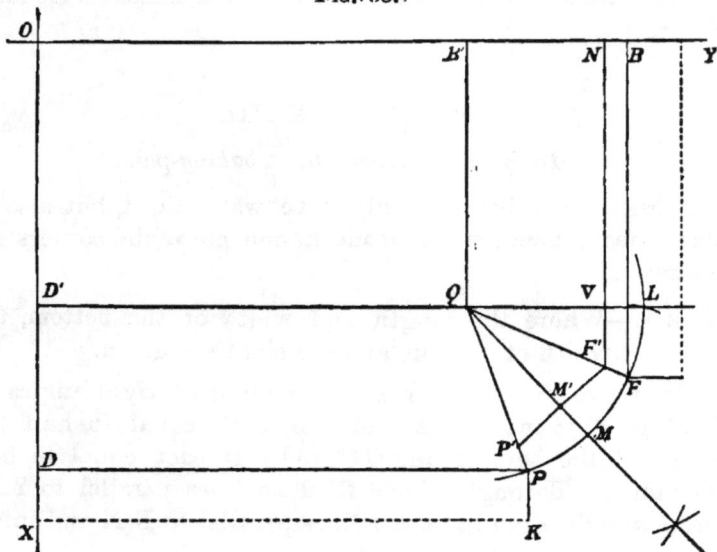

(QUARTER OF FIG. 37, ENLARGED.)

secting the lines drawn through C' and D'; we get by this a rectangular figure, which is the shape of the bottom. Make A' A, B' B, C' C, and D' D each equal to the given slant; through B and A draw E F and G H parallel to X X; through C and D draw S T and R P parallel to Y Y; and make B E, B F, A H, and A G each equal to half the given width of the top. Join Q F and with Q, which is one of the angular points of the bottom, as centre and radius Q F describe an arc cutting P R in P. Join Q P (the working can here be best followed in Fig. 38); bisect the angle F Q P by Q M (Problem VIII., Chap. II.); and draw a line Q L making with F Q an angle equal to the angle F Q M. The readiest way of doing this is by continuing the arc P F to L, then setting off F L equal to F M, and joining L Q. Now on B B' set off B N equal to the thickness of the wire to be used for wiring, and through N draw N F' parallel to E F (Fig. 37) and cutting Q L and Q F in V and F'; make Q P' equal to Q F', and Q M' equal to Q V; and join P' M' and M' F'. Repeat this construction for the other three corners and the pattern will be completed. This is not done in Fig. 37, for a reason that will appear presently. We shall quite understand the corners if we follow the letters of the Q corner in Fig. 38. These are B F F' M' P' P. The dotted lines drawn outside the pattern (Fig. 37) parallel to E F, P R, G H, and S T show the allowance for fold for wiring. It is important that the ends of this allowance for fold shall be drawn, as, for instance, P K (both figs.), perpendicular to their respective edges.

Now as to bending up to form the pan. When the end adjacent and the side adjacent to the corner Q come to be bent up on B' Q and D' Q so as to bring F Q and P Q into junction, it is evident that as F Q and P Q are equal we shall obtain a true corner. To bring F Q and P Q into junction it is likewise manifest that the pattern will have also to be bent on the lines F' Q, P' Q, and M' Q. This fold on each other of P' Q M' and F' Q M' is generally still further bent round against Q V.

(39.) The truth of the pan when completed, and the ease with which its wiring can be carried out, depend entirely on the accuracy of the pattern at the corners. This must never be forgotten. In marking out a pattern, only one corner need be drawn, as the like to it can be cut out separately and used to mark the remaining corners by. The points R, S, and T, the fixing of which will aid in this marking out, can be readily found, thus:—For the R corner to come up true it is clear that D R will have to be equal to D P, from which we learn that D P is half the length of the top. Having then determined the point P we have simply to set off D R, C S, and C T each equal to D P.

If the pattern Fig. 37 were completed with the corners as at R, S, and T, that would be the solution of Problem XXII., and the ends and sides being bent up, we should get an oblong equal-tapering body with top and bottom parallel and having flat sides and ends.

CASE II.—Where the length and width of the bottom, the length of the top, and the slant are given.

The only difference between this case and the preceding is that D P (Fig. 37), half the length of the top, is known instead of B F the half-width of top. To find the half-width of top; with Q as centre and Q P as radius describe an arc cutting B F in F. Then B F will be half the width of the top, just as (§ 39, previous case) we saw that D P was half the length of the top. The remainder of the construction is now as in Case I.

(40.) It will be evident that, in this problem, choice of dimensions is not altogether arbitrary. The lines Q F and Q P the meeting of which forms the corner, must always be of the same length. This limits the choice; for with the dimensions of bottom given, and the slant, and the width of the top, the length of the top cannot be fixed at pleasure, but must be such as will bring Q F and Q P into junction; and *vice versâ* if the length of the top is given. It is unnecessary to follow the limit with other data.

Case III.—Where the length and width of the top, and the
slant and the height are given.

The data in this case and the next are the usual data when
a pan has to be made to order. The difference between this
case and the preceding is that the size of· the bottom is not
given, but has to be determined from the data. Excepting
as to finding the dimensions of this, the case is the same as
Cases I. and II. All that we have now to do is therefore to
find the dimensions of the bottom, thus :—

Fig. 39. Fig. 40.

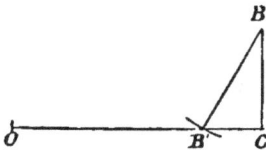 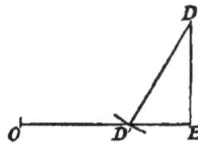

Draw O C (Fig. 39) equal to half the given length of the
top, and through C draw C B perpendicular to O C and equal
to the given height. With B as centre and radius equal to
the given slant describe an arc cutting O C in B'. Then O B'
is the required half-length of the bottom.

Next draw O E (Fig. 40) equal to half the given width of
the top, and through E draw E D perpendicular to O E and
equal to the given height. With D as centre and radius
equal to the given slant describe an arc cutting O E in D'.
Then O D' is the required half-width of the bottom.

Case IV.—Where the length and width of the top and the
length and inclination of the slant are given.

This is a modification of Case III.

Let B B' (Fig. 39) be the length of the given slant, and
the angle that B B' make with B' C be the inclination of the
slant. Through B draw B C perpendicular to B' C. Then
half the given length of the top less B' C will be the required
half-length of the bottom ; and similarly half the given
width of the top less B' C will be the required half-width of
the bottom.

G

PROBLEM XXIV.

To draw the pattern for an equal-tapering body with top and base
parallel, and having flat sides and ends (same as Problem
XXII.), but with bottom, sides, and ends in separate pieces;
the length and width of the bottom, the width of the top, and
the slant being given.

To draw the pattern for the end.

Draw B B' (Fig. 41) equal to the given slant, and through
B and B' draw lines C D and C' D' perpendicular to B B'.
Make B C and B D each equal to half the given width of the
top; also make B' C' and B' D' each equal to half the given
width of the bottom. Join C C' and D D'; then C' C D D'
will be the end pattern required.

FIG. 41.   FIG. 42.

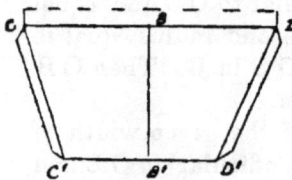

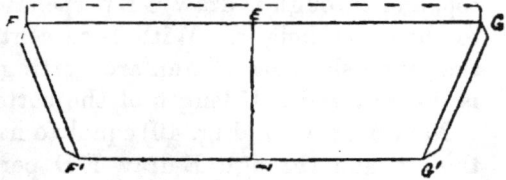

END PATTERN.   SIDE PATTERN.

To draw the pattern for the side.

Draw E E' (Fig. 42) equal to the given slant, and through
E and E' draw F G and F' G' perpendicular to E E'. Make
E' F' and E' G' each equal to half the given length of the
bottom, and with F' and G' as centres and radius D' D
(Fig. 41) describe arcs cutting F G in points F and G. Join
F' F and G' G; then F' F G G' will be the side pattern
required.

It should be noted that, as in the preceding problem, the
width of the top determines the length of the top and *vice
versâ*, also that lines such as D D', G G' must be equal.

(41.) If the ends of the body are seamed ('knocked up') on to the sides, as is usual, twice the allowance for lap shown in Fig. 41, must be added to the figure $C\,C'\,D'\,D$. Similarly if the sides are seamed on to the ends, a like double allowance for lap must be added to $F\,F'\,G'\,G$, instead of to the end pattern.

CHAPTER VII.

PATTERNS FOR EQUAL-TAPERING ARTICLES OF FLAT AND CURVED SURFACES COMBINED.

(CLASS I. *Subdivision c.*)

From what has been stated about the plans of equal-tapering bodies, and from *d*, p. 55, it will be evident that the curved surfaces of the articles now to be dealt with, are portions of frusta of cones.

PROBLEM XXV.

To draw the pattern for an equal-tapering body with top and bottom parallel, and having flat sides and equal semicircular ends (an 'equal-end' pan, for instance), the dimensions of the top and bottom of the body and its height being given.

Four cases will be treated of; three in this problem and one in the problem following.

CASE I.—Patterns when the body is to be made up of four pieces.

We may suppose the article to be a pan. Having drawn (Fig. 43) the plan A R C S A' D' C' B' by the method of Problem XVI. (as well as the plan, the lines of its construction are shown in the fig.), let us suppose the seams are to be at A, B, C, and D, where indeed they are usually placed. Then we shall require one pattern for the flat sides and another for the curved ends.

To draw the pattern for the sides.

Anywhere in A D (Fig. 43) and perpendicular to it draw

F F', and make F H equal to the given height; join F' H; then F' H is the length of the slant of the article, and therefore the width of the pattern for the sides.

Fig. 43.

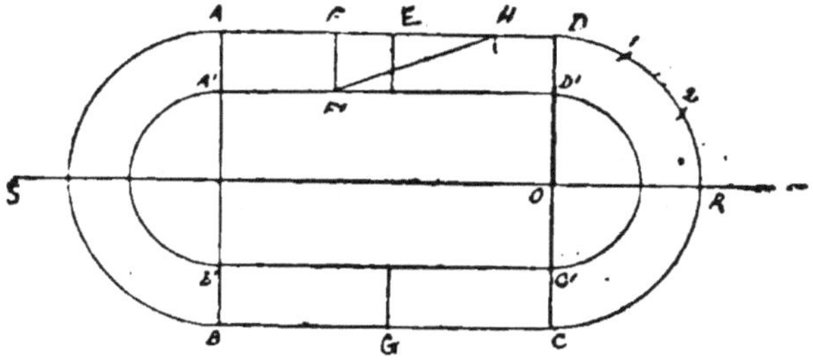

Draw A A' (Fig. 44) equal to F'H (Fig. 43). Through A and A' draw lines perpendicular to A A'. Make A D (Fig. 44) equal to A D (Fig. 43), and draw D D' parallel to

Fig. 44.

Side Pattern.

A A'. The rectangle A D D' A' will be the side pattern required. The fig. also shows extras for lap.

To draw the pattern for the ends.

Draw (Fig. 45a) two lines D A, O A, perpendicular to one another and meeting in A; make D A equal to O D (Fig. 43), the larger of the radii of the semicircular ends; and on A O set off A G equal to the given height. Draw G D' perpendicular to A G and equal to O D' (Fig. 43), the smaller of the radii of the semicircular ends; join D D' and produce it, meeting A O in O. Then with any point O' (Fig. 45b), as centre, and O D (Fig. 45a) as radius, describe an arc D C, and with the same centre and radius O D' (Fig. 45a) describe an arc D' C'. Draw any line D O', cutting the arcs in points D and D'. Divide D R (Fig. 43) into any number of equal parts, say three. From D (Fig. 45b) mark off these three

Fig. 45a.

Fig. 45b.

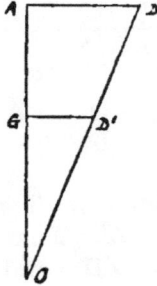
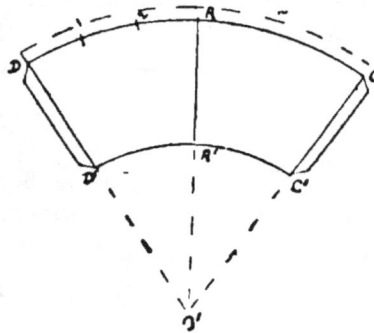

dimensions to R, make R C equal to D R, and join C O'; then D C C' D' will be the end pattern required.

If R be joined to O', then D D' R' R will be half the pattern. The centre line R R' is very useful, because, in making up the article, the point R' must meet the line S R (axis) of Fig. 43, otherwise the body will be twisted in consequence of the bottom not being true with the ends.

CASE II.—Pattern when the body is to be made up of two pieces.

Secondly, suppose the article can be made in two pieces (halves), with the seams at E and G (Fig. 43); the line E G (part only of it shown) being the bisecting line of the plan. It will be seen by inspection of the fig. that we require one pattern only, namely, a pattern that takes in one entire end of the article with two half-sides attached.

Draw the end pattern D D' C' C (Fig. 46) in precisely the same manner that it is drawn in Fig. 45b. Through C and C' draw indefinite lines C G C' G', perpendicular to C C', and through D and D' draw indefinite lines D E, D' E' perpendicular to D D'. Make D E, D' E' each equal to D E (Fig. 43)

FIG. 46.

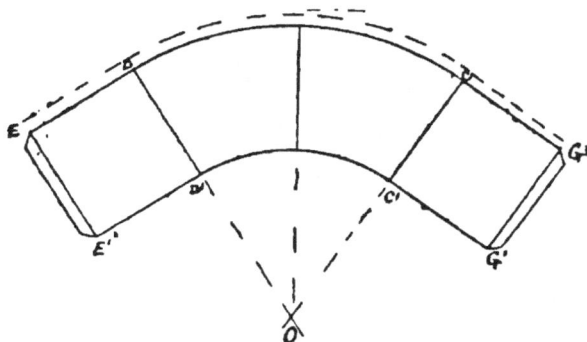

and join E E'; also make C G, C' G', each equal to C G (Fig. 43), and join G G'. Then E E' G' G will be the pattern required.

CASE III.—Pattern when the body can be made of one piece.

Thirdly, suppose the article can be made in one piece, with the seam at S (Fig. 43). Then evidently the pattern will be made up of the pattern of one entire end, side patterns attached to this, and half an end pattern attached to each side pattern.

Draw D D' C' C (Fig. 47) the end pattern. Through C and C' draw lines perpendicular to C C', and each equal to C B (Fig. 43); and join B B'. Produce B B', and with B as centre and O D (Fig. 45a), the larger of the radii for the end pattern, as radius, describe an arc cutting the produced line

in O. With O as centre and O B and O B' as radii respectively, draw arcs B S, B' S'. Make B S equal to D R (Fig. 47)

FIG. 47.

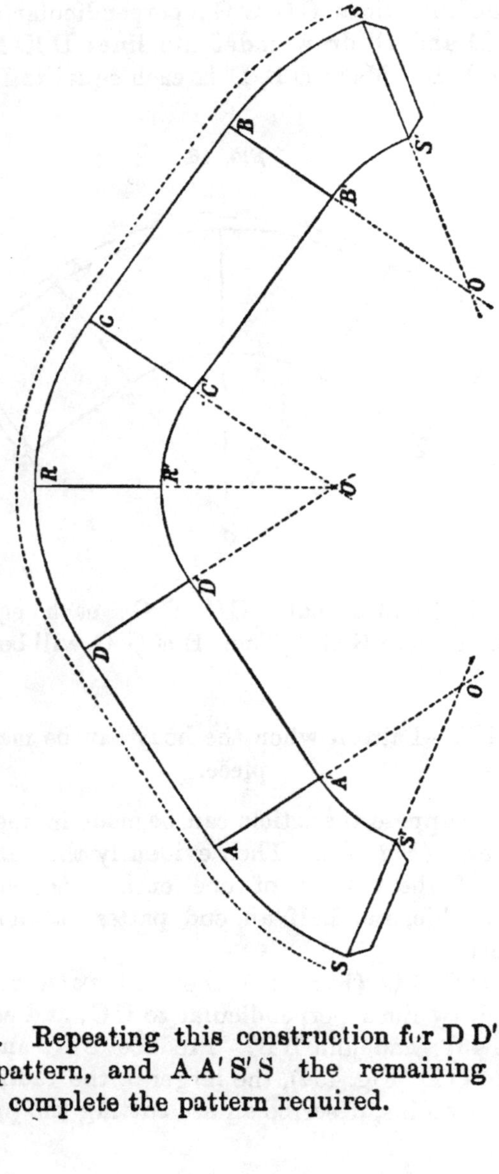

and join S O. Repeating this construction for D D' A' A the
other side pattern, and A A' S' S the remaining half-end
pattern, will complete the pattern required.

PROBLEM XXVI.

To draw, **without long radii,** *the pattern for an equal-tapering body with top and bottom parallel, and having flat sides and equal semicircular ends ; the dimensions of the top and bottom of the body and its height being given.*

This problem is a fourth case of the preceding, and exceedingly useful where the work is so large that it is inconvenient to draw the whole of the plan, and to use long radii.

Draw half the plan (Fig. 48a). Divide D C into six or more equal parts, and join 1, 2, &c., to O, by lines cutting D' C' in 1', 2', &c., and join D 1'. Draw D E perpendicular to

Fᴵɢ. 48*a.*   Fᴵɢ. 48*b.*

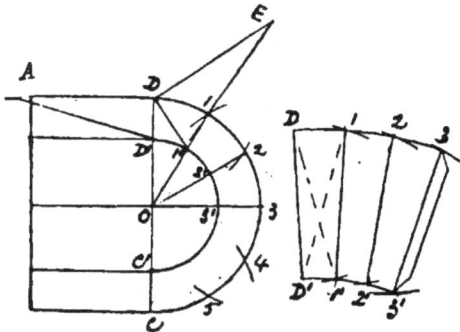

D 1' and equal to the given height, and join E 1'. (The line 1 to 1' appears to, but does not, coincide with E 1'.) Then E 1' may be taken as the true length of D 1'. Next, producing as necessary, make D A equal to the given height. Joining D' to A gives the true length of D D'.

To draw the end pattern. Draw Fig. (48b) D D' equal to D'A (Fig. 48a). With D and D' as centres, and radii respectively D 1 and E 1' (Fig. 48a) describe arcs intersecting

in point 1 (Fig. 48*b*). With D' and D as centres and radii
respectively D' 1' and E 1' (Fig. 48*a*) describe arcs intersecting in point' 1' (Fig. 48*b*). By using points 1 and 1'
(Fig. 48*b*) as centres, instead of D and D', and repeating the
construction, the points 2 and 2' can be found. Next, using
points 2 and 2' as centres and repeating the construction,
find points 3 and 3', and so on for the remaining points
necessary to complete the end pattern, which is completed
by joining the various points, as 3 to 3' by a straight line ;
D, 1, 2, and 3, by a line of regular curve ; and D', 1', 2' and
3', also by a line of regular curve. Only half the end
pattern is shown in Fig. 48*b*. The side pattern can be
drawn as shown in Fig. 44.

PROBLEM XXVII.

To draw the pattern for an equal-tapering body with top and
   bottom parallel, and having flat sides and ends, and round
   corners (an oblong pan with round corners, for instance) ;
   the height and the dimensions of the top and bottom being
   given.

Again four cases will be treated of; three in this problem,
and one in the problem following.

Case I.—Pattern when the body is to be made up of four
pieces.

The plan of the article, with lines of construction, drawn
by the method of Problem XVII., is shown in Fig. 49. We
will suppose the seams are to be at P, S, Q, and R, that is, at
the middle of the sides and ends. The pattern required is
therefore one containing a round corner, with a half-end and
a half-side pattern attached to it. The best course to take
is to draw the pattern for the round corner first, which, as
will be seen from the plan, is a quarter of a frustum of a
cone.

Draw two lines (Fig. 50a) O A, C A, meeting perpendicularly to one another in A ; make A C equal to O A (Fig. 49) the radius of the larger arc of the plan of one of the corners,

FIG. 49.

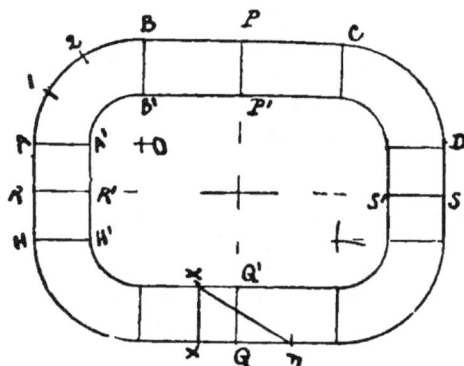

and make A D equal to the given height of the body.   Draw D E perpendicular to A D and equal to the radius of the smaller arc of the plan of a corner, and join C E and produce

FIG. 50a.   FIG. 50b.

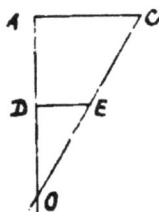 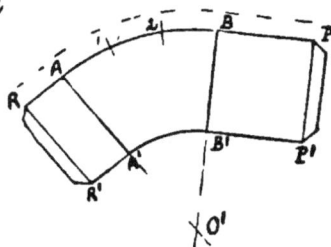

it to meet A O in O.   With any point O' (Fig. 50b) as centre and radius equal to O C (Fig. 50a), describe an arc A B, and with the same centre and radius equal to O E (Fig. 50a) describe an arc A' B'.   Draw any line A O' cutting the arcs in the points A and A', and make A B equal to A B (Fig. 49) by marking off the same number of equal parts along A B (Fig. 50b) that we divide (arbitrarily) A B (Fig. 49) into. Join B O', cutting in B' the arc A' B'. Through B and B' draw B P, B' P' perpendicular to B B', make B P equal to B P (Fig. 49) and draw P P' perpendicular to B P. Through A and A' draw A R and A' R' perpendicular to A A', make A R equal to A R (Fig. 49) and draw R R' perpendicular to A R. Then R R' P' P will be the pattern of one round corner, with a half-end and a half-side pattern attached left and right.

CASE II.—Pattern when the body is to be made up of two pieces.

Now suppose the seams are to be at the middle of each end, at R and S (Fig. 49). The pattern required will then be of twice the amount shown in Fig. 50b. It will be found best to commence with the side pattern.

Draw in the plan (Fig. 49), any line X X perpendicular to the Q side-line; then X and X' will be plans of corresponding points (§ 26). Make X F equal to the given

FIG. 51.

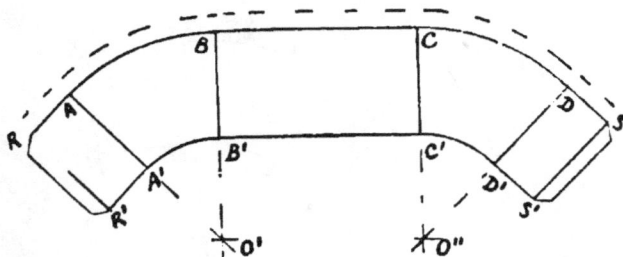

height of the body and join X' F. Then X' F is the length of a slant of the body. Draw B B' (Fig. 51) equal to X' F (Fig. 49), and through B and B' draw B C and B' C' perpendicular to B B'. Make B C equal to B C (Fig. 49) and draw

C C′ perpendicular to B C ; then B C C′ B′ will be the pattern for the side.

We have now to join on to this, at B B′ and C C′, the patterns for the round corners, which can be done thus. Produce B B′ and C C′, and make B O′ and C O″ each equal to O C (Fig. 50a) the radius of the larger arc of the corner pattern. With O′ and O″ as centres and O′B as radius, describe arcs B A and C D, and with the same centres and O′ B′ as radius, describe arcs B′ A′ and C′ D′. Make B A and C D equal each to B A in the plan (Fig. 49), and join A C, cutting the arc B′ A′ in A′, and D O″ cutting the arc C′ D′ in D′. Through A and A′ draw A R and A′ R′ perpendicular to A A′ ; make A R equal to A R (Fig. 49), and draw R R′ perpendicular to A R. Next through D and D′ draw D S and D′ S′ perpendicular to D D′ ; make D S equal to D S (Fig. 49), and draw S S′ perpendicular to D S. Then R R′ S′ S will be the pattern required.

It should be noted that O′ B′ and O″ C′ should be each equal to O E (Fig. 50a), or the pattern will not be true. This gives a means of testing its accuracy.

CASE III.—Pattern when the body is to be made up of one piece.

When the body is made in one piece it is usual to have the seam at S (Fig. 49). It is now best to commence with the end pattern.

Draw A A′ (Fig. 52) equal to X′ F (Fig. 49), which, the body being equal-tapering, is the length of its slant anywhere, and draw A H and A′ H′ perpendicular to A A′. Make A H equal to A H (Fig. 49) and draw H H′ perpendicular to A H ; then A A′ H′ H will be the end pattern. Next produce A A′, and make A O′ equal to C O (Fig. 50a). Then O′ will be the centre for the arcs of the corner pattern, which, drawn by Case II. of this problem, can now be attached at A A′. Through B and B′ draw B C and B′ C′ perpendicular to B B′ ; make B C equal to B C (Fig. 49), and draw C C′ perpendicular to B C. That completes the addition of the side pattern

B B' C' C to the corner pattern. Next produce C C', mark the necessary centre, and attach to C C' the corner pattern C D D' C', just as A B B' A' was attached to A A'. Then through D and D' draw D S and D' S' perpendicular to D D'; make D S equal to D S (Fig. 49) and draw S S' perpendicular

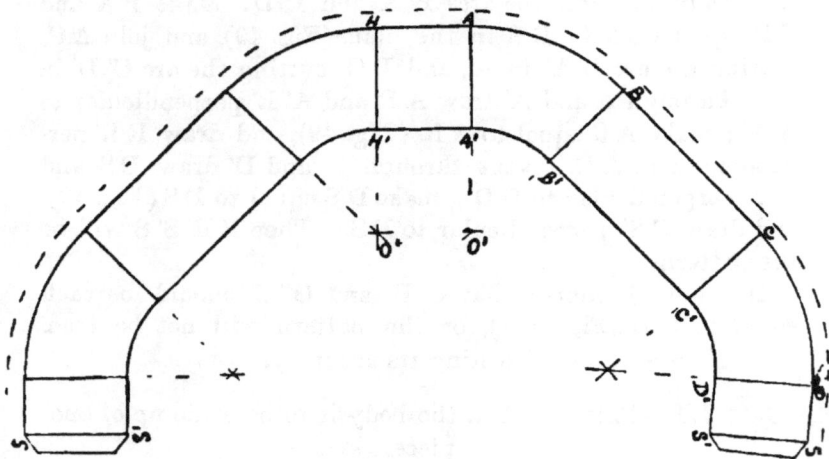

to D S; this adds half an end-pattern to D D'. By a repetition of the foregoing working on the H H' side of the end-pattern H H' A' A, the portion H H' S' S on the H H' side may be drawn, and the one-piece pattern S H A S S' A' H' S' of the body we are treating of completed.

PROBLEM XXVIII.

To draw, **without long radii,** *the pattern for an equal-tapering body with top and bottom parallel, and having flat sides and ends, and round corners; the height and the dimensions of the top and bottom being given.*

This is a fourth case of the preceding problem, and we will apply it to the first case of that problem, that is, when the body is to be made up of four pieces ; now, however, assuming that the pieces are so large that their patterns cannot be conveniently drawn by the method there given.

Draw (Fig. 53a) P S S' P', one-quarter of the plan of the body. Divide C D the larger arc of the plan of the round corner, into three or more equal parts, and join each division-point to O the centre of the arcs of the plan. Draw D F perpendicular to D D' and equal to the given height of the body, and join D' F ; then D' F will be the true length of D D'. Next join D 2'; draw D E perpendicular to it and

FIG. 53a.

FIG. 53b.

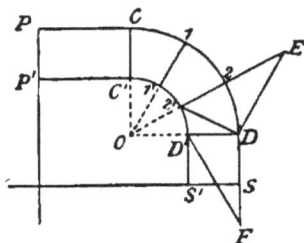

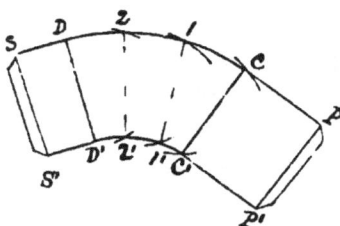

equal to the given height, and join 2' E; then 2' E may be taken as the true length of D 2'. (In the fig. a portion of the line 2' E coincides with the line from 2 to 2'. This is, however, merely a coincidence of the particular case.)

Now draw (Fig. 53b) D D' equal to D' F (Fig. 53a), and with D and D' as centres and radii respectively D 2 and 2' E (Fig. 53a) describe arcs intersecting in point 2. Also with D' and D as centres and radii respectively D' 2' and 2' E describe arcs intersecting in point 2'. Repeating this construction, but using points 2 and 2' just found as centres instead of D and D', the points 1 and 1' can be found. Similarly using points 1 and 1' as centres and repeating the construction we can find the points C and C'. Join C C'; draw a regular curve from C through 2 and 1 to D, and another regular curve from D' through 2' and 1' to C'; then in D C C' D' we have the pattern of the corner. To this corner pattern, to get at the pattern we require, we have to attach a half-end and a half-side pattern.

Draw D S perpendicular to D D' and equal to D S (Fig. 53a). Also draw S S' perpendicular to D S, and through D' draw D' S' parallel to D S. The half-end pattern is now attached.

Through C and C' draw C P and C' P' perpendicular to C C'. Make C P equal to C P (Fig. 53a), and draw P P' perpendicular to C P. This attaches the half-side pattern, and completes the pattern required.

If the body is to be made in more than four pieces it will still be generally possible to have a complete corner in one of the pieces. The corner should always be marked out first, and whatever has to be attached should be added as described.

PROBLEM XXIX.

To draw the pattern for an oval equal-tapering body with top and bottom parallel, the height and the dimensions of the top and bottom being given.

Again we deal with four cases; three in this problem and one in the following.

Case I.—Patterns when the body is to be made up of four pieces.

The plan of the body with lines of construction, drawn by Problem XVIII., is shown in Fig. 54. It will be evident from the plan and from d, p. 55, that the body is made up of two round equal-tapering bodies (frusta of cones); the ends C D D' C', F E E' F', being equal portions of a frustum of one cone, and the sides, E C C' E', D F F' D', equal portions of a frustum of another cone. Such being the case, it is clear that we require two patterns, one for the two ends, and the other for the two sides; also that the seams should be at E, C, D, and F, where the portions meet.

To draw (Fig. 55b) the ends' pattern.

From any point A draw two lines O A, B A (Fig. 55a)

Fig. 54.

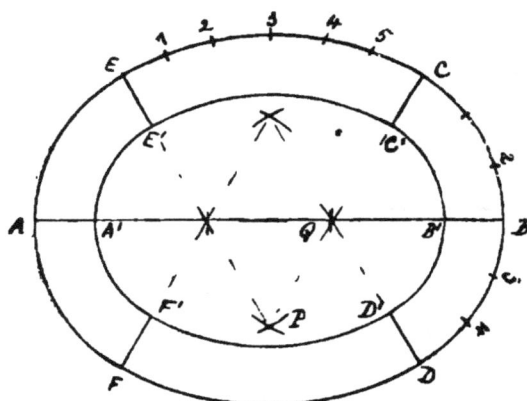

Fig. 55a.

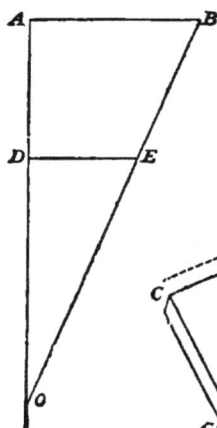

Fig. 55b.

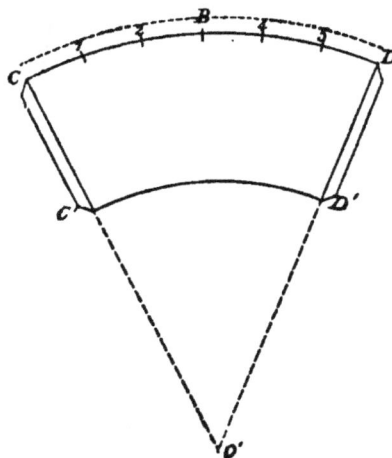

FIG. 56a.

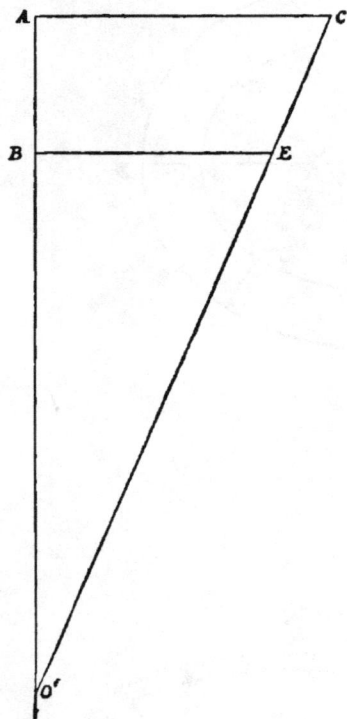

FIG. 56b.

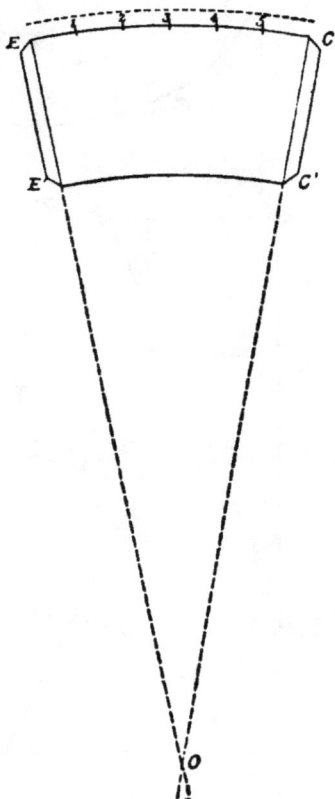

perpendicular to each other ; make A B equal to Q B (Fig. 54) the radius of the larger end-curve in the plan, and A D equal to the height of the body. From D draw D E perpendicular to A O and equal to Q B' (Fig. 54) the radius of the smaller end-curve of the plan ; join B E and produce it, meeting A O in O. Then with any point O' (Fig. 55b) as centre, and O B (Fig. 55a) as radius, describe an arc C D ; and with same centre and O E (Fig. 55a) as radius describe an arc C' D'. Join any point C in the arc C D to O' by a line cutting the arc C' D' in C' ; divide the arc C D (Fig. 54) into any number of equal parts, and set off from C (Fig. 55b) along C D parts equal to and as many as C D (Fig. 54) is divided into. Join D O' cutting the arc C' D' in D' ; then D D' C' C will be the pattern required.

To draw the pattern for the sides.

From any point A draw two lines A C, A O perpendicular to each other, and make A C equal to P C (Fig. 54), the radius of the larger side-curve and A B equal to the given height of the body. From B draw B E perpendicular to A B and equal to P C', (Fig. 54) the radius of the smaller side-curve. Join C E and produce it, meeting A O in O. With any point O' (Fig. 56b) as centre and O C (Fig. 56a) as radius describe an arc C E ; and with same centre and O E (Fig. 56a) as radius describe an arc C' E'. Now join C O', and, proceeding exactly similarly as in drawing the ends' pattern, complete E E' C' C the pattern for the sides.

CASE II.—Patterns when the body is to be made up of two pieces.

In this case the seams are usually put at the ends A and B (Fig. 54). It is evident that one pattern only is now required, and that this is made up of a side pattern, with, right and left attached to it, a half-end pattern.

Draw (Fig. 57) a side pattern E C C' E' in the way described in Case I. Then with C as centre and the radius of the ends' pattern, that is B O (Fig. 55a) as radius, set off the distance C O. With O as centre and radii O C and O C' describe arcs C B and C' B'. Make C B equal to C B (Fig. 54)

H 2

and join B O, cutting C' B' in B'.   Then C B B' C' is a half-end
pattern attached to C C'.   The half-end pattern E A A' E' is

Fig. 57.

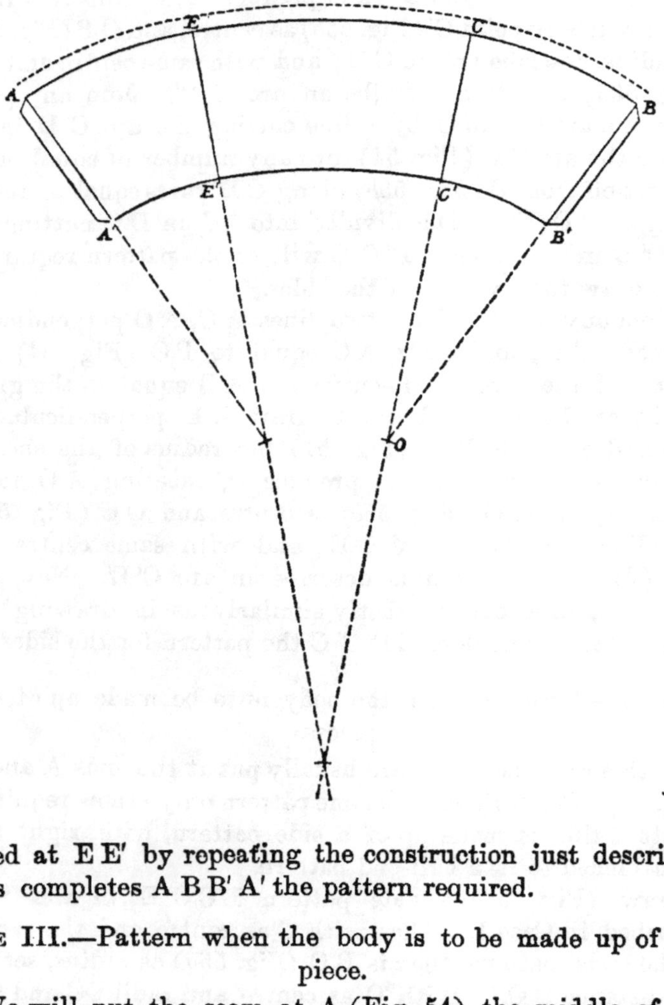

added at E E' by repeating the construction just described.
This completes A B B' A' the pattern required.

CASE III.—Pattern when the body is to be made up of one
piece.

We will put the seam at A (Fig. 54), the middle of one
end.   Draw C D D' C' (Fig. 58), an end pattern in the way
described in Case I.   Produce C C' and make C O equal to

C O (Fig. 56a), the radius for a side pattern. With O as centre and radii O C and O C′ describe arcs C E, C′ E′. Make

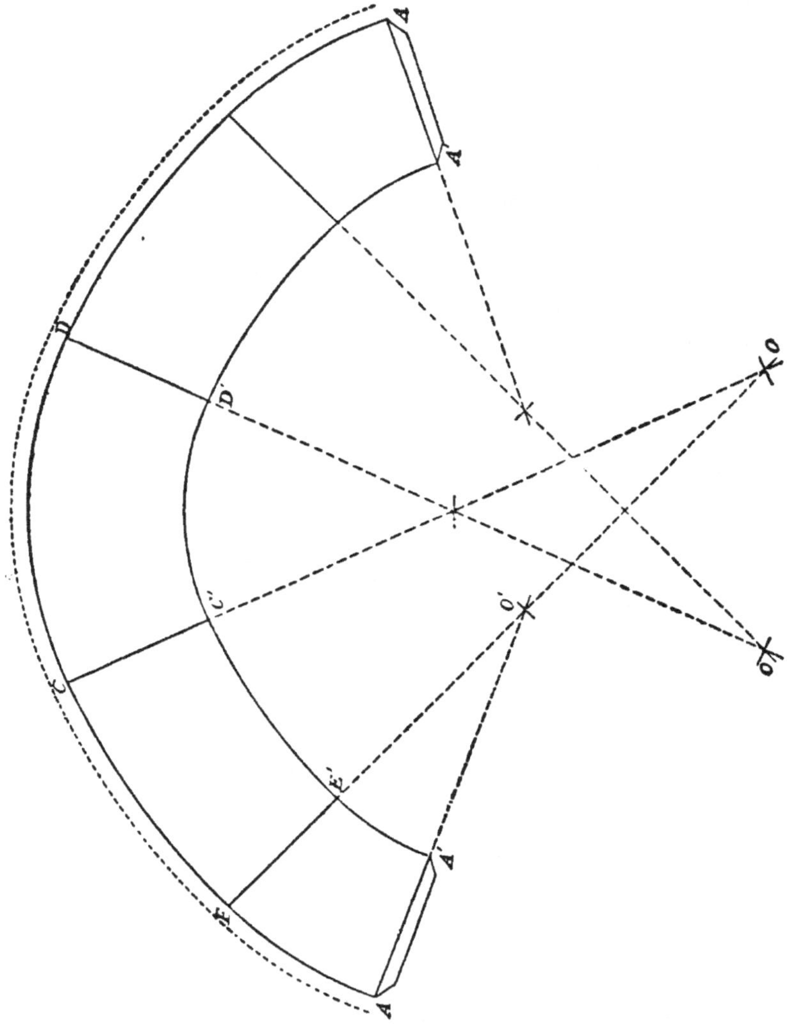

Fig. 58.

C E equal to C E (Fig. 54) and join E O cutting C′ E′ in E′. Make E O′ equal to B O (Fig. 55a) and with O′ E as radius

describe an arc E A. With same centre and O' E' as radius
describe an arc E' A'. Make E A equal to E A (Fig. 54) and
join A O', cutting E' A' in A'. The remainder D A A' D' of
the pattern can be drawn by repeating the foregoing construction. The figure A C D A A' D' C' A' will be the pattern
required.

PROBLEM XXX.

To draw, **without long radii,** *the pattern for an oval equal-
tapering body with top and bottom parallel, the height and
the dimensions of the top and bottom being given.*

This is a problem which will be found very useful for
large work, especially with the pattern for the sides, the
radius for which is often of most inconvenient length.

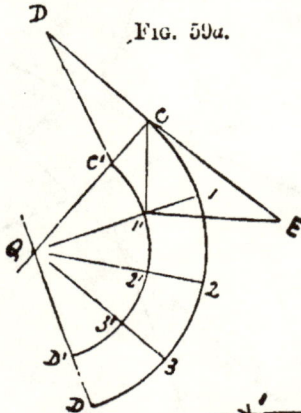

Fig. 59*a*.

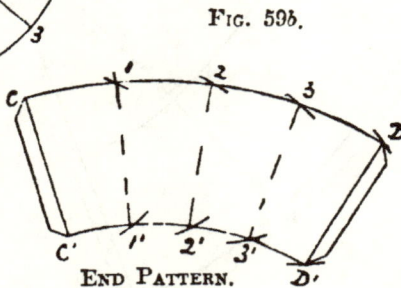

Fig. 59*b*.

END PATTERN.

To draw the ends' pattern.

Draw (Fig. 59*a*) the plan C D D' C' of the end of the body.
Divide C D into four or more equal parts, and join, 1, 2, &c.,

to Q, by lines cutting C' D' in 1', 2', &c. From C draw C D
perpendicular to C C' and equal to the given height. Join
C' D, then C' D is the true length of C C'. Join C 1'; draw
1' E perpendicular to it and equal to the given height; and
join C E. Then C E may be taken as the true length of C 1'.
Draw C C' (Fig. 59b) equal to C' D (Fig. 59a). With C and
C' as centres and radii respectively equal to C 1 and C E
(Fig. 59a) describe arcs intersecting in point 1. With C' and
C as centres and radii respectively equal to C' 1' and C E
(Fig. 59a) describe arcs intersecting in point 1'. By using
points 1 and 1' as centres and repeating the construction,
points 2 and 2' can be found. Similarly find points 3 and 3',
and D and D'. Join D D', draw a regular curve from C
through 1, 2, and 3, to D, and another regular curve from C'
through 1', 2', and 3', to D'. The figure C C' D' D is the
pattern required.

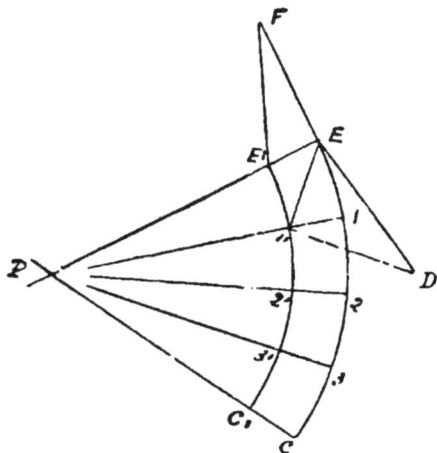

Fig. 60a.

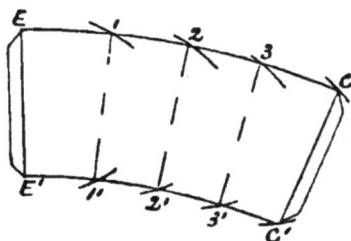

Fig. 60b.

SIDE PATTERN.

To draw the sides' pattern.

Draw (Fig. 60a) the plan E C C' E' of the side of the body.
Divide E C into four or more equal parts, and join 1, 2, &c.,
to P, by lines cutting E' C' in 1', 2', &c. From E draw E F

perpendicular to E E' and equal to the given height. Join E' F ; then E' F is the true length of E E'. Next join E 1' ; draw 1' D perpendicular to E 1' and equal to the given height, and join E D ; then E D may be taken as the true length of E 1'. Now draw (Fig. 60b) E E' equal to E' F (Fig. 60a), and with E and E' as centres and radii respectively equal to E 1 and D E (Fig. 60a) describe arcs intersecting in point 1. With E' and E as centres and radii respectively equal to E' 1' and D E (Fig. 60a) describe arcs intersecting in point 1'. Repeating the construction with points 1 and 1' as centres, the points 2 and 2' can be found ; and similarly the points 3 and 3' and the points C and C'. Join C C' ; draw a regular curve from E through 1, 2, and 3, to C, and another from E' through 1', 2', and 3', to C'. The figure E E' C' C will be the pattern required.

Book II.

CHAPTER I.

PATTERNS FOR ROUND ARTICLES OF UNEQUAL TAPER OR INCLINATION OF SLANT.

(CLASS II. *Subdivision a.*)

(42.) WE stated at the commencement of the preceding division of our subject that it was advisable, in order that the rules for the setting out of patterns for articles having equal slant or taper should be better understood and remembered, to consider the principles on which the rules were based; and the remark is equally true and of greater importance in respect of the rules for the setting out of patterns for articles having unequal slant or taper. We have shown that the basis of the rules for the setting out of patterns for equal tapering bodies is the right cone; we purpose showing that the basis of the rules for the setting out of patterns for articles of unequal slant or taper is what is called the *oblique* cone. The consideration of the cone apart from its species, that is, apart from whether it is right or oblique, which becomes now necessary, immediately follows.

DEFINITION.

(43.) *Cone.*—A cone is a solid of which one extremity, the base, is a circle, and the other extremity is a point, the apex. The line joining the apex and the centre of the base is the axis of the cone.

(44.) Given a circle and a point *in* the line passing through the centre of the circle at right angles to its plane. If an indefinite straight line, passing always through the given point, move through the circumference of the given circle, there will be thereby generated between the point and the circle, a solid; this solid is the *right* cone.

(45.) If, all other conditions remaining the same, the given point is *out of* the line passing through the centre of the circle at right angles to its plane, the solid then generated will be the *oblique* cone.

(46.) Figs. 1 and 2 represent oblique cones. The lines

FIG. 1. FIG. 2.

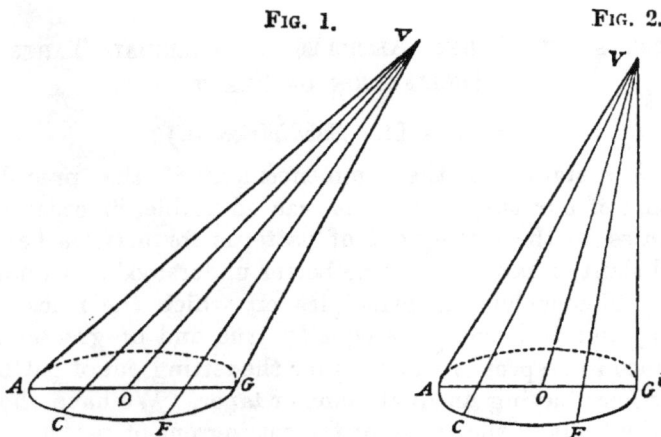

V A, V C, V F, and V G (either fig.) drawn from the apex V to the circumference of the base of the cone, are portions of the generating line at successive stages of its revolution. It is but a step from this and will be a convenience, to regard these lines, first, as each of them part of an independent generating line, and then as each of them a complete generating line. Shaded representations of various oblique cones will be found later on.

(47.) Comparing now the right and oblique cones. A right cone may be said to be made up of an infinite number of equal generating lines, and an oblique cone of an infinite number of unequal generating lines.

(48.) If a right cone is placed on a horizontal plane, the apex is vertically over the centre of the base.

(49.) If when a cone is placed on a horizontal plane the apex is not vertically over the centre of the base, the cone is oblique. Hence all cones not right cones are oblique cones. Hence also a cone is oblique if its axis is not at right angles to every diameter of its base.

(50.) In the right cone, any plane containing the axis is perpendicular to the base of the cone, and contains two generating lines (see Figs. 1a and 2a, p. 25).

(51.) In the oblique cone, only *one plane* containing the axis is perpendicular to the base of the cone, and this plane contains its longest and shortest generating lines. In Fig. 1 or 2 the lines V A and V G are respectively the longest and shortest generating lines, and the plane that contains these lines contains also V O the axis of the cone.

(52.) The obliquity of a cone is measured by means of the angle that its axis makes with that radius of the base that terminates in the extremity of the shortest generating line. Thus the angle V O G gives the obliquity of either cone V A G. The angle V O G also gives the inclination of the axis. As the angle V A G is in the same plane as V O G and smaller than that angle, it will be seen that not only is V A the longest generating line, but it is also the line of greatest inclination on the cone. Similarly as the angle that V G makes with A G produced is greater than V O G, the line V G is not only the shortest generating line, but also the line of least inclination on the cone.

(53.) The plane that contains the longest and shortest generating lines bisects the cone; consequently the generating lines of either half are, pair for pair, equal to one another.

(54.) If the elevation (see § 21, p. 48) of an oblique cone be drawn on a plane parallel to the bisecting plane, the elevations of the longest and shortest generating lines will be of the same lengths respectively as those lines. Thus if the triangle V A G (either fig.) be regarded as the elevation of the cone represented, then V A will be the true length of the longest generating line of the cone, and V G of the shortest. In speaking in the pages that follow, of elevation with regard to the oblique cone, we shall always suppose it to be on a plane parallel to the bisecting plane.

(55.) If the hypotenuse of a right-angled triangle represent

a generating line of any cone, right or oblique, then one of
the sides containing the right angle is equal in length to the
plan of that line, and the other side is equal to the height
(distance between the extremities) of any elevation of it.
See, p. 25, the triangles O B A, O G A, O E A, O D C, O K C,
O F C, and, p. 109, the triangles V' A V, V' B V, V' C V, &c.
See also § 22a.

(56.) As the generating lines of the oblique cone vary in
length, the setting out of patterns of articles whose basis is
the oblique cone (that is to say, the development of the
curved surfaces of such articles) differs from that appertaining to articles in which the right cone is involved.　The
principles, however, are the same in both cases.　In developments of the right cone, if a number of its generating lines
are laid out in one plane, then as they are all equal and have
a point (the apex) in common, the curve joining their
extremities is an arc of a circle of radius equal to a generating
line; while in developments of the oblique cone, as the
generating lines, though they still have a point in common,
vary in length, the individual lengths of a number of them
must be found, as well as the distances apart of their
extremities, before, by being laid out in one plane at their
proper distances apart, a curve can be drawn through their
extremities, and the required pattern ascertained.

The problem which follows gives the working of this in
full.

PROBLEM I.

*To draw the pattern (develop the surface) of an oblique cone, the
inclination of the axis (§ 52), its length and the diameter
of the base of the cone being given.*

First (Fig. 3) draw V' A G the elevation of the cone, and
A d G half the plan of its base; thus.　Draw any line X X,
and at any point O in it, make the angle V' O X equal to
the given inclination of the axis (a line O V' is omitted in
the fig. to avoid confusion), and make O V' equal to the
given length of the axis.　With O as centre and half the

given diameter of the base as radius describe a semicircle
A *d* G cutting X X in A and G.   Join A V', G V', then
V' A G is the elevation required, and A *d* G is the half-plan
of base.   Next divide A *d* G into any convenient number of
parts, equal or unequal.   The division here is into six, and
the parts are equal; to make them so being an advantage.
Now let fall V' V perpendicular to X X; and join V to the
division points *b*, *c*, *d*, *e*, *f* (to save multiplicity of lines this
is only partially shown in the fig.).   The lines V A, V *b*,
V *c*, &c., will be the lengths in plan of seven lines from apex

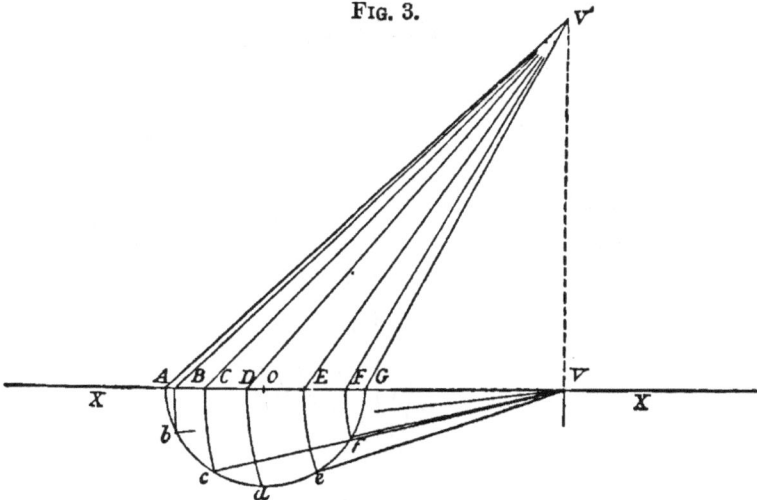

Fig. 3.

to base of cone, that is, of seven generating lines.   Now
with V as centre and radii successively V *b*, V *c*, V *d*, V *e*, and
V *f*, describe arcs respectively cutting X X in B, C, D, E, and
F.   Join V' B, V' C, &c.; then as V' V is the height of any
elevation of either of the seven generating lines (see § 22*a*,
p. 48, and § 55, p. 107) and as V A, V B, &c., are their
plan lengths, we have in V' A, V' B, V' C, &c., their true
lengths.   V' G is not only so however, it is (§ 54) the shortest
of all the generating lines of the cone.
     To draw the pattern of the cone with the seam to corre-

spond with V' G the shortest generating line.   Draw (Fig. 4)
V' A equal to V' A (Fig. 3), and with V' as centre and
V' B, V' C, V' D, V' E, V' F, and V' G (Fig. 3) successively
as radii describe arcs, respectively, *b b*, *c c*, *d d*, *e e*, *f f*, and *g g*.
With A as centre and radius equal to A *b* (A *b* being the
distance apart on the round of the cone at its base of the
generating lines V' A and V' B') (Fig. 3) describe an arc
cutting the arc *b b* right and left of V' A in B and B.   With

FIG. 4.

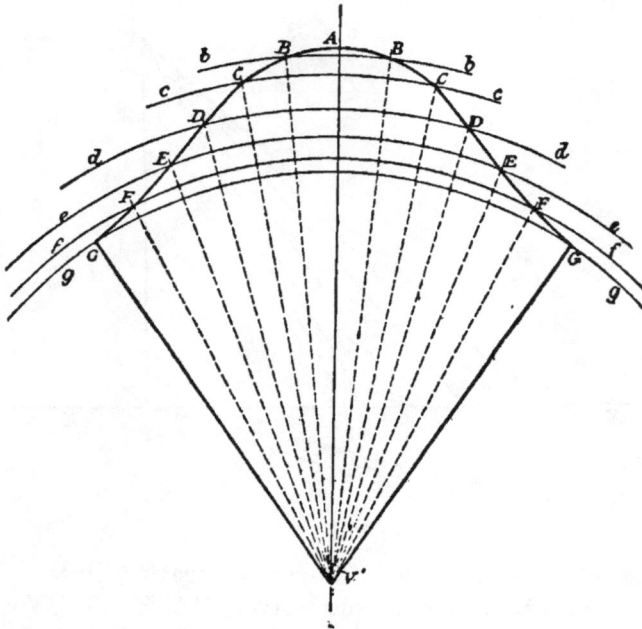

these points B and B as centres and radius as before (the
distances apart of the extremities of the generating lines at
the base of the cone being all equal), describe arcs cutting
the arc *c c* right and left of V' A in C and C.   With same
radius and the last-named points as centres describe arcs
cutting *d d* right and left of V' A in D and D.   With
D and D as centres and same radius describe arcs cutting *e e*
right and left of V' A in E and E, and with E and E as

centres and same radius describe arcs cutting ff right and left of V' A in F and F. Similarly with same radius and F and F as centres find points G and G. Draw through A and the points B, C, D, E, F and G, right and left of V' A an unbroken curved line. Also join G V' right and left of V' A, then G A G V' will be the pattern required.

The dotted lines V' B, V' C, V' D, &c., are not necessary for the solution of the problem, and are only drawn to show the position of the seven generating lines on the developed surface.

(57.) As the plane of the elevation V' A G bisects the cone (§§ 53 and 54), it is clear that the seven generating lines found, correspond with seven other generating lines on the other half of the cone ; so that in finding them for one half of the cone, we find them for the other, as the halves necessarily develop alike.

(58.) Round articles of unequal slant or taper having their ends parallel are portions (frusta) of oblique cones.

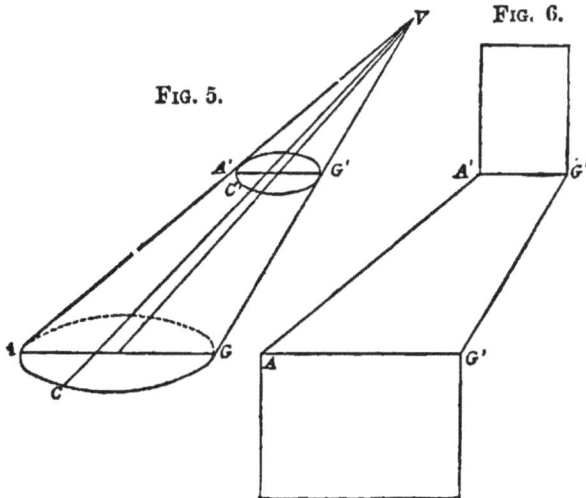

Fig. 5.

Fig. 6.

DEFINITION.

(59.) *Frustum.*—If an oblique cone be cut by a plane parallel to its base, then the part containing the apex is still an oblique cone as V A' C' G' (Fig. 5), and the

part A′ A C G G′ containing the base is a *frustum* of an
oblique cone. A shaded representation of such frustum
will be found in Fig. 15. This frustum, however, differs
from that of the right cone (§ 12, p. 33), in that, though
having circular ends, its sides are not of equal, but of unequal slant. Conversely, a round unequally tapering body,
having its top and base parallel, is a frustum of an oblique
cone. A tapering piece of pipe A′ A G G′ (Fig. 6) joining
two cylindrical pieces which are not in line with each other
is a frustum of an oblique cone.

(60.) Referring to Fig. 5, it is evident that if the pattern
for the larger cone, V A G, and the pattern for the smaller
cone, V A′ G′, be drawn from a common centre V′ (Fig. 9),
the figure G′ G A G G′ (Fig. 9) will be the pattern for the
portion A G G′ A′ (Fig. 5) of the cone V A G. The line
C C′ shows a generating line of the frustum (*b*, p. 126).

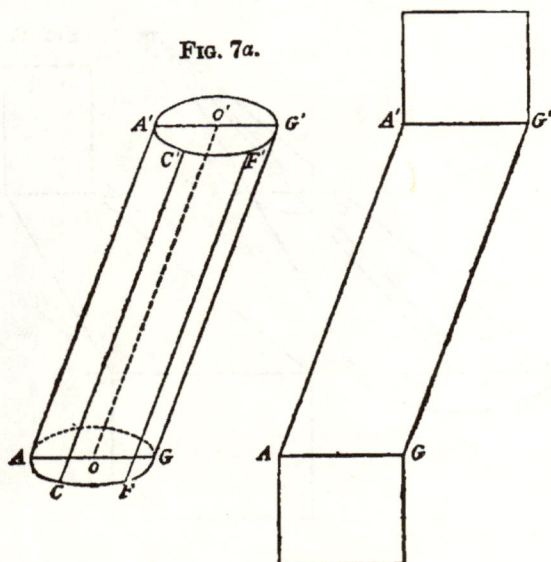

Fig. 7*b*.

Fig. 7*a*.

(61.) A case that not unfrequently occurs needs mention
here. Suppose the diameters of the top and base of a round

unequal-tapering body of parallel ends are very nearly equal, then it is evident that the apex of the oblique cone of which the body is a portion will be a long distance off, and, if the diameters be equal, then the body becomes what is called an oblique cylinder, (a cylinder is a round body without any taper at all). Of course this is an extreme case, but it is quite an admissible one. For, if the diameters of the ends of the body differ by only $\frac{1}{10000}$ inch, then, clearly, however short the body may be, we are dealing with a frustum of an oblique cone, although so nearly a cylinder that, for almost any purpose occurring in practice, it could be treated as a cylinder. Later on the advantages of looking upon the oblique cylinder as a special case of frustum of an oblique cone, and considering its generating lines as parallel, will be seen. Such a frustum is represented in A A' G' G in Figs. 7a and 7b, the line O O' being the axis of the frustum and C C' and F F' generating lines. A construction easily dealing with its development will be given presently.

PROBLEM II.

To draw the pattern of a round unequal-tapering body with top and base parallel (frustum of an obliquecone, as in Fig. 5), the diameters of top and base, the height, and the inclination of the longest generating line being given.

First (Fig. 8) draw the elevation and half the plan of the body, thus: Draw a line X X', and at any point A in it, make the angle X' A A' equal to the given inclination of the longest generating line. At a distance from X X' equal to the given height, draw a line A' G' parallel to X X'. From the point A' where A' G' cuts A A', make A' G' equal to the diameter of the top, also make A G equal to the diameter of the base; and join G G'. Then A A' G' G is the elevation of the body. Now on A G describe a semicircle A d G, this will be half the plan of the base. Produce A A', G G', to intersect in V'; this point will be the elevation of

I

the apex of the cone of which the body is a portion. Through
V' draw V' V perpendicular to X X'; divide the semicircle
into any convenient number of equal parts, here six, in the

FIG. 8.

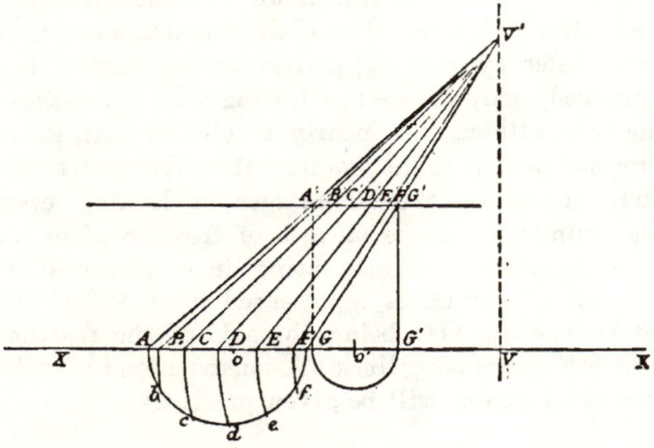

points *b, c, d, e,* and *f*; with V as centre and radii successively
V *b*, V *c*, V *d*, &c., describe arcs respectively cutting X X in
B, C, D, &c.; and join B, C, D, &c., to V' by lines cutting A' G'
in B', C', D', E', and F'. Then B B', C C', D D', &c., will be
(see construction of last problem) the true lengths of various
generating lines of the frustum (*b.* p. 126).

Now to draw the pattern (Fig. 9) so that the seam shall
correspond with G G' (Fig. 8) the shortest generating line.
Draw (Fig. 9) V' A equal to V' A (Fig. 8) and with V'
(Fig. 9) as centre and radii successively equal to V' B, V' C,
V' D, V' E, V' F, and V' G (Fig. 8), describe, respectively, arcs
b b, c c, d d, e e, f f, and *g g*. With A as centre and radius
equal A *b* (Fig. 8) (see preceding problem for the reason of
this), describe arcs cutting the arc *b b* right and left of V' A
in B and B. With these points, B and B as centres and
radius as before, describe arcs cutting the arc *c c* right and
left of V' A in C and C. With same radius and the last-

named points as centres, describe arcs cutting *d d* right
and left of V' A in D and D. With D and D as centres and
same radius, describe arcs cutting *e e* right and left of V' A in
E and E, and with E and E as centres and same radius
describe arcs cutting *f f* in F and F. Similarly, with same
radius and F and F as centres find points G and G. Join the
points B, C, D, E, &c., right and left of V' A to V'. With
V' as centre and V' A' (Fig. 8) as radius describe an arc

FIG. 9.

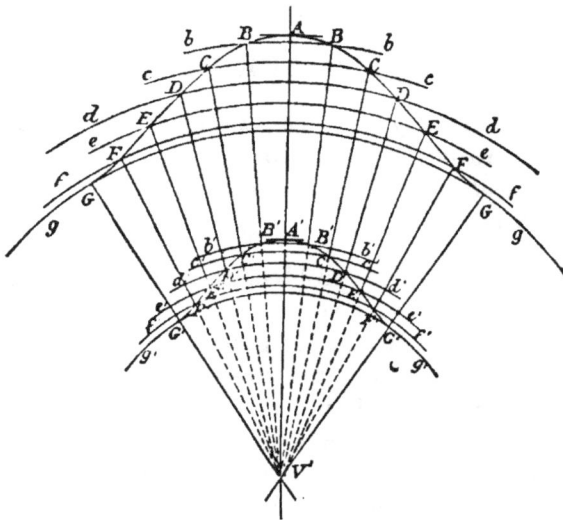

cutting V' A in A'. With same centre and V' B' (Fig. 8) as
radius describe an arc *b' b'* cutting V' B right and left of V' A
in B'. With same centre and V' C' (Fig. 8) as radius
describe an arc *c' c'* cutting V' C right and left of V' A in C'.
Similarly, with the same centre and V' D', V' E', V' F', and
V' G' (Fig. 8) successively as radii describe, respectively, arcs
d' d', *e' e'*, *f' f'*, and *g' g'*, cutting V'D, V' E, V' F, and V'G,
right and left of V' A respectively in D', E', F', and G'.
Draw through A and the points B, C, D, E, F, and G, right

I. L

and left of V' A an unbroken curved line. Also draw through
A' and the points B', C', D', E', F', and G', right and left of
V' A an unbroken curved line.   Then G A G G' A' G' will
be the pattern required.

The small semicircle (Fig. 8) and perpendicular lines from
its extremities are not needed in this problem, but are
introduced in illustration of that next following.

(62.) In the applications of the oblique cone, it is generally
in the form next following that the problem presents itself.

PROBLEM III.

*To draw the pattern of a round unequal-tapering body with top
and base parallel (frustum of oblique cone), its plan and the
perpendicular distance between the top and base (the height of
the body) being given.*

The working of this problem should be carefully noted,
for the reason just above stated.

Let *a g a' g'* (Fig. 10) be the given plan of the body.   The
side lines of the plan are not drawn, but only the circles of
its top and base, as we do not make use of the side lines.   In
fact, all that we really make use of is, as will be presently
seen, the halves of the plan-circles of the top and base of the
body, not, however, placed anyhow, but *in their proper positions
relatively to one another in plan*.   This is the all-essential point.
If the plan-circle *a' g'* (Fig. 10) of the top of the frustum
were further removed than represented from the plan-circle
a g of its base, we should, by the end of our working, get at
the pattern of some other oblique-cone frustum than that the
pattern of which we require.   Through O O' the centres of
the circles draw an indefinite line *a* V, and draw any line
X X parallel to O O'.   Through *a* and *g* draw *a* A and *g* G
perpendicular to X X, and through *a'* and *g'* draw indefinite
lines *a'* A" and *g'* G" perpendicular to X X, and cutting it in

points A', G'.   Make A' A'' and G' G'' each equal to the given height or perpendicular distance that the top and base are apart.   Join A A'' and G G'', and produce these lines to intersect in V'.   The problem can now be completed by Problem II.

Fig. 10.

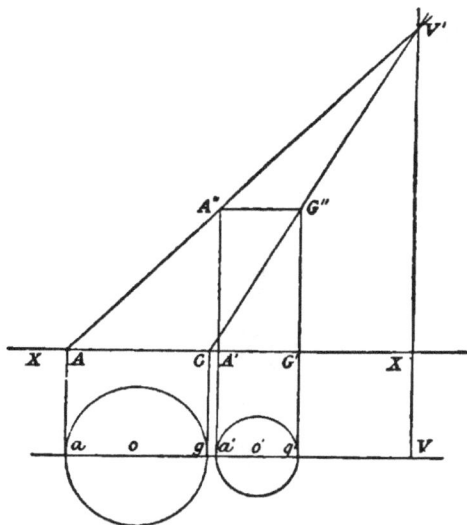

It should be noted that it is only necessary to draw half the plans of the top and base as shown by the semicircles A G and A' G' in Fig. 8.

The case of large work, where long radii would be inconvenient, we treat as a separate problem, and we will suppose the dimensions given to be those of Problem II.   The method is also suitable where there is little difference between the diameters of top and base of the body.

PROBLEM IV.

To draw **without long radii** *the pattern of a round unequal-tapering body with top and base parallel, the diameters of the top and base, the height, and the inclination of the longest generating line being given.*

Draw any line X X' (Fig. 11) and at point A in it make the angle X' A A" equal to the given inclination of the longest generating line. At a distance from X X' equal to the given height draw a line A" F parallel to X X'. Make A E equal to the longer of the given diameters, and on it describe the semicircle A *c* E. From the point A" where

Fig. 11.

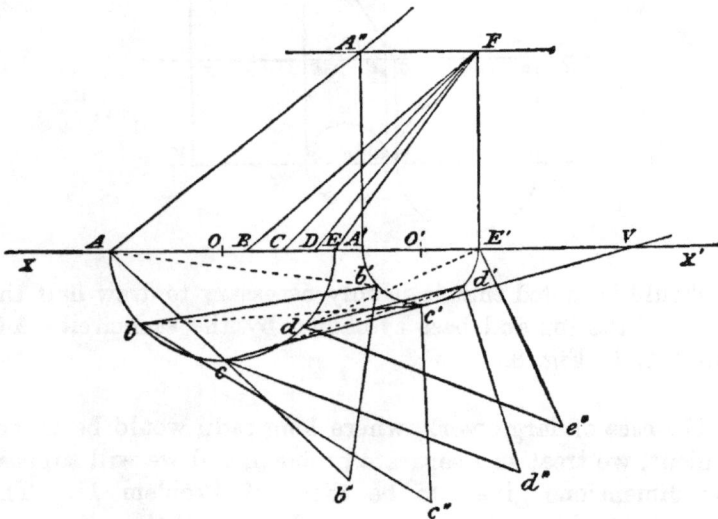

A" F cuts A A" let fall A" A' perpendicular to X X. Make A' E' equal to the shorter of the given diameters, and on it describe the semicircle A' *c'* E'. Next find *c* and *c'* the midpoints of the semicircular arcs. These are easily found by drawing lines (not shown in the fig.) through O and O' (the centres from which the semicircles are described) perpendicular to X X' and cutting the semicircular arcs in the required points c and c'. Join c c' and produce it meeting X X' in V; this point will be the plan of the apex of the cone of which our tapering body is a frustum. This point is by no means always within what may be termed workable reach, as for instance where the two semicircles are nearly equal the lines X X' and c c' therefore very nearly parallel, and the producing c c' to V impracticable. We will work under both suppositions.

(63.) If V is accessible, then divide the larger semicircle into any convenient number of parts (four parts only are taken in the figure in order to keep it clear), as A b, b c, c d, d E. Join b and d to V (c is already thus joined, the lines from b and d to V are only drawn in the fig. as far as the smaller semicircle) by lines cutting the smaller semicircle in b' and d'. Then A A', b b', c c', &c., are the plans of generating lines of the frustum (tapering body), and in order to draw its pattern their true lengths must be found.

(64.) If V is inaccessible, then divide the smaller semicircle into four parts (the same 'convenient number' of parts that the larger semicircle was divided into), in the points b', c', and d', and join b b', c c', and d d'.

The true lengths of b b', c c', &c., are found as follows : From E' draw a line perpendicular to X X' and cutting A" F in F, and join E F. Then E F is the true length of E E'. Now make E' D equal to d d' and join D F; then D F is the true length of d d'. Next set off E' C equal to c c', and E' B equal to b b'; and join C F and B F. Then C F and B F are the true lengths respectively of c c' and b b'. The true length of A A' we already have in A A", and as this is the longest generating line of the frustum, E F will be the shortest.

We proceed now to find the distance the points A and b', b and c', c and d', &c , are apart, which we do by finding the true lengths of the lines A b', b c', c d', and d E', joining the points. Through b' draw b' b" perpendicular to A b', and equal to the given height. Join A b"; then A b" may be taken as the true length of A b'. Similarly, through c', d', and E' draw lines equal to the given height, and perpendicular to $b c'$, $c d'$, and d E' respectively. Join $b c''$, $c d''$, and $d e''$; then $b c''$, $c d''$, and $d e''$ may be taken as the true lengths required.

To draw the pattern (Fig. 12) the seam to correspond with the shortest generating line. Draw A A' equal to A A'' (Fig. 11) and with A and A' as centres and radii respectively A b'' and A' b' (Fig. 11) describe arcs intersecting in b'. Next with b' and A as centres and radii respectively B F and A b (Fig. 11) describe arcs intersecting in b. Then A, b, A', b', are points in the curves of the pattern. With b and b' as

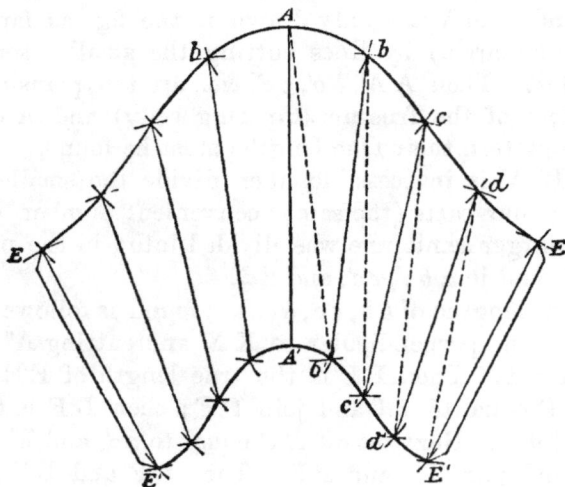

FIG. 12.

centres and radii respectively $b c''$ and $b' c'$ (Fig. 11) describe arcs intersecting in c'. With c' and b as centres and radii respectively C F and $b c$ (Fig. 11) describe arcs intersecting in c; and with c and c' as centres and radii respectively $c d''$ and $c' d'$ (Fig. 11) describe arcs intersecting in d'. With d' and c as centres and radii respectively D F and $c d$ (Fig. 11) describe arcs intersecting in d. Similarly find E' and E. Draw unbroken curved lines through A $b c d$ E and A' b' c' d' E and join E E'; that will give us half the pattern. By like procedure we find the other half of the pattern, that to the left of A A'.

(65.) The lines $b\,b'$, $c\,c'$, &c., and the dotted lines A b', $b\,c'$, &c., are drawn in Fig. 12 simply to show the position that the lines which correspond to them in Fig. 11 ($b\,b'$, A b', &c.) take upon the developed surface of the tapering body. It is evident that it is not a necessity to make distinct operations of the two halves of the pattern; for as the points b', b, c', c, &c., are successively found, the points on the left of A A' corresponding to them can be set off.

PROBLEM V.

To draw the pattern of an oblique cylinder (inclined circular pipe for example), the length and inclination of the axis and the diameter being given.

Draw (Fig. 13) any line A'G', and at the point A' in it

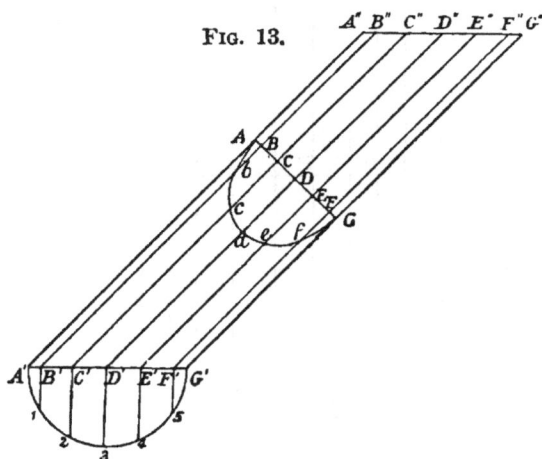

FIG. 13.

make the angle G' A' A" equal to the given inclination of the axis. Make A' G' equal to the given diameter, and draw a line G' G" parallel to A' A". Make A' A" and G' G" each equal to the length of the cylinder (the length of a cylinder is the length of its axis) and join A″ G″. Then A′ A″ G′ G″ is the elevation of the cylinder. Now on A′ G′ describe a semicircle, and divide it into any number of equal parts, in the points 1, 2, 3, &c.; through each point draw lines perpendicular to A′ G′, meeting it in points B′, C′, D′, &c., and through B′, C′, D′, &c., draw lines parallel to A′ A″. Draw any line A G perpendicular to A′ A″ and G′ G″, cutting the lines B′ B″, C′ C″, &c., in points B, C, &c. Next make B b equal to B′ 1, C c equal to C′ 2, D d equal to D′ 3, E e equal to E′ 4, and F f equal to F′ 5, and draw a curve from A through the points b, c, d, &c., to G. It is necessary to remark that this curve is not a semicircle, but a semi-ellipse (half an ellipse).

Fig. 14.

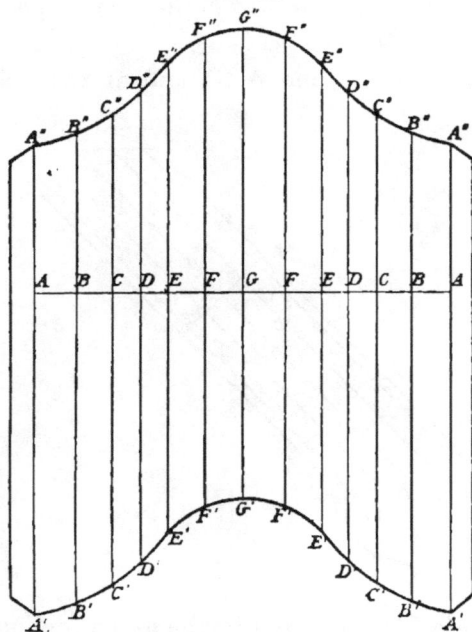

To draw the pattern (Fig. 14). Draw any line A A, and at about its centre draw any line G″ G′ perpendicular to it and cutting it in G. From G, right and left of it, on the line A A mark distances G F, F E, E D, D C, C B, and B A equal respectively to the distances G*f*, *f e*, *e d*, *d c*, &c. (Fig. 13). Through the points F, E, D, &c., right and left of G, draw lines parallel to G″ G′. Make G G′, G G″ equal to G G′, G G″ (Fig. 13) respectively. Similarly make F F′, F F″, E E′, E E″, D D′, D D″, &c., right and left of G′ G″ equal respectively to F F′, F F″, E E′, E E″, D D′, D D″ &c. (Fig. 13). Draw an unbroken curved line from G″ through F″, E″, D″, &c., right and left of G″ and an unbroken curved line through F′, E′, D′, &c., right and left of G′. The figure A″ G″ A″ A′ G′ A′ will be the pattern required. The two parts G″ A″ A′ G′ of the pattern are alike in every respect.

CHAPTER II.

Unequal-tapering Bodies, of which Top and Base are Parallel, and their Plans.

(66.) Before going into problems showing how to draw the patterns of unequal-tapering bodies with parallel ends, bodies which are (as the student will realise as he proceeds) partly or wholly portions of oblique cones, it will be necessary to enter into considerations in respect of the plans of frusta of such cones (see § 58, p. 111), similar to those appertaining to the plans of frusta of right cones treated of in Chap. V., Book I.; but to us of greater importance, because the constructions in problems for the setting out of patterns of bodies having unequal taper or inclination of slant are a little more difficult than those in problems for patterns of equal-tapering bodies. The chapter referred to may be now again read with advantage.

As much use will be hereafter made of the terms *Proportionate Arcs* and *Similar Arcs* we now define them. We also extend our explanation of *Corresponding Points*.

Definitions.

(67.) Proportional Arcs: Similar Arcs.—Arcs are *proportional* when they are equal portions of the circumferences of the circles of which they are respectively parts; they are *similar* when they are contained between the same generating lines. Similar arcs are necessarily proportional. In Fig. 15 the arcs A D and A' D' are proportional because each is a quarter of the circumference of the circle to which it belongs. They are similar because the generating lines V' A and V' D contain them both.

(68.) Corresponding Points.—Points on the same generating line are *corresponding points* (compare § 24); thus,

the points A and A' are corresponding points, because they
are on the same generating line V' A ; also the points C and
C' on the generating line V' C.   The point A' on V A is the

FIG. 15.

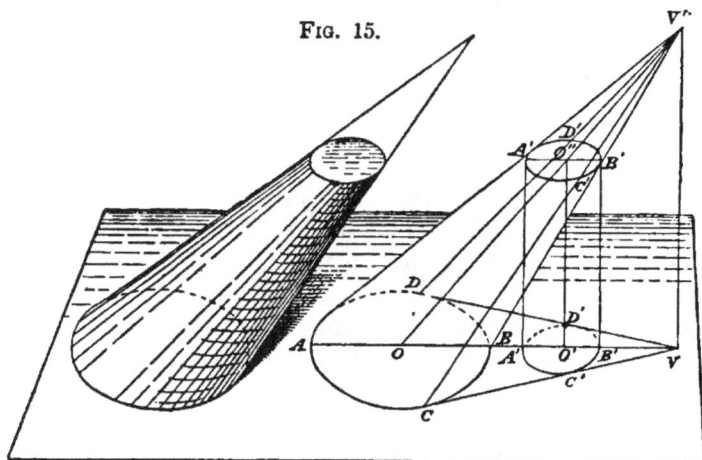

plan of A on V' A ; the point C' on V C is the plan of C' on
V' C ; and so on.

(69.) From the figure, which shows a frustum of an oblique
cone standing on a horizontal plane, it will be seen that the
plan of a round unequal-tapering body (frustum of oblique
cone) consists mainly of two circles C A D B, C' A' D' B', the
plans of the ends of the body.   In Fig. 16 is shown a
complete plan of an oblique-cone frustum.   With the con-
necting lines of the two sides we are not concerned, but may
simply mention that they are *tangents* (lines which touch
but do not cut) to the circles.   Further from Fig. 15 it will
be seen that, completing the cone of which the tapering
body is a portion.

a. The plan of the axis (line joining the centres of the
ends) of a frustum is a line joining the centres of the circles
which are the plans of its ends ; thus, the line O O' is the
plan of the axis O O" (see also O O', Fig. 16).

Similarly, the plan of the axis of the complete cone, is the plan, produced, of the axis of its frustum; thus, O V is the plan of O V'.

b. The plan, produced both ways, of the axis of a frustum contains the plans of the lines of greatest and least inclination on the frustum (see § 52); that is to say, of the longest and shortest lines on it. Thus, O O', produced both ways, contains the plans of A A' and B B'. It is convenient to regard lines joining corresponding points of a frustum (corresponding points of a frustum are points on one and the same generating line of the complete cone) as generating lines of the frustum. (See in connection with this, § 46). Then lines A A', B B', for instance, may be spoken of as generating lines of the frustum represented.

Similarly, the plans of the longest and shortest generating lines of a cone are contained in the plan, produced, of the axis of its frustum; thus, the plans of V'A and V'B are contained in O O' produced, both ways.

c. The line, produced, which joins the centres of the plans of the ends of a frustum, contains the plan of the apex of the cone; thus, O O', produced, contains V the plan of V'.

d. The line, produced, which joins the plans of corresponding points of a frustum (see definition, § 68) contains the plan of the apex of the complete cone, and, produced only as far as the plan of the apex is the plan of a generating line of the cone; thus, C and C' being corresponding points on the cone, the plan, C C', produced, of the joining C and C', contains V; and C V is the plan of the generating line C V'.

e. The plans produced of all generating lines of a frustum intersect the plan of its axis produced, in one point, and that point is the plan of the apex of the complete cone; for example, the plans produced, of the generating lines C C' and D D' of the frustum intersect O O' produced in V.

(70.) It follows from *e* that the plan of the apex of the complete cone of which a given frustum is a portion can easily be found if we have given the plans of the ends of the

frustum and the plans of two corresponding points not in
the line passing through the centres of the plans of its ends.
This is a valuable fact for us, as it spares us elevation
drawing which in many cases is very troublesome, and
indeed, sometimes practically impossible, as, for instance,
where an unequal-tapering body is frustum of an exceedingly high cone the axis of which is but little out of the
perpendicular. This is a case in which although the apex
cannot be found in elevation because of the great length of
the necessary lines, it can readily be found in plan, because,
in plan, the requisite lines are short. An oblique cone may
of course not only be exceedingly long, but also very greatly
out of the perpendicular. In this case it is impracticable
anyhow to find the plan of the apex. Problem IV., just
solved, meets both cases. It was by e that we there found
the plan of the apex when accessible, that is, where the lines
of the plan are not unduly long (see Fig. 11) by joining the
plans of corresponding points c and c' (c and c' are corresponding points in that they are mid-points on the half-
plans of the ends of the frustum, and therefore necessarily
on one and the same generating line), and producing $c c'$ to
intersect $O O'$, the line joining the centres of the plans of
the ends, that is to intersect the plan produced of the axis
of the frustum.

(71.) Passing the foregoing under review, it will be seen
that if we have two circles which are the plans of the ends
of a round unequal-tapering body (frustum of an oblique
cone) standing on a horizontal plane, and the circles are in
their proper relative positions as part plan of the frustum,
then the line produced, one or both ways as may be necessary,
which joins the centres of the circles, contains:—

The PLAN of the AXIS of the frustum (see a, p. 125).

The PLAN of the AXIS of the cone of which the frustum is
a part (see a, p. 125).

The PLAN of the APEX of the cone (see c, p. 126).

The PLANS of the LONGEST and SHORTEST GENERATING lines of
the frustum (see b, p. 126).

The PLANS of the LONGEST and SHORTEST GENERATING
LINES of the cone, of which the frustum is a part (see *b*,
p. 126).

The PLANS of the LINES of GREATEST and LEAST INCLINATION
of the frustum (see *b*, p. 126).

The PLANS of the LINES of GREATEST and LEAST INCLINA-
TION of the cone, of which the frustum is a part (see
§ 52, p. 107).

And this is a matter of very great practical importance, as
will be seen later on.

(72.) As with circles under the conditions stated, so exactly
with arcs which form the plans either of the ends or of
portions of the ends, of an unequal-tapering body.

(73.) Further, referring to *c* and *d* of p. 55 as to round
equal-tapering bodies we are now in a position to deduce (see
Fig. 15) the following as to round unequal-tapering bodies.

f. The plan of a round unequal-tapering body with top
and base parallel (frustum of oblique cone) consists essentially
of two circles, not concentric, definitely situate relatively to
one another.   See Fig. 16.

FIG. 16.

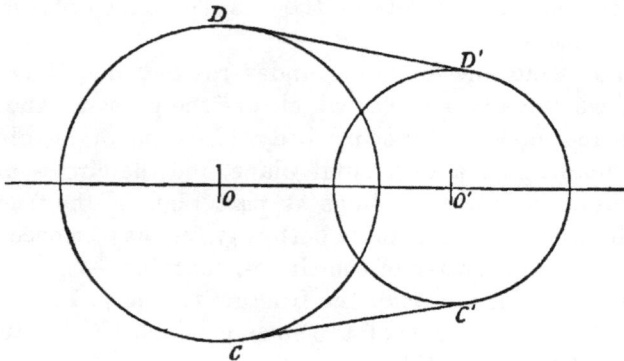

Similarly the plan of a *portion* of such round unequal-
tapering body (frustum of oblique cone) consists essentially
of two arcs definitely situate relatively to one another, and

not concentric. See Fig. 17. O O' is the axis of the complete frustum.

g. Conversely.—If two circles, not having the same centre, definitely situate relatively to one another, form essentially the plan of a tapering body having parallel ends, that body is a round unequal-tapering body (frustum of oblique cone).

In Fig. 16, if the two circles represent essentially the plan of a tapering body having parallel ends, then the body, of which the circles are the essential plan, is a round unequal-body (frustum of oblique cone).

Similarly if two arcs definitely situate relatively to one another, and not having a common centre, form the essential part of the plan either of a tapering body or of a portion of

FIG. 17.

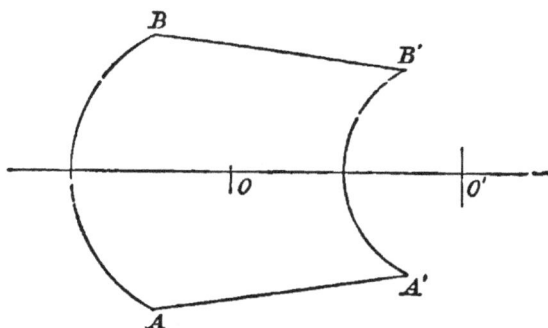

a tapering body having parallel ends, then that body or portion is a portion of a round unequal-tapering body (frustum of oblique cone).

In Fig. 17 if the arcs form the essential part of the plan, either of a tapering body or of a portion of a tapering body having parallel ends; then the body or portion of body, of which that fig. is the plan, is a portion of a round unequal-tapering body (frustum of oblique cone). In the particular plan represented the arcs are similar; the points B and B', and A and A' are therefore corresponding points.

K

We will now proceed to draw the plans of some unequal-tapering bodies, of which patterns will be presently set out.

PROBLEM VI.

To draw the plan of an unequal-tapering body with top and base parallel and having straight sides and semicircular ends (an " equal-end " bath with semicircular ends), from given dimensions of top and bottom.

Draw A'B'C'D' (Fig. 18) the plan of the bottom by Problem XVI., p. 20. Bisect A'B' in O, and through O draw C D perpendicular to A'B'. Make O A and O B each

FIG. 18.

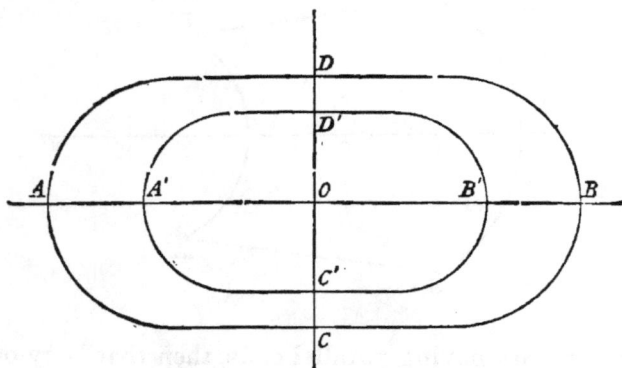

equal to half the length, and O C and O D each equal to half the width of the top. The plan of the top can now be drawn in the same manner as that of the bottom, completing the plan required.

PROBLEM VII.

To draw the plan of an oval unequal-tapering body with top and base parallel (an oval bath), from given dimensions of top and bottom.

Draw (Fig. 19) A′ B′ C′ D′ the plan of the bottom, the given length and width, by Problem XII., p. 13; and make

FIG. 19.

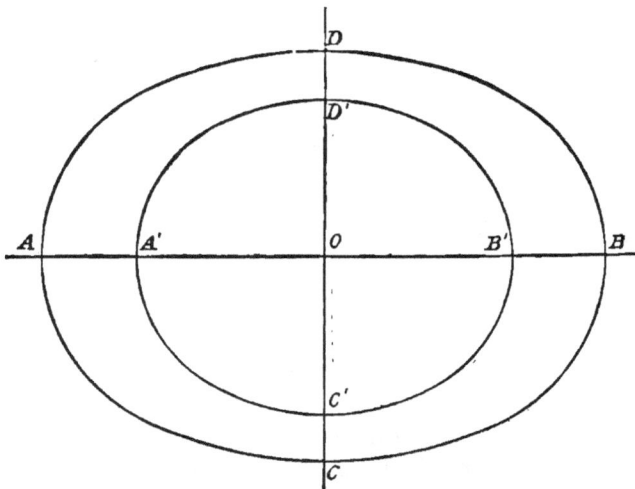

O A and O B each equal to half the length, and O C and O D each equal to half the width of the top. The plan of the top can now be drawn in the same manner as that of the bottom; this completes the plan required.

PROBLEM VIII.

To draw the plan of a tapering body with top and base parallel and having oblong bottom with semicircular ends and circular top (tea-bottle top), from given dimensions of top and bottom.

Draw (Fig. 20) the plan of the oblong bottom by Problem XVI., p. 20, and with O the intersection of the axes of the oblong as centre and half the diameter of the top as radius, describe a circle. This completes the plan.

FIG. 20.

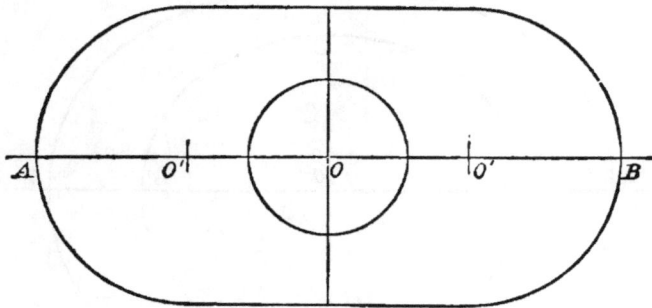

We here, for the first time, extend the use of the word 'axes' (see Problems XII. and XIV., pp. 13 and 15). It is convenient to do so, and the meaning is obvious.

PROBLEM IX.

To draw the plan of a tapering body with top and base parallel, the top being circular and the bottom oval (oval canister-top), from given dimensions of top and bottom.

Draw (Fig. 21) the plan of the oval bottom by Problem XII., p. 13, and with O the intersection of the axes as centre,

and half the given diameter of the top as radius, describe a
circle. This completes the plan.

Fig. 21.

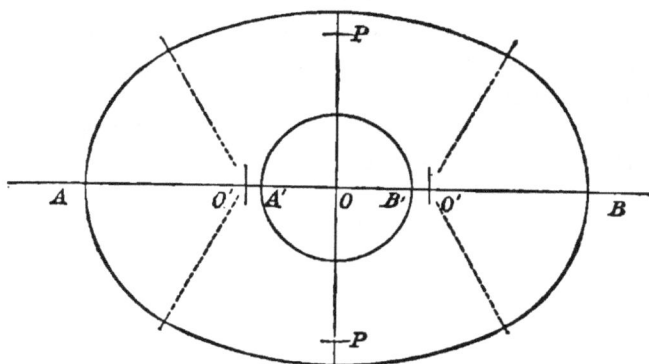

PROBLEM X.

*To draw the plan of a tapering body with top and base parallel
and having oblong base with round corners and circular top,
from given dimensions of the top and bottom.*

Draw (Fig. 22) the plan of the oblong bottom by Problem
XV., p. 19, and with O the intersection of the axes of the

Fig. 22.

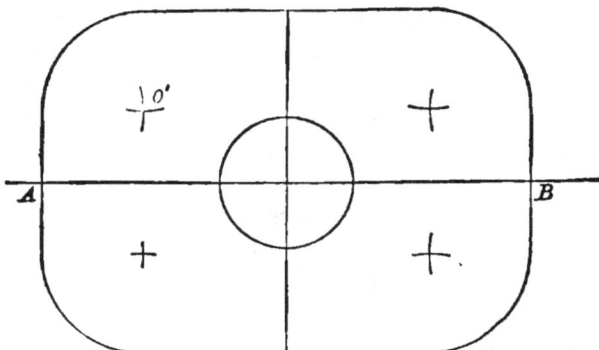

oblong bottom as centre and half the diameter of the top as
radius, describe a circle. This completes the plan.

PROBLEM XI.

To draw the plan of an Oxford hip-bath.

Fig. 23 is a side elevation of the bath, drawn here only to make the problem clearer, not because it is necessary for the working. No method that involves the drawing of a full-size side elevation is practical, on account of the amount of space that would be required.

FIG. 23.

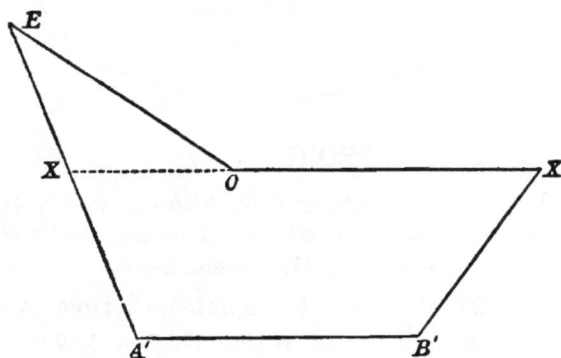

The bottom of an Oxford hip-bath is an egg-shaped oval. The portion $O X'$ of the top is parallel to the bottom $A' B'$, and the whole $X X'$ top, the portion $O X E$ of the bath being removed, is also an egg-shaped oval. In speaking of the plan of the 'bath,' we mean the plan of the $X X' B' A'$ portion of it, as the plan of this portion is all that is necessary to enable us to get at the pattern of the bath.

We will first suppose the following dimensions given :— The length and width of the bottom, and the length of the $X X'$ top, the height of the bath in front, and the inclination of the slant at back.

First draw (Fig. 25) the plan of the bottom $A' D' B' C'$ by Problem XIII., p. 14. To draw the horizontal projection of

the X X' top, make (Fig. 24a) the angle A A' E equal to the
given inclination of the slant at the back. Through A' draw
A' H perpendicular to A A', and equal to the height of the
bath in front; through H draw H X parallel to A A' and
cutting A' E in X; and draw X A perpendicular to X H;
then A A' will be the distance, in plan, at the back, between
the curve of the bottom and the curve of the X X' top
(Fig. 23). Make A' A (Fig. 25) equal to A A' (Fig. 24a), and
make A B equal to the length of the X X' top. With O as
centre and O A as radius describe a semicircle; the remainder
of the oval of the X X' top can now be drawn as was that of
the bottom. This completes, as stated above, all that is
necessary of the plan of the bath to enable its pattern to be
drawn.

FIG. 24a.   FIG. 24b.

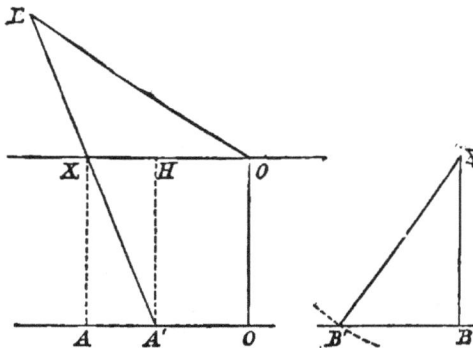

If the width of the X X' top is given, and not the inclination of the bath at back, make O A (Fig. 25) equal to half
that width, and proceed as before. The seam in an Oxford
hip-bath, at the sides, is on the lines of which C C' and
D D' are the plans.

If the length of the X X' top (Fig. 23) is not given, it can
be determined in the following manner:—

Let the angle X B' B (Fig. 24b) represent the inclination
of the slant of the front, and B' X its length. Through X

draw X B perpendicular to B' B; then B B' will be the distance in plan, at the front, between the curve of the bottom and the curve of the X X' top (Fig. 23), and this distance, marked from B' to B (Fig. 25), together with the distance A' A at the back end of the plan of the bottom, fixes the length required.

FIG. 25.

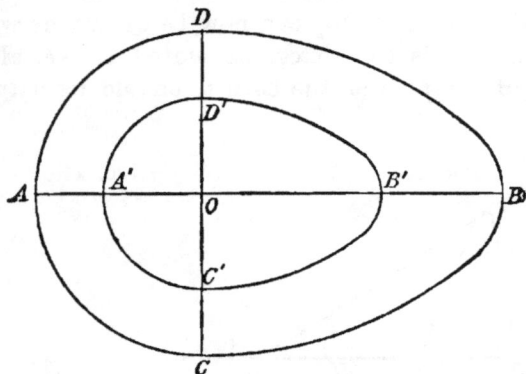

If the lengths only of the slants of the bath at back and front are given and not their inclinations, the plan of the X X' top can be drawn as follows :—

Draw two lines X B, B' B (Fig. 24b) perpendicular to one another and meeting in B; make B X equal to the given height of the bath in front, and with X as centre and radius equal to the given length of the slant at the front; describe an arc cutting B' B in B'. Make B' B (Fig. 25) equal to B B' (Fig. 24b), and B A equal to the given length of the X X' top; this will give the distance A' A. Now make A A' (Fig. 24a) equal to A A' (Fig. 25), draw A X and A' H perpendicular to A A' and equal to the given height of the bath in front; join H X and draw A' E, through X, equal to the length of the slant at the back; the remainder of the plan of the X X' top can then be drawn as already described.

It will be useful to show here in this problem how to

complete the back portion already commenced in Fig 24a
of the side elevation of the bath.   Make A' E equal to the
slant (§ 4, p. 24) at back, which must of course be given,
and make X O equal to half the given width of the X X'
top ; join O E, and draw O O perpendicular to A A' produced ;
then A' E O O is the elevation required.

PROBLEM XII.

To draw the plan of an Athenian hip-bath or of a sitz-bath.

Fig. 26 is a side elevation of the bath, drawn for the reason
mentioned in the preceding problem.

FIG. 26.

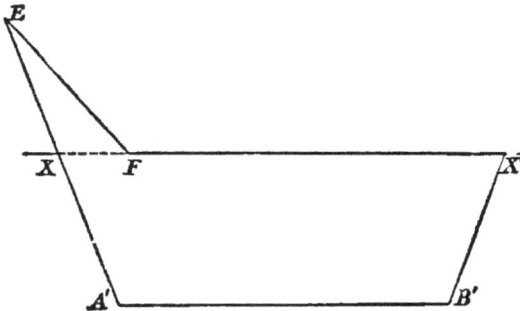

The bottom of an Athenian hip-bath or a sitz-bath is an
ordinary oval.   The portion X' F of the top is parallel to the
bottom A' B', and the whole X X' top, the portion F X E
of the bath being removed, is also an ordinary oval.   Similarly as with the bath of the last problem ; we mean by plan
of the 'bath,' the plan of the X X B' A' portion of it ; no .
more being required for the drawing of the pattern of the
bath.

We will first suppose the given dimensions to be those of

the bottom and the X X' top of the bath, also height of the
bath in front.

First draw A' D' B' C' (Fig. 27) the plan of the bottom by
Problem XII., p. 13. To draw the plan of the X X' top
(Fig. 26) set off O A and O B each equal to half the given
length of that top, and O C and O D each equal to half its
given width. The plan of the X X' top can now be drawn

FIG. 27.

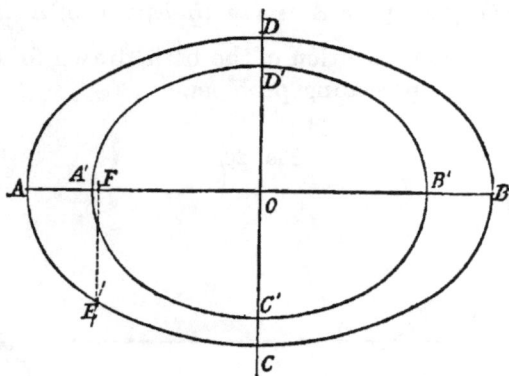

as was that of the bottom. This completes, as stated above,
all that is necessary of the plan of the bath to enable its
pattern to be drawn.

If the length of the X X' top (Fig. 26) is not given but the
inclination of the slant at front and back, these inclinations
being the same, the required length can be determined in the
following manner:—

Make the angle A A' E (Fig. 28a) equal to the given
inclination. Through A' draw A' H perpendicular to A A'
and equal to the given height of the bath in front; through
H draw H X parallel to A A' and cutting A' E in X, and
draw X A perpendicular to A A'; then A A' will be the
distance in plan, at back and front, between the curve of the

bottom and the curve of the X X' top. Make A A' (Fig. 27) and B B' each equal to A A' (Fig. 28a) ; then A B will be the length required.

FIG. 28a.   FIG. 28b.

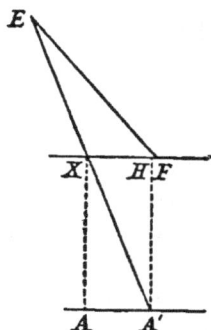

If the length of the X X' top of the bath (Fig. 26) is not given, nor the inclination of the slant at front and back, but only the length of the slant at front, the required length can be thus ascertained.

Draw two lines X B, B' B (Fig. 28b) perpendicular to one another and meeting in B; make B X equal to the given height of the bath in front, and with X as centre, and radius equal to the length of the slant at the front, describe an arc cutting B B' in B'. Make A' A and B' B (Fig. 27) each equal to B B' (Fig. 28b), then A B is the length wanted. The remainder of the plan can be drawn as described above.

By a little addition to Fig. 28a we get at the back portion of the side elevation of the bath. It will be useful to do this. Produce A' X and make A' E equal to the slant at back, which must of course be given. Then, on the plan (Fig. 27), E being the meeting point of the end and side curves of the oval A D B C, draw E F perpendicular to A B. Make X F (Fig. 28a) equal to A F (Fig. 27); join F E; this completes the elevation required.

PROBLEM XIII.

To draw the plan of an oblong taper bath, the size of the top and bottom, the height, and the slant at the head being given.

To draw D E F C (Fig. 30) the plan of the top. Draw A B equal to the given length of the top, and through A and B draw lines perpendicular to A B. Make A E and A D each equal to half the width of the top at the head of the bath, and B F and B C each equal to half the width of the top at the toe; and join E F and D C. Next from E mark off along E F and E D equal distances, E G and E H, according to the size of the round corner required at the head. (It will be useful practice for the student to work this problem, commencing with the plan of the bottom, and its smaller corners, for the reason given in § 27*a*, p. 63). Through G and H draw

Fig. 29*a*.   Fig. 29*b*.

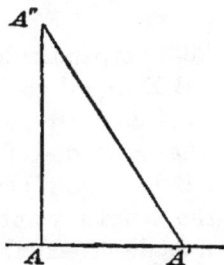

lines perpendicular to E F and E D respectively, intersecting in O; and with O as centre and O G as radius describe an arc H G to form the corner. The round corners at D F C, &c., are drawn in like manner.

To draw the plan of the bottom. Let the angle A″ A′ A (Fig. 29*a*) be the angle of the inclination of the slant at the head, and A′ A″ the length of the slant. Through A″ draw

A″A perpendicular to A A′, then A A′ will be the distance
between the lines, in plan, of the top and bottom at the
head.   Make A A′ (Fig. 30) equal to A A′ (Fig. 29a), and
A′B′ equal to the length of the bottom.   Through A′ and
B′ draw lines each perpendicular to A B; make A′E′ and
A′D′ each equal to half the width of the bottom at the head,
and B′F′ and B′C′ each equal to half the width of the
bottom at the toe.   Join E′ F′ and D′C′.   The round corner
of the bottom at the head must be drawn in proportion to
the round corner of the top at the head, and this is done in
the following manner.   Join E E′ and produce it, to meet

FIG. 30.

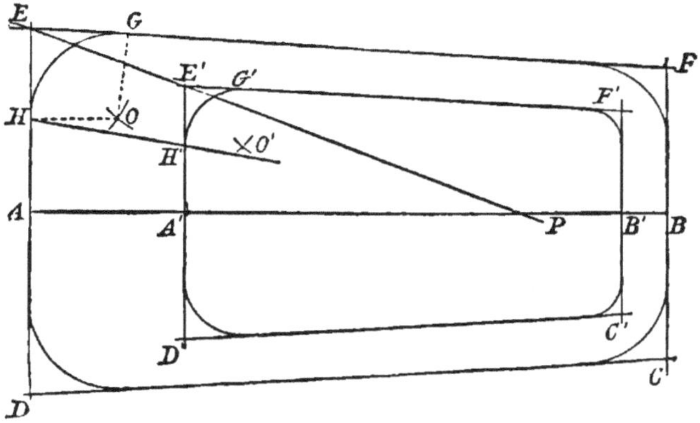

A B in P, and join H P by a line cutting D′E′ in H′; make
E′G′ equal to E′H′, and complete the corner from centre O′
obtained as was the centre O.   Draw the other corners in
similar way, and this will complete the plan required.   The
D corner is like the E corner; the corners also at F and C
correspond.   Similarly with the E′ and D′, and F′ and C′
corners.

If the length of the bath is given and the length of slant at
(but not its inclination) head or toe, the distance A A′ can be

found by drawing two lines A″ A, A′ A (Fig. 29a) perpendicular to one another and meeting in A, and making A A″ equal to the given height; then, with A″ as centre and A″ A′, the given length of the slant at the head, as radius, describe an arc cutting A A′ in A′. Then A A′ is the distance required. Similarly (Fig. 29b) the distance B B′ can be found.

CHAPTER III.

PATTERNS FOR ARTICLES OF UNEQUAL TAPER OR INCLINATION OF SLANT, AND HAVING FLAT (PLANE) SURFACES.

(CLASS II. *Subdivision b.*)

ARTICLES of unequal taper or inclination of slant, and having plane or flat surfaces (hoppers, hoods, &c.), are frequently portions (*frusta*) of oblique pyramids, or parts of such *frusta*.

DEFINITIONS.

(74.) Oblique Pyramid : Frustum of Oblique Pyramid :— Oblique pyramids have not yet been defined. For our purpose it will be sufficient to define an oblique pyramid negatively, that is, as a pyramid which is not a right pyramid ; and when cut by a plane parallel to its base (that

FIG. 31.

is, when truncated), to define its frustum (§ 33, p. 69) as the frustum of a pyramid which is not a right pyramid. In the oblique pyramid the faces are not all equally inclined. Articles of which the faces are not all equally inclined are not necessarily portions of oblique pyramids. One such case will be given later on. The problems immediately following deal with articles of which the faces are not all equally inclined, but which are portions of oblique pyramids.

(75.) Further, an oblique pyramid, when it has a base through the angular points of which a circle can be drawn, can be inscribed in an oblique cone like as a right pyramid in a right cone, and this property gives constructions for solving most of our oblique-pyramid problems, somewhat similar to those in Book I., Chapter VI., where the right pyramid is concerned. Fig. 31 represents an oblique hexagonal pyramid inscribed in an oblique cone. This fig. should be compared with Fig. 31, p. 67. The edges of the oblique pyramid are generating lines of the cone.

This fig. should be compared with Fig. 31, p. 67.

Fig. 32.

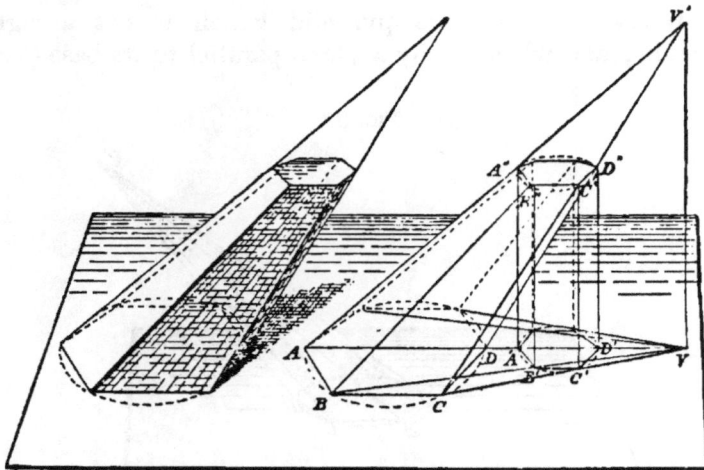

(76.) Also from Fig. 32 it will be seen that the plan of a frustum of an oblique hexagonal pyramid standing on a horizontal plane consists of two hexagons A B C D and A' B' C' D' (the plans of the ends), whose similarly situated sides, A B and

A' B', B C and B' C', C D and C' D' for instance, are parallel, and whose corresponding points (§ 68, p. 124) A, A' and B, B', for instance, are joined by lines A A', B B', which are the plans of the edges of the frustum. Just as in the case of the frustum of the oblique cone (see d and e, p. 126), if a line joining corresponding points in plan be produced, it will contain the plan of the apex of the complete pyramid of which the frustum is a portion; and if another such line be produced to intersect the first line, the point of intersection will be that plan of apex. For example, the lines A A', B B' and C C' produced meet in a point which is the plan of the apex of the pyramid of which the frustum A B D D″B″A″ is a portion.

(77.) From this it follows that if the plan of a tapering body with top and base parallel and having plane or flat surfaces be given, we can at once determine whether the tapering body is or not a frustum of an oblique pyramid by producing the plans of the edges. If these meet in one point, then the given plan is that of a frustum of an oblique pyramid.

PROBLEM XIV.

To draw the pattern of an oblique pyramid.

CASE 1.—Given the plan of the pyramid and its height.

Let A B C D E F V (Fig. 33) be the plan of the pyramid (here a hexagonal pyramid), V being the apex, and O V the plan of the axis. Draw X X parallel to O V, and through V draw V V' perpendicular to X X, and cutting it in v; make v V' equal to the given height of the pyramid. Next make v a, v b, v c, v d, v e, and v f equal respectively to V A, V B, V C, V D, V E, and V F, the plans of the edges of the pyramid. Joining V' a, V' b, V' c, &c., will give the true lengths of these edges.

To draw the pattern of the pyramid with the seam at the edge V A. Draw V A (Fig. 34) equal to V' a (Fig. 33); with

L

V as centre and V' b (Fig. 33) as radius describe arc b, and with
A as centre and A B (Fig. 33) as radius describe an arc intersecting the arc b in B.   The other points C, D E, F, A, are

FIG. 33.

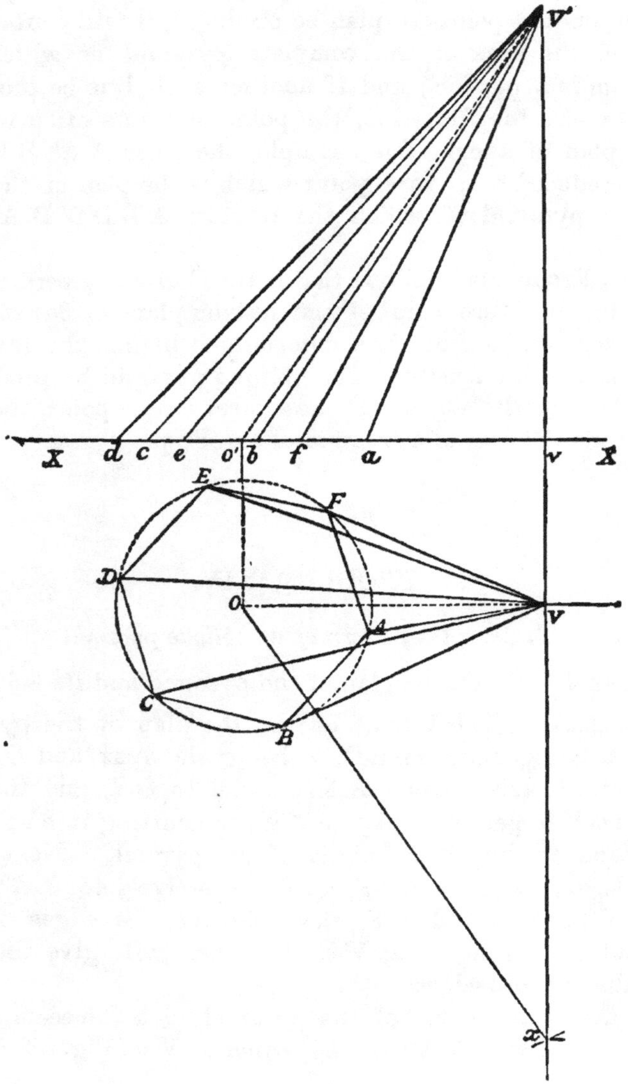

found in similar manner. Thus, with V (Fig. 34) as centre
and V' c, V' d, V' e, V' f, and V' a (Fig. 33) successively as
radii, describe arcs c, d, e, f, and a (Fig. 34). Next, with
B (Fig. 34) as centre and B C (Fig. 33) as radius, describe an
arc intersecting arc c in C ; with C D (Fig. 33) as radius and
C (Fig. 34) as centre describe an arc cutting arc d in D ; with
D (Fig. 34) as centre and D E (Fig. 33) as radius describe an
arc intersecting arc e in E ; and so on for points F and A.
Join A B, B C, C D, DE, E F, F A and A V, and this will
complete the pattern required.

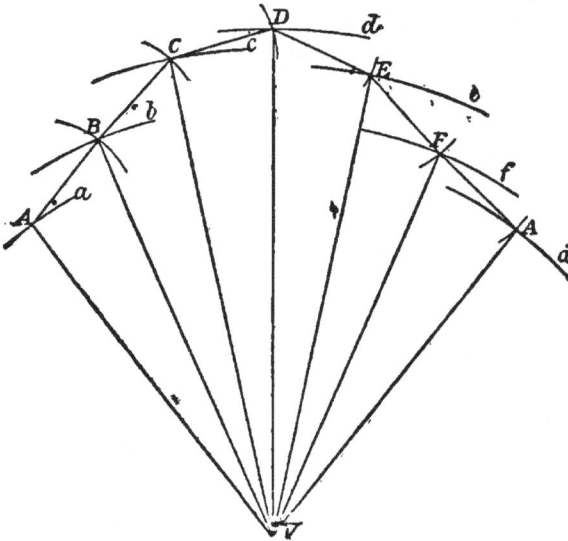

Fig. 34.

Joining the points B, C, D, &c., to V, it will be seen that
the pattern is made up of a number of triangles, each triangle
being of the shape of a face of the pyramid, also that the
construction of the pattern is very similar to the construction
of that of an oblique cone.

Should it be inconvenient to draw X X in the position
shown in Fig. 33, the true lengths of the edges of the pyramid

can be found in the following manner. Draw X X quite apart from the plan of the pyramid, and from any point *v* in it draw *v* V' perpendicular to X X, and equal to the height of the pyramid, and proceed as just described.

CASE II.—Given the plan of the pyramid and the length of its axis.

Draw X X (Fig. 33) parallel to O V, the plan of the axis; through V draw V V' perpendicular to X X, and through O draw O O' perpendicular to X X. With O' as centre and the given length of the axis as radius describe an arc cutting V V' in V'; then *v* V' will be the height of the pyramid, and we now proceed as in Case 1.

Or, draw V *x* perpendicular to O V, and with O as centre and radius equal to the length of the axis describe an arc cutting V *x* in *x*; V *x* will be the height of the pyramid.

PROBLEM XV.

To draw the pattern of a frustum of an oblique pyramid.

CASE I.—Given the plan of the frustum and its height.

Let A B C D D' A' B' C' (Fig. 35) be the plan of the frustum (here of a square pyramid). Produce A A', B B', &c., the plans of the edges to meet in a point V; this point is the plan of the apex of the pyramid of which the frustum is a part. Join O, the centre of the square which is the plan of the large end of the frustum, to V. The line O V will pass through *o'*, the centre of the plan of the small end; O O' will be the plan of the axis of the frustum, and O V the plan of the axis of the pyramid of which the frustum is a portion.

Draw X X parallel to O V; through V draw V V' perpendicular to X X, and cutting it in *v*. Make *v x* equal to the given height of the frustum, and through *x* draw *x x* parallel to X X; through O draw O Q perpendicular to X X and meeting it in Q and through O' draw O'Q' perpendicular to X X and cutting xx in e'. Join Q e' and produce it to intersect v V' in V'. Next make $v\,a, v\,b, v\,c, v\,d$ equal to V A, V B, V C, V D respectively; join a, b, c, and d to V' by lines cutting xx in points a', b', c', and d'; $a\,a', b\,b'$, &c., are the lengths of the edges of the frustum.

FIG. 35.

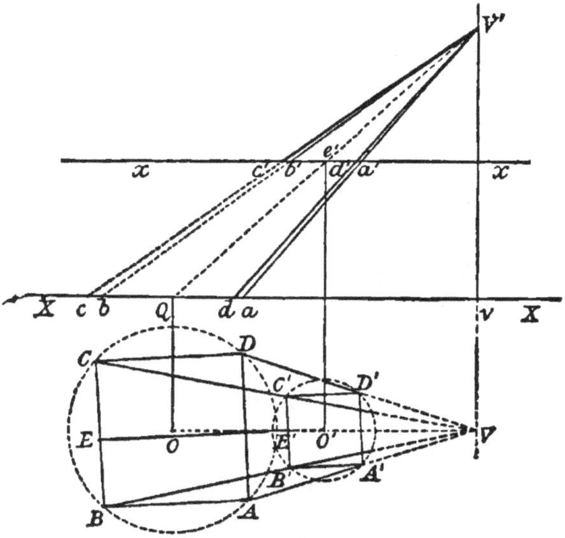

To draw the pattern with the seam at A A'. Draw V A (Fig. 36) equal to V' a (Fig. 35); with V as centre and V' b (Fig. 35) as radius describe an arc b, and with A as centre and A B (Fig. 35) as radius describe an arc intersecting arc b in B; with V' c (Fig. 35) as radius and V as centre describe arc c, and with B C (Fig. 35) as radius and B as centre describe an arc intersecting the arc c in C. Next with V' d and V' a (Fig. 35) as radii and V as centre describe arcs d and a; with C as centre and radius C D (Fig. 35) describe an arc intersecting arc d in D; and with D A (Fig. 35) as radius and D as centre describe an arc intersecting the arc a in A. Join A, B, C, D,

and **A** to **V**; make **A A'**, **B B'**, **C C'**, **D D'** respectively equal
to *a a'*, *b b'*, *c c'*, *d d'* (Fig. 35), and join **A B**, **B C**, **C D**, **D A**,
A' B', **B' C'**, **C' D'**, &c.   Then **A B C D A A' D' C' B A'** is the
pattern required.

FIG. 36.

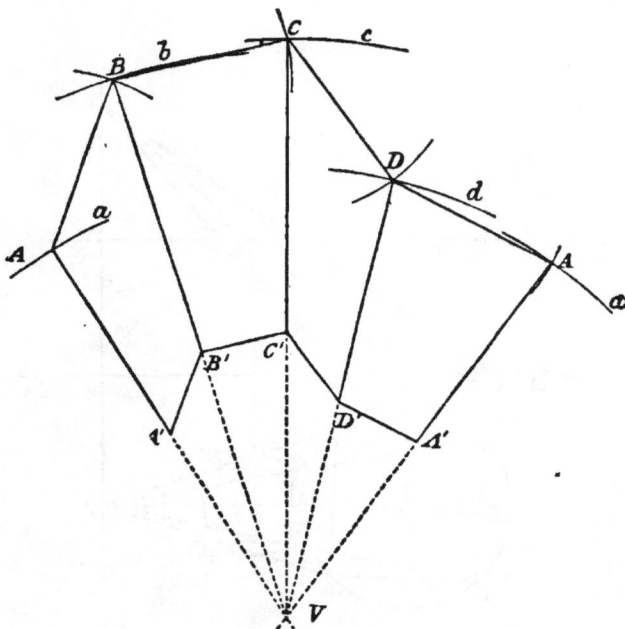

(78.) The dotted circles (Fig. 35) through the angular
points of the plans of the ends show the plans of the ends of
the frustum of the oblique cone which would envelop the
frustum of the pyramid. From the similarity of the construction above to that for the pattern of a frustum of an
oblique cone, it will be evident that we have treated the
edges of the frustum as generating lines (see *b*, p. 126) of the
frustum of the oblique cone in which the frustum of the
pyramid could be inscribed.

Should it be inconvenient to draw **X X** in conjunction with
the plan of the pyramid draw **X X** quite apart, and from any

point v in it draw v V' perpendicular to X X; make $v x$ equal to the height of the frustum and draw $x x$ parallel to X X. Make $v a, v b, v c, v d$ equal to V A, V B, V C, V D (Fig. 35) respectively; and make $x a', x b', x c', x d'$ equal to V A', V B', V C', V D' (Fig. 35) respectively. Join $a a', b b', c c'$, and $d d'$ by lines produced to meet v V' in V', and proceed as stated above.

CASE II.—Given the dimensions of the two ends of the frustum, the slant of one face and its inclination (the slant of the face of a frustum of a pyramid is a line meeting its end lines and perpendicular to them).

Draw (Fig. 37) a line E E" equal to the given slant, make the angle E" E E' equal to the given inclination, and let fall E" E' perpendicular to E E'. Draw A B C D (Fig. 35), the plan of the large end of the frustum, and let B C be the plan of the bottom edge of the face whose slant is given. Bisect B C in E and draw E E' perpendicular to B C and equal to

FIG. 37.

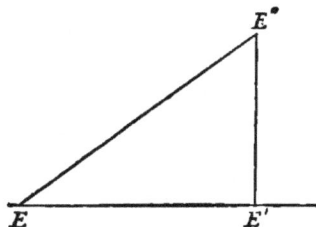

E E' (Fig. 37). Through E' draw B' C' parallel to B C; make E' C' and E' B' each equal to half the length of the top edge of the B C face, through C' and B' draw C' D' and B' A' parallel to C D and B A; make C' D' and B' A' each equal to B' C'; join D' A', also A A', B B', C C', and D D'; this will complete the plan of the frustum. E' E" (Fig. 37) is the height of the frustum. The remainder of the construction is now the same as that of Case I.

For large work and where the ends of a frustum are of nearly the same size, it would be inconvenient to use long radii. For unequal-tapering bodies which are not portions of oblique pyramids, as in Problem XVII., the method now given, or a modification of it, must be used.

PROBLEM XVI.

To draw, **without long radii,** *the pattern for a frustum of an oblique pyramid. The plan of the frustum and its height being given.*

Let A B C D D' A' B' C' (Fig. 38) be the plan of the frustum. From any point E in B C draw E E' perpendicular to B C and B' C', the plans of the bottom and top edges of the face B C B' C' of the frustum. Draw E' E'' perpendicular to E E' and

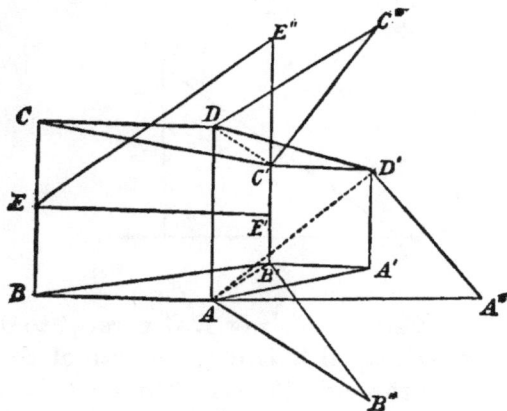

equal to the height (which either is given or can be found as in Case II. of last problem), and join E E'', then E E'' is the true length of a slant of the face B C B' C', of the frustum. Join D C', and find its true length (D C'') by drawing C' C'' perpendicular to D C' and equal to the height of frustum and joining D C". Next join D' A and B' A; through D' and B' draw lines D' A", B' B" perpendicular to D' A and B' A respectively, and make D' A" and B' B" each equal to the given height of the frustum; join A A" and A B", then A A" and A B" are the true lengths of D' A and B' A respectively.

To draw the pattern of the face B C B' C', draw E E' (Fig. 39) equal to E E" (Fig. 38), and through E and E' draw B C and B' C' perpendicular to E E'. Make E C, E B, E' C', and E' B' equal to E C, E B, E' C', and E' B' (Fig. 38) respectively; join C C' and B B'; this completes the pattern of the face. The patterns of the other faces are found in the following manner :—With C' (Fig. 39) and C as centres and

FIG. 39.

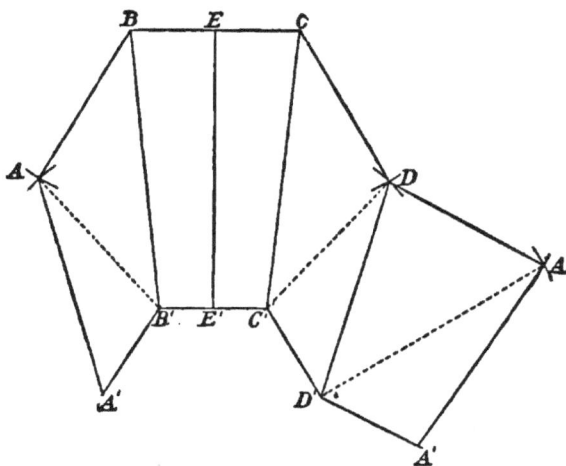

D C" and C D (Fig. 38) as radii respectively, describe arcs intersecting in D; join C D, draw C' D' parallel to C D and equal to C' D' (Fig. 38); and join D D'. With D' and D (Fig. 39) as centres and A A" and D A (Fig. 38) as radii respectively describe arcs intersecting in A ; join D A, draw D' A' parallel to D A and equal to D' A' (Fig. 38), and join A to A'. Next, with B' and B as centres and A B'' and B A (Fig. 38) respectively as radii, describe arcs intersecting in A ; join B A and draw B' A' parallel to B A and equal to B' A' (Fig. 38). Join A A', and this will complete the pattern required.

PROBLEM XVII.

To draw the pattern for a hood.

The plan of the hood is necessarily given, or else dimensions from which to draw it. Also the height of the hood, or the slant of one of its faces. The hood is here supposed to be a body of unequal taper with top and base parallel, but not a frustum of an oblique pyramid.

Let A B C D A' B' C' D' (Fig. 40) be the given plan of the

FIG. 40.

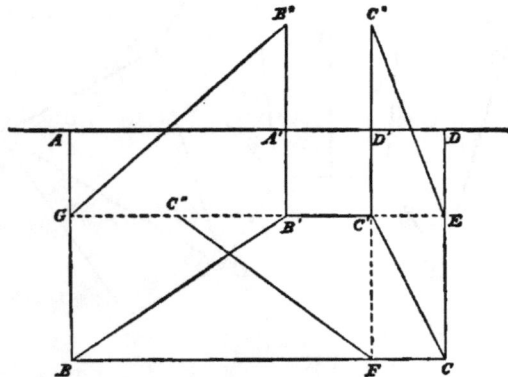

hood (a hood of three faces), A D being the 'wall line,' A B and D C perpendicular to A D and B C parallel to it, also let the length of F C'', a slant of face B B' C' C, be given. Draw C' F perpendicular to B C and through C' draw C' C'' perpendicular to C' F, and with F as centre and radius equal to the given length, describe an arc cutting C' C" in C". Join F C"; then C' C" is the height of the hood, which we need. If the height of the hood is given instead of the length F C", make C' C" equal to the height and join F C", which will be the true length of F C'. Next, through C' draw C' E perpendicular to C D; draw C' C" perpendicular to C' E, make C' C" equal to the height and join E C". Now produce C' B' to meet A B in G; draw B' B" perpendicular to B'G and equal to the height, and join G B".

To draw the pattern of the hood. Draw F C' (Fig. 41) equal to F C" (Fig. 40); through F and C' draw B C and B' C', each perpendicular to F C'; make F B equal to F B (Fig. 40); make F C equal to F C (Fig. 40), and C' B' equal to C' B' (Fig. 40). Join B B' and C C', then B B' C' C will be the pattern of the face of which B B' C' C (Fig. 40) is the plan. To draw the pattern of the face C' D' D C (Fig. 40). With C' and C (Fig. 41) as centres and E C" and C E

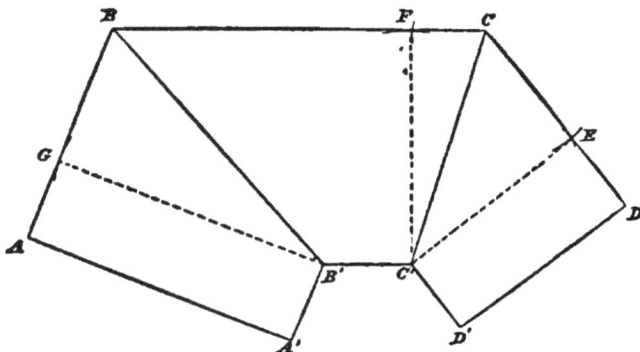

FIG. 41.

(Fig. 40) as radii respectively, describe arcs intersecting in E. Join C E and produce it, making C D equal to C D (Fig. 40), and through C' draw C' D' parallel to C D and equal to C' D' (Fig. 40). Join D D', then C' C D D' is the pattern of the face of which C' C D D' (Fig. 40) is the plan. With B' and B as

centres and radii respectively equal to B″ G and B G (Fig. 40),
describe arcs intersecting in G.   Join B G and produce it,
making B A equal to B A (Fig. 40), and through B′ draw
B′ A′ parallel to B A and equal to B′ A′ (Fig. 40).   Join A A′;
and the pattern for the hood is complete.

CHAPTER IV.

PATTERNS FOR UNEQUAL-TAPERING ARTICLES OF FLAT AND CURVED SURFACE COMBINED.

CLASS II. (*Subdivision c.*)

FROM what has been stated about the plans of unequal-tapering bodies and from *g*, p. 129, it will be evident that the curved surfaces of the articles now to be dealt with are portions of frusta of oblique cones.

(79.) The advantages referred to in § 61 of looking upon the oblique cylinder as frustum of an oblique cone will be evident in this chapter. For there is to each of the problems a Case where the plan arcs of the curved portions of the body treated of have equal radii. To deal with these as problems exceptional to a general principle would be most inconvenient. As extreme cases, however, of the one principle that the curved portions of the bodies before us are portions of frusta of oblique cones, their solution presents no difficulty. It will be sufficient to take one such Case in connection with only one of the bodies. This we shall do in Case IV. of the next problem.

PROBLEM XVIII.

To draw the pattern for an unequal-tapering body with top and base parallel and having flat sides and semicircular ends (an 'equal-end' bath, for instance), the dimensions of top and bottom of the body and its height being given.

Five cases will be treated of; four in this problem and one in the problem following.

CASE I.—Patterns when the body is to be made up of four
pieces.

Draw (Fig. 42) the plan of the body (see Problem VI.,
p. 130), preserving of its construction the centres O, O' and the
points A, A' in which the plan lines of the sides and curves

FIG. 42.

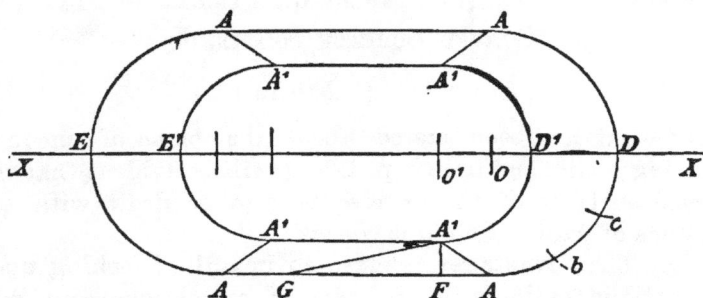

of the ends meet each other.   Join A A', as shown (four
places) in the fig.   The ends A D A A' D' A' and A E A A' E' A'
of the body (see g, p. 129) are portions of frusta of oblique
cones.   Let us suppose that the seams are to be at the four A
corners where they are usually placed, and to correspond
with the four lines A A'.   Then we shall require one pattern
for the flat sides, and another for the semicircular ends.

To draw the end pattern.

(80.) Draw A b D D' A' (Fig. 43) the A b D D' A' portion of
Fig. 42 separately, thus.   Draw any line X X and with any
point O (to correspond with O, Fig. 42) in it as centre and
O D (Fig. 42) as radius describe an arc (here a quadrant) D b A
equal to the arc D b A (Fig. 42).   Make D O' equal to D O'
(Fig. 42), and with O' as centre and O' D' (Fig. 42) as radius
describe an arc (here a quadrant) D' A' equal to the arc D' A'
(Fig. 44).   Joining A A' completes the portion of Fig. 42
required.   Now divide D A into any number of equal parts,
here three, in the points b and c.   From D' draw D' D" perpendicular to ˒X X and equal to the given height.   Then
D, D" are, in elevation, the corresponding points of which

D, D' are the plans. Being corresponding points, they are
in one and the same generating line (§ 68). Join D D" and
produce it indefinitely, then somewhere in that line will lie
the elevation of the apex of the cone of a portion of which

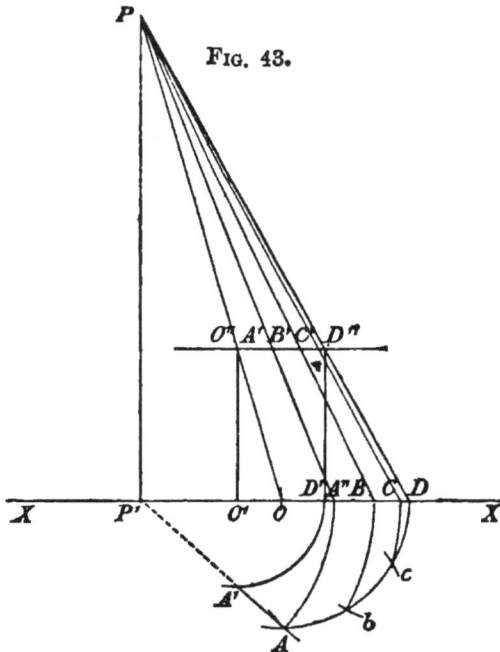

Fig. 43.

A b D D' A' is the plan. Now from O' draw O' O" perpendicular to X X and equal to the given height. Then O, O"
are, in elevation, the centres of the ends of the frustum in
the same plane that D, D" are represented in, that is in the
plane of the paper; O, O' being the plans of these centres.
Join O O" and produce it indefinitely; then in this line lies
the elevation of the axis of the cone of a portion of which
A b D D' A' is the plan, and necessarily therefore the elevation
of the apex. That is, the intersection point P of these two
lines is the elevation of the apex. Next, from P let fall P P'
perpendicular to X X; then P' will be the plan of the apex.
Join D" O". With P' as centre and P' c, P' b, and P' A suc-

cessively as radii, describe arcs cutting X X in C, B, and A″. Join these points to P by lines cutting O″ D″ in C′, B′, and A′.

Next draw a line P D (Fig. 44) equal to P D (Fig. 43), and with P as centre and P C, P B, and P A″ (Fig. 43) successively as radii describe arcs $c\,c$, $b\,b$, and $a\,a$. With D as centre and radius equal D c (Fig. 43) describe arcs cutting arc $c\,c$ in C and C right and left of P D. With same radius and these points C ·and C successively as centres describe arcs cutting arc $b\,b$ in B and B right and left of P D. With B and B

FIG. 44.

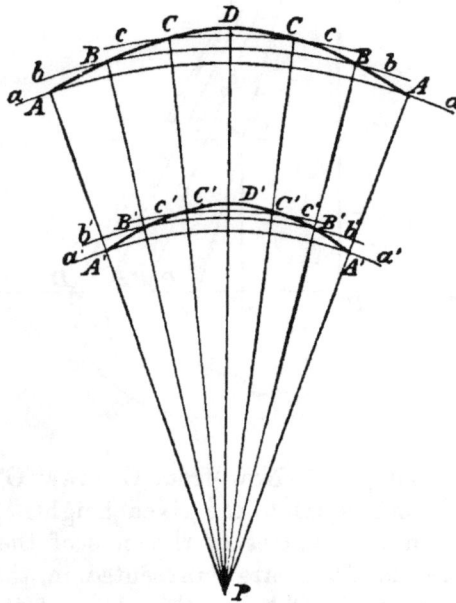

successively as centres and same radius describe arcs cutting arc $a\,a$ in A and A right and left of P D. Join the points C, B, and A right and left of P D to P. With P as centre and P D″ (Fig. 43) as radius describe an arc cutting P D in D″. With same centre, and P C′ (Fig. 43) as radius, describe arc $c'\,c'$ cutting lines P C right and left of P D in C′ and C′. With same centre, and P B′ (Fig. 43) as radius, describe arc

b' b' cutting lines P B right and left of P D in B' and B'.
Similarly find points A' and A'. Through the successive
points A, B, C, D, C, B, A, draw an unbroken curved line.
Also through the successive points A', B', C', D', C', B', A,
draw an unbroken curved line. Then A D A A' D' A' will
be the required pattern for ends of the body.

To draw the pattern for the sides.

Through A' (Fig. 42) draw A' F perpendicular to A' A'
make F G equal to the given height and join A' G. Then
A' G is the slant of the body at the side. Next draw
(Fig. 45) A A equal to A A (Fig. 42), and make A F equal

FIG. 45.

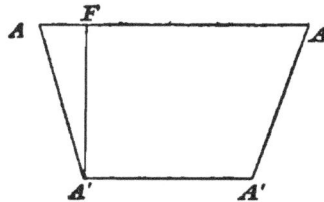

to A F (Fig. 42); through F draw F A' perpendicular to A A
and equal to A' G (Fig. 42), and through A' draw A' A'
parallel to A A. Make A' A' equal to A' A' (Fig. 42). Join
A A', A A', then A A' A' A is the pattern for the sides.

CASE II.—Pattern when the body is to be made up of two
pieces.

We will take it that the seams are to be at D D' and E E'
(Fig. 42). It is evident that we want but one pattern, which
shall include a side of the body and two half-ends.

First draw as just explained A' A F A A' (Fig. 46) a side-
pattern of the body. Produce one of the lines A A' of this
pattern, and make A P' equal to A" P (Fig. 43). With P' as
centre and P B, P C, and P D (Fig. 43) successively as radii
describe arcs b, c, and d, and with A as centre and A b
(Fig. 43) as radius describe an arc cutting arc b in B. With
same radius and B as centre describe an arc cutting arc c
in C; similarly with C as centre and same radius find D.

M

Join B P′, C P′, D P′. Now with P B′ (Fig. 43) as radius and P″ as centre describe an arc b′ cutting P B in B′, and with P C′, P D″ (Fig. 43) successively as radii describe arcs c′ and d cutting P′ C and P′ D in C′ and D′. Through the points

FIG. 46.

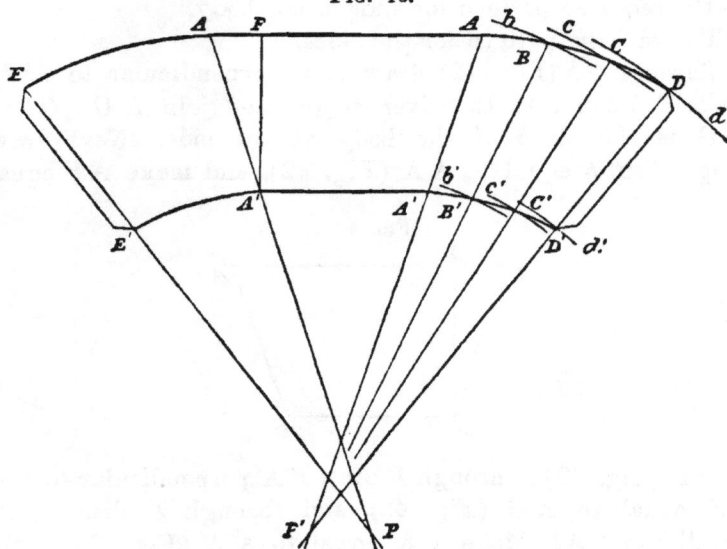

A, B, C, and D draw an unbroken curved line. Also through the points A′, B′, C′, and D′ draw an unbroken curved line. Then A D D′ A′ will be a half-end pattern attached to the right of the side pattern. Draw the other half-end pattern A E E′ A′ in the same manner; then E F D D′ A′ A′ E′ will be the complete pattern required.

CASE III.—Pattern when the body is to be made up of one piece.

In this case we will put the seam to correspond with D D′ (Fig. 42).

First draw A E A A′ E′ A′ (Fig. 47) an end pattern of the body in the same manner that A D A A′ D′ A′ (Fig. 44) was drawn. With A′ and A (right of E E′) as centres and A′ G and A F (the small length A F) (Fig. 42) respectively as radii describe arcs intersecting in F; join A F and produce it, making A A equal to A A (Fig. 42). Through A' (extremity of F A') draw A' A' parallel to A A and equal to

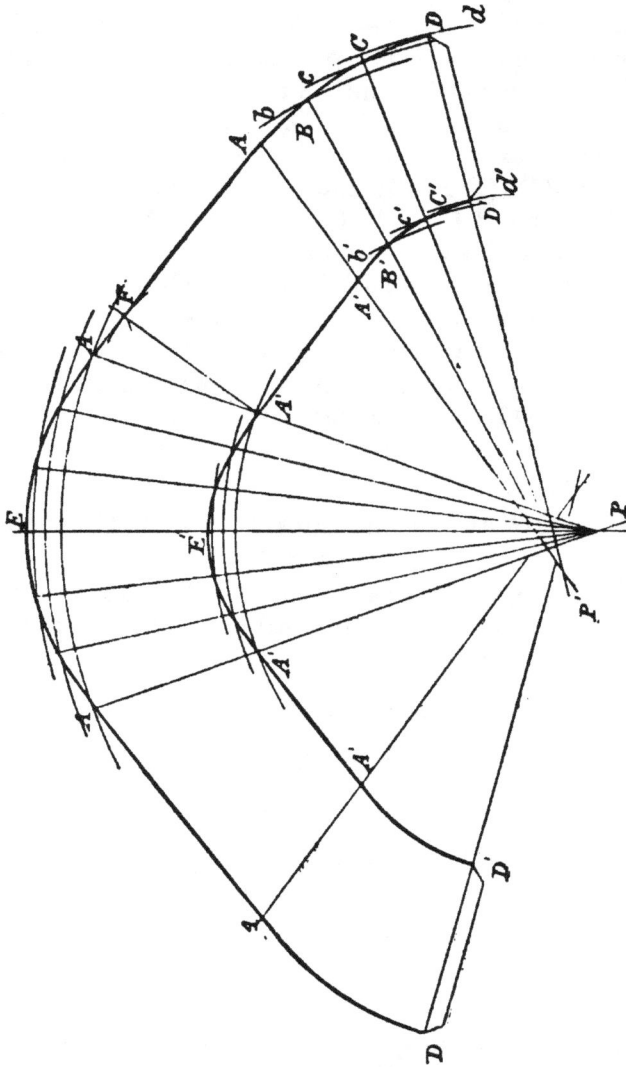

FIG. 47

A′ A′ (Fig. 42). Join A A′, the extremities of the lines just drawn, and produce it indefinitely; and make A P′ equal A″ P (Fig. 43). With P′ as centre and P B, P C, and P D (Fig. 43) successively as radii describe arcs b, c, and d, and with A (of A P′) as centre and A b (Fig. 43) as radius describe an arc cutting arc b in B. With same radius and B as centre describe an arc cutting arc c in C, and similarly with C as centre and same radius find D. Join B P′, C P′, and D P′. Now with P′ as centre and radii successively equal to P B′, P C′, and P D″ (Fig. 43) describe arcs b′, c′, and d′ cutting P′B, P′C, and P′D in B′, C′, and D′. Through the points A, B, C, and D draw an unbroken curved line. Also through the points A′, B′, C′, and D′ draw an unbroken curved line. We have now in D F A A′ A′ D′ attached to the right of the end pattern we started with, a side pattern and a half-end pattern. By a repetition of the foregoing construction we can attach A A D D′ A′ A′ to the left of the end pattern we started with. The figure D E D D′ E′ D′ will be the complete pattern required.

CASE IV.—Where the plan arcs D A, D′ A′ (Fig. 42) have equal radii.

This is the extreme case above (§ 79, p. 157) referred to, where the cone becomes cylindrical. Problem V., p. 121, may

FIG. 48.

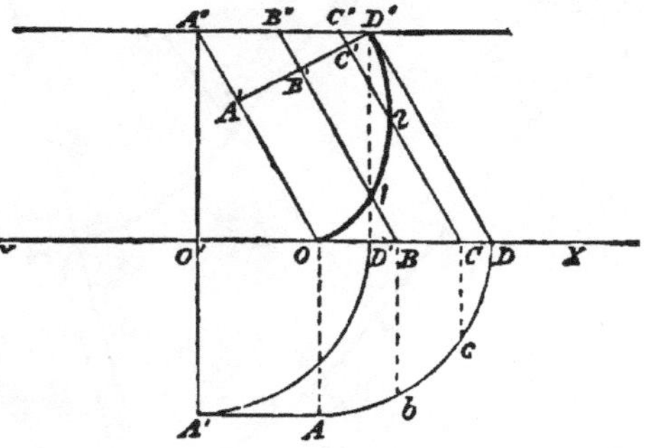

advantageously be compared with the work now given. The arcs (Fig. 48) D A and D′ A′ (here quadrants) being equal,

their radii O D, O' D' are also equal. Through D' and O' draw
D' D", and O' A" perpendicular to X X and each equal to the
given height of the body. Join D D", A" D"; divide the arc
D A into any number of equal parts, here three, in the
points *b* and *c*; and through *c*, *b*, and A draw *c* C, *b* B, and
A O each perpendicular to X X and cutting it in C, B, and O
respectively. The arc D A being here a quadrant the point
where the line from A perpendicular to X X cuts X X is
necessarily O, the centre whence the arc is drawn. Through
C, B, and O draw C C", B B", and O A" parallel to D D" and
cutting A" D" in points C", B", and A". Also through D"
draw a line D" A' perpendicular to D D" and cutting the
lines just drawn in C', B' and A'. Make *c* 2 equal to C *c*;
B' 1 equal to B *b*, and A' O equal to O A. From D" through
2, 1, to O draw an unbroken curved line.

To draw the pattern.

Draw D D" (Fig. 49) equal to D D" (Fig. 48) and through
D" draw an indefinite line A' A' perpendicular to D D".

FIG. 49.

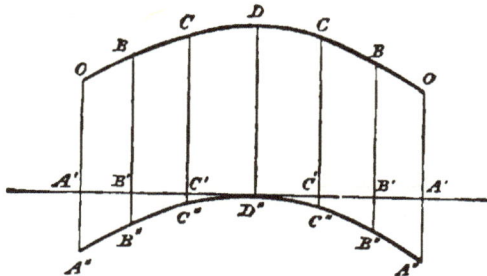

Mark off on A' A', right and left of D", D" C', C' B', and B A'
respectively equal to the distances between D" and 2, 2 and
1, and 1 and 0 (Fig. 48); and through C', B', and A', right
and left of D" draw indefinite lines each parallel to D D".
Make C' C right and left of D D" equal to C' C (Fig. 48); and
make B' B right and left of D D" equal to B' B (Fig. 48).
Also make A' O right and left of D D" equal to A' O (Fig. 48).

Next make $C'C''$ right and left of DD'' equal to $C'C''$ (Fig. 48). Similarly find points B'', A'' right and left of DD'' by making $B'B''$, $A'A''$ respectively equal to $B'B''$, and $A'A''$ (Fig. 48). Through the points O, B, C, D, C, B, O, draw an unbroken curved line. Also through the points A'', B'', C'', D'', C'', B'', A'', draw an unbroken curved line. Then $O D O A'' D'' A''$ will be the pattern required.

PROBLEM XIX.

To draw, **without long radii,** *the pattern for an unequal-tapering body with top and base parallel and having flat sides and equal semicircular ends (an ' equal-end' bath, for instance). The dimensions of the top and bottom of the body and its height being given.*

This problem is a fifth case of the preceding, and is exceedingly useful where the work is so large that it i inconvenient to draw the whole of the plan, and to use long radii.

To draw the pattern.

FIG. 50.

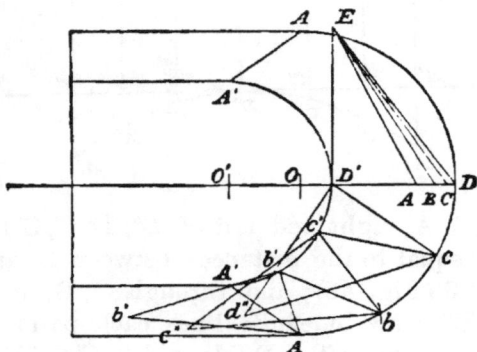

First draw half the plan (Fig. 50). It is evident that the drawing of the side pattern presents no difficulty, as long

radii are not involved. It can be drawn as in Case I. of
preceding problem. Divide the quadrants D A, D' A', each
into the same number of equal parts, here three, in the points
c, b, c', b'; join $c c', b b'$. Through D' draw D' E perpendicular
to D' D and equal to the given height of the body. From
D' along D' D mark off D' A, D' B, and D' C respectively
equal to A' A, $b' b$, and $c' c$; and join points D, C, B, and A to
E, then E A, E B, E C, and E D, will be the true lengths of
A' A, $b' b$, $c' c$, and D' D respectively. Next join c D', and
draw D' d'' perpendicular to D' c and equal to the given
height. Join $c d''$, then $c d''$ may be taken as the true length
of D' c. Similarly join $b c'$ and A b', and through c' and b'
draw $c' c''$ and $b' b''$ perpendicular to $c' b$ and b' A respectively,
and each equal to the given height. Join $b c''$ and A b'', then
$b c''$ and A b'' may be taken as the true lengths of $b c'$ and A b'
respectively.

Now draw (Fig. 51) a line D D' equal to D E (Fig. 50) and
with D' and D as centres and radii respectively equal to $d'' c$

FIG. 51.

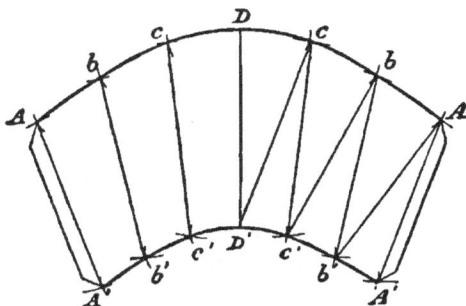

and D c (Fig. 50) describe arcs right and left of D D', intersecting in c and c. With c, right of D, and D' as centres, and
radii respectively equal to C E and D' c' (Fig. 50) describe
arcs intersecting in c', right of D'. With c, left of D, and D'
as centres, and radii as before, describe arcs intersecting in c',
left of D'. With successively c' and c right and left of D D',

as centres and radii respectively equal to $c'' b$ and $c\, b$ (Fig. 50) describe arcs intersecting in b and b right and left of D D'. With successively b and c' right and left of D D' as centres and radii respectively equal to B E and $c'\, b'$ (Fig. 50) describe arcs intersecting in b', and b' right and left of D D', and with successively b' and b right and left of D D' as centres and radii respectively equal to A b'' and b A (Fig. 50) describe arcs intersecting in A and A right and left of D D'. Similarly with A E and b' A' (Fig. 50) as radii and centres respectively A and b' describe intersecting arcs to find points A' and A', right and left of D' D. Through the points A, b, c, D, c, b, A draw an unbroken curved line. Also through the points A', b', c D', c', b', A', draw an unbroken curved line. Join A A', A A', right and left of D D', then A D A A' D' A' will be the pattern required.

The lines $c\, c'$, c D', $b\, b'$, &c., are not needed in the working; they are drawn here to aid the student by showing him how the pattern corresponds with the plan, line for line of same lettering (see also § 65, p. 121).

PROBLEM XX.

To draw the pattern for an oval unequal-tapering body with top and base parallel (an oval bath, for instance). The height and dimensions of the top and bottom of the body being given.

Four cases will be treated of, three in this problem, and one in the problem following (see also § 79, p. 157).

Draw (Fig. 52) the plan of the body (see Problem VII., p. 131), preserving of its construction, the centres O, O', P, P' and the several points d and d' in which the side and end curves meet each other. Join $d\, d'$, as shown (four places) in the fig. From the plan we know (see g, p. 129) that d G d d' G' d', d B $d\, d'$ B' d', the ends of the body are like portions of the frustum of an oblique cone; we also know that

d A d d' A' d', d E d d' E' d', the sides of the body are like portions of the frustum of an oblique cone.

In Plate I. (p. 181), is a representation of the oval unequal-

Fig. 52.

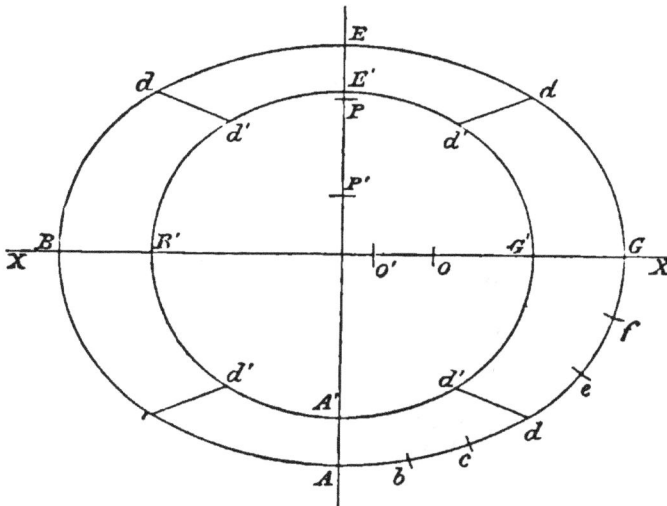

tapering body for which patterns are required, also of two oblique cones (x and Z). The oblique cones show (except as to dimensions) to what portions of their surfaces the several portions of the surface of the oval body correspond. Thus the sides, A', of the body correspond to the portion A of cone x, and the ends, B', B', correspond to the B portion cone Z. The correspondence will be more fully recognised as we proceed with the problem. The difference of obliquity between B' and B is seeming only, not real ; and arises simply from Z being turned round so that the whole of the d B d d' B' d' (Fig. 52) of the cone shall be seen. If the representation of Z showed its full obliquity, then the line on it from base to apex would be the right-hand side line of the cone, and only half of the B portion could be seen.

F_IG_. 53

CASE I.—Patterns when the body is to be made up of four pieces.

It is clear that we require two patterns; one for the two ends, and one for the two sides; also that the seams should correspond with the four lines $d\,d'$, where the portions of the respective frusta meet each other.

To draw the pattern for the ends.

Draw separately (Fig. 53) the $G'\,G\,f\,d\,d'$ portion of Fig. 52, thus. Draw any line X X, Fig. 53, and with any point O (corresponding to O, Fig. 52) in it as centre and O G (Fig. 52) as radius, describe an arc G d equal to G d of Fig. 52. Make G O' equal to G O' (Fig. 52), and with O' as centre and O' G' (Fig. 52) as radius describe an arc G' d' equal to G' d' of Fig. 52. Joining $d\,d'$ completes the portion of Fig. 52 required. Now divide the arc G d into any number of equal parts, here three, in the points f and e. At G' and O' draw G' G'', O' O'' perpendicular to X X, and each equal to the given height of the body. Join O O'', G G''; produce them to their intersection in S (§ 80, p. 158); and from S let fall S S' perpendicular to X X. Join O'' G''. With S' as centre and S' f, S' e, and S' d successively as radii, describe arcs cutting X X in F, E, and D. Join these points to S by lines cutting O'' G'' in F', E', and D'.

Next draw S G (Fig. 54) equal to S G (Fig. 53), and with S as centre and S F, S E, and S D (Fig. 53) successively as radii describe arcs $f\!f$, $e\,e$, and $d\,d$. With G as centre and radius equal to Gf (Fig. 53) describe arcs cutting arc $f\!f$ right and left of S G in F and F. With each of these points, F and F as centre and same radius describe arcs cutting arc $e\,e$ right and left of S G in E and E. With same radius and each of the last-obtained points as centre describe arcs cutting $d\,d$ right and left of S G in D and D. Join all the points right and left of S G to S. With S as centre and S G'' (Fig. 53) as radius, describe an arc cutting S G in G'. With same centre and S F' (Fig. 53) as radius, describe an arc $f'\!f'$ cutting the lines S F right and left of S G in F' and F'. With same centre and S E' (Fig. 53) as radius, describe an

Fig. 54.

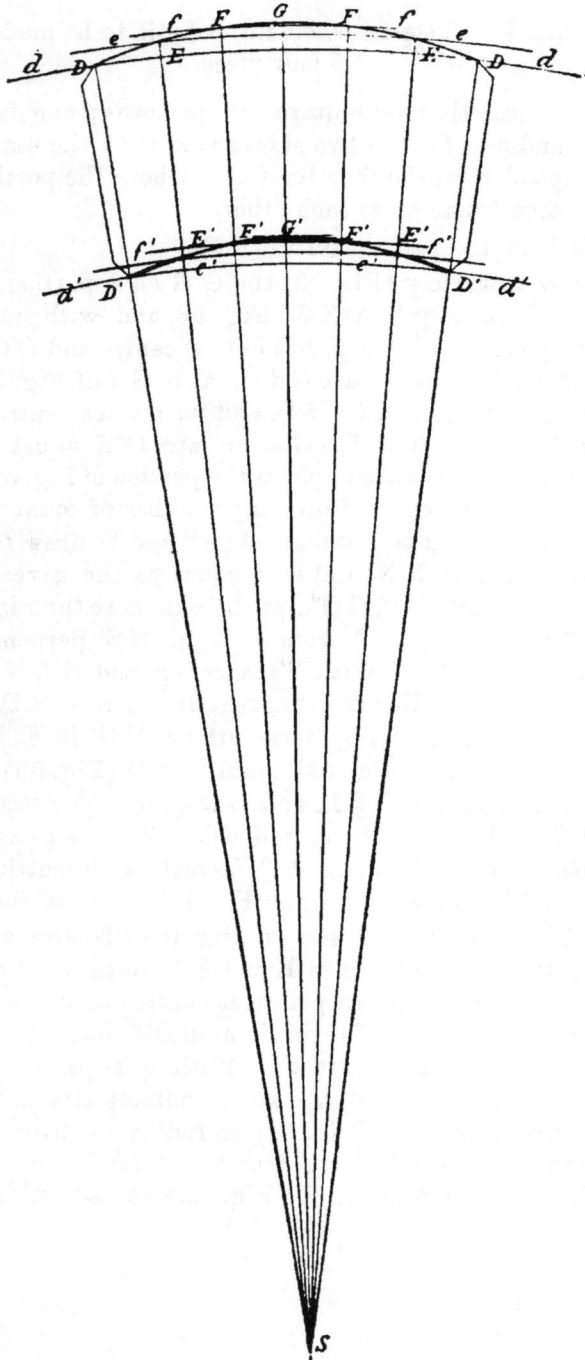

arc e' e' cutting the lines S E right and left of S G in E' and E'.
Similarly by arc d' d' obtain points D' and D'. Through the
points D, E, F, G, F, E, D, draw an unbroken curved line.
Also through points D', E', F', G', F', E', D', draw an unbroken
curved line. Then D G D D' G' D' will be the required pattern
for the ends of the body. It is in fact the development of
the B portion of cone Z of Plate I.

To draw the pattern for the sides.

Draw separately (Fig. 55) the A' A b d d' portion of Fig. 52,
:hus. Draw any line X X, Fig. 55. and with any point P

FIG. 55.

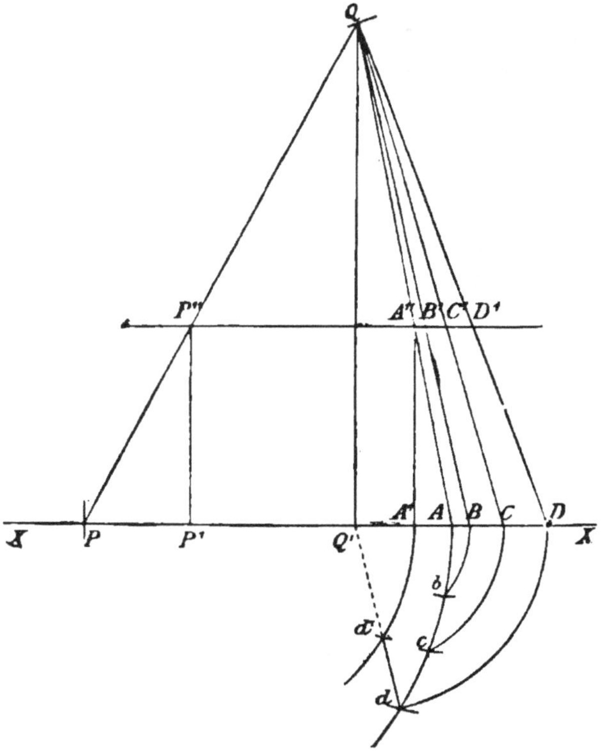

(corresponding to P, Fig. 52) in it as centre and P A (Fig. 52)
as radius, describe an arc A d equal to A d of Fig. 52. Make

A P' equal to A P' (Fig. 52) and with P' as centre and P' A'
(Fig. 52) as radius describe an arc A' d' equal to A' d' of
Fig. 52. Joining d d' completes the portion of Fig. 52 required.
Now divide the arc A d into any number of equal parts, here
three, in the points b and c. At A' and P' draw A' A", P' P"
perpendicular to X X, and each equal to the given height
of the body. Join P P", A A", produce them to their intersection in Q (§ 80, p. 158); and from Q let fall Q Q' perpendicular to X X. Draw a line through P" A". With Q' as
centre and Q' b, Q' c, and Q' d successively as radii describe
arcs cutting X X in B, C, and D. Join these points to Q by
lines cutting that through P" A", in B', C', and D'.

Next draw Q A (Fig. 56) equal to Q A (Fig. 55), and with
Q as centre and Q B, Q C, and Q D (Fig. 55) successively as

FIG. 56.

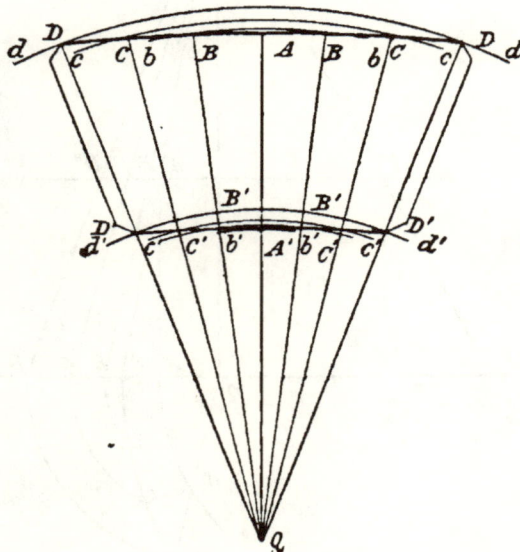

radii describe arcs b b, c c, and d d. With A as centre and
radius equal to A b (Fig. 55) describe arcs cutting arc b b
right and left of Q A in B and B. With each of these points

B and B as centre and same radius describe arcs cutting arc
c c right and left of Q A in C and C.   With same radius and
each of the last-named points as centre describe arcs cutting
arc d d right and left of Q A in D and D.   Join all the points
right and left of Q A to Q.   With Q as centre and Q A''
(Fig. 55) as radius, describe an arc cutting Q A in A'.   With
same centre, and Q B' (Fig. 55) as radius, describe an arc b' b
cutting the lines Q B right and left of Q A in B' and B'.
With same centre and Q C' (Fig. 55) as radius, describe arc
c' c' cutting the lines Q C right and left of Q A in C' and C'.
Similarly by arc d' d' obtain points D' and D'.   Through the
points D, C, B, A, B, C, D, draw an unbroken curved line.
Also through the points D', C', B', A', B', C', D', draw an unbroken curved line.   Then D A D D' A' D' will be the required
pattern for the sides of the body, and is in fact the development of the A portion of cone x of Plate I.

CASE II.—Pattern when the body is to be made up of two
pieces.

In this case the seams are usually made to correspond with
B B' and G G' (Fig. 52).   It is evident that only one pattern
is now required, made up of a pattern for the side A' of the
body (Plate I.) with right and left a half-end (B', B', Plate I.)
pattern attached.

Draw (Fig. 57) a side pattern D A D D' A' D' as described
in Case I.   Produce D Q and make D S equal to D S (Fig. 53).
With S as centre and S E, S F, and S G (Fig. 53) successively
as radii describe arcs e, f, and g, and with D as centre and
d e (Fig. 53) as radius describe an arc cutting arc e in E.
With same radius and E as centre describe an arc cutting
arc f in F, and similarly with F as centre and same radius
find G   Join E S, F S, G S.   Now with S E' (Fig. 53) as
radius and S as centre describe an arc e' cutting S E in E',
and with S F' and S G'' (Fig. 53) successively as radii describe
arcs f' and g' cutting S F and S G in F' and G'.   From points
D to G draw an unbroken curved line.   Also from points D'
to G' draw an unbroken curved line.   Draw the other half-end

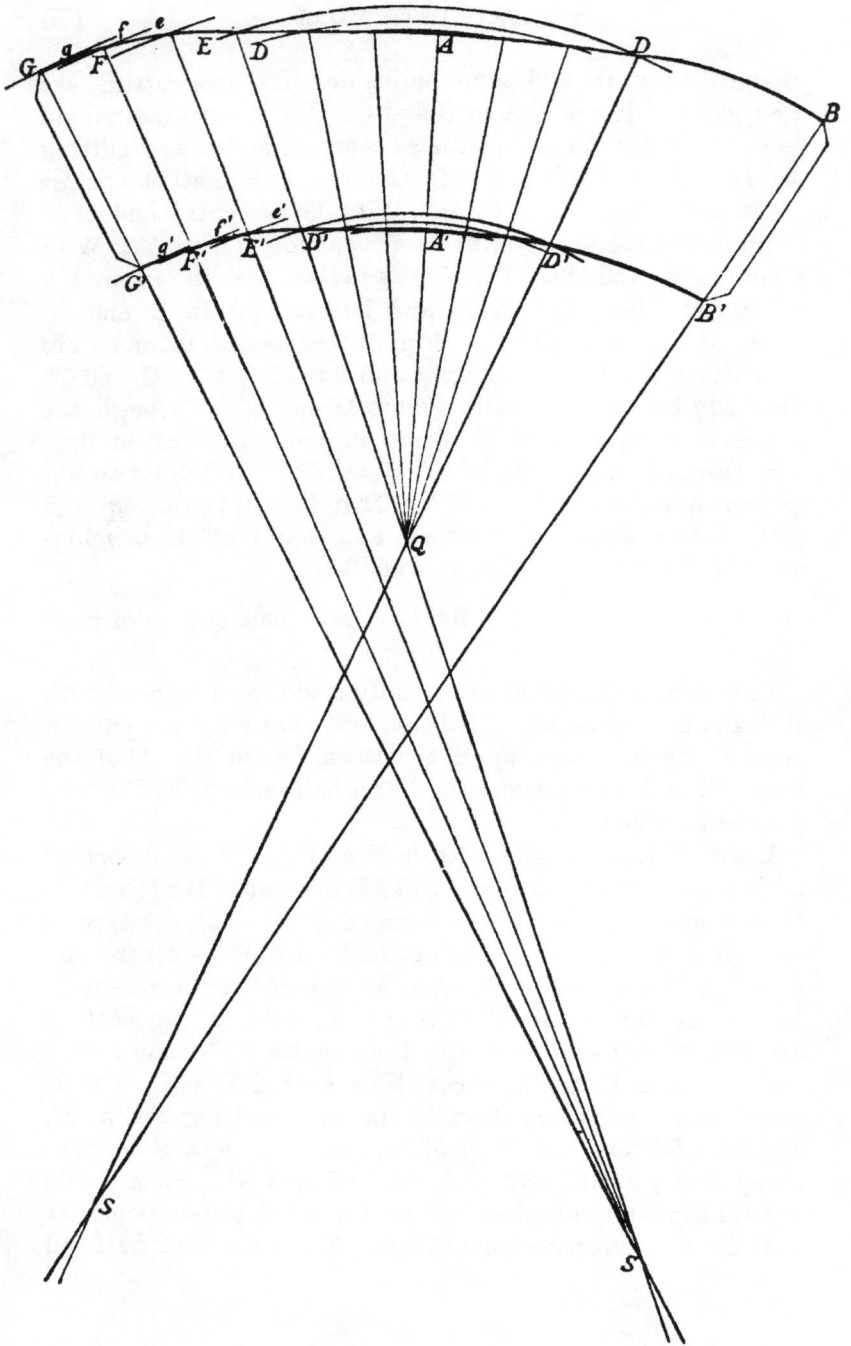

pattern D B B′ D′ in the same manner; then G A B B′ A′ G′ will be the pattern required.

CASE III.—Pattern when the body is to be made up of one piece.

We will put the seams at the middle of one end of the body, say, to correspond with B B′ (Fig. 52). We now need an end pattern (the end *d* G *d d′* G′ *d′* in plan), with side pattern attached right and left (*d* E *d d′* E′ *d′*, *d* A *d d′* A′ *d′* in plan), and attached to each of these a half-end pattern (*d* B B′ *d′*, *d* B B′ *d′* in plan). For want of space we do not give the pattern, but it is evident from what has just been stated, that the pattern will be double that shown in Fig. 57. It will be a useful exercise and should present no difficulties to the student, to himself draw the complete pattern, first drawing an end pattern (see Case I. and Fig. 54) and attaching right and left, a side pattern and a half-end pattern.

PROBLEM XXI.

To draw, without long radii, *the pattern for an oval unequal-tapering body, with top and base parallel (an oval bath, for instance). The height and the dimensions of the top and bottom of the body being given.*

This problem is a fourth case of the preceding, and will be found very useful for both the end and side patterns, the radii of which are often of a most inconvenient length.

To draw the end pattern.

First draw (Fig. 58) the plan of the end of the body; that is the *d* G *d d′* G *d′* portion of Fig. 52. Divide the arcs G *d*, G′ *d′* each into the same number of equal parts, here three, in the points *f*, *e*, *f′*, *e′*; join *f f′*, *e e′*. Through G draw G H perpendicular to G G′ and equal to the given height of the body. From G along G G′ mark off G F, G E, and G D respectively equal to *f f′*, *e e′*, and *d d′*; join G′, F, E and D to H; then G′ H, F H, E H, and D H will be the true lengths of

N

GG', ff', ee', and dd' respectively. Next join fG'; draw $G'g''$ perpendicular to fG' and equal to the given height and join fg''; then fg'' may be taken as the true length of $G'f$. Similarly join ef', de', and through f' and e' draw $f'f''$ and

FIG. 58..

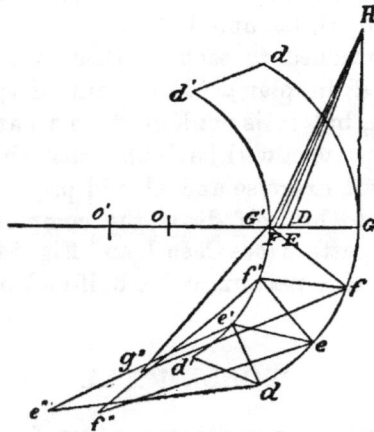

$e'e''$ perpendicular to $f'e$ and $e'd$ respectively, and each equal to the given height, and join ef'' and de''; then ef'' and de'' may be taken as the true lengths of ef' and de' respectively.

Next draw (Fig. 59) a line GG' equal to $G'H$ (Fig. 58) and with G' and G as centres and radii respectively equal to fg'' and Gf (Fig. 58) describe arcs right and left of GG', intersecting in f and f. With f, right of G, and G' as centres, and radii respectively equal to FH and $G'f'$ (Fig. 58) describe arcs intersecting in f', right of G'. With f, left of G, and G' as centres, and radii respectively as before, describe arcs intersecting in f', left of G'. With successively f' and f, right and left of GG', as centres and radii respectively equal to ef'' and fe (Fig. 58) describe arcs intersecting in e and e. With successively e and f' right and left of $G'G$ as centres and radii respectively equal to EH and $f'e'$ (Fig. 58) describe arcs intersecting in e' and e'; and with successively e' and e right and left of G G' as centres, and radii respectively equal to $d\,e''$ and $e\,d$ describe arcs intersecting in d and d. Also with successively d and e' as centres and D H and $e'\,d'$ respectively as radii describe arcs intersecting in d' and d'.

FIG. 59.

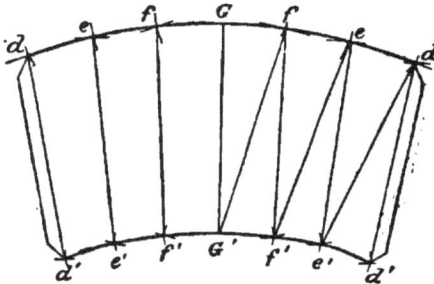

END PATTERN.

Through d, e, f, Gf, e, d, draw an unbroken curved line. Also through d', e', f', G', f', e', d', draw an unbroken curved line. Join $d\,d'$, right and left of G G', then d G d d' G' d' will be the end pattern required.

The lines ff', $e\,e'$, G'f, $f'\,e$, &c., are not needed in the working, they are drawn for the reason stated in § 65, p. 121.

To draw the side pattern.

First draw (Fig. 60) the plan of the side of the body; that is the d A d d' A' d' portion of Fig. 52. Divide the arcs A d, A' d' each into the same number of equal parts, here three, in the points b, c, b', c'; join $b\,b'$, $c\,c'$. Through A draw A E perpendicular to A A' and equal to the given height of the body. From A along A A' mark off A B, A C, and A D respectively equal to $b\,b'$, $c\,c'$, and $d\,d'$. Join A', B, C and D to E; then A' E, B E, C E, and D E will be the true lengths of A A', $b\,b'$, $c\,c'$, and $d\,d'$ respectively. Next join b' A and draw $b'\,b''$ perpendicular to b' A and equal to the given height.

N. 2

Join A b''; then A b'' may be taken as the true length A b'. Similarly join $c'b$ and $d'c$; through c' and d' draw $c'c''$, and $d'd''$ perpendicular to $c'b$ and $d'c$ respectively, and each equal to the given height; join $b\,c''$, $c\,d''$, then $b\,c''$ and $c\,d''$ may be taken as the true lengths of $b\,c'$ and $c\,d'$ respectively.

FIG. 60.

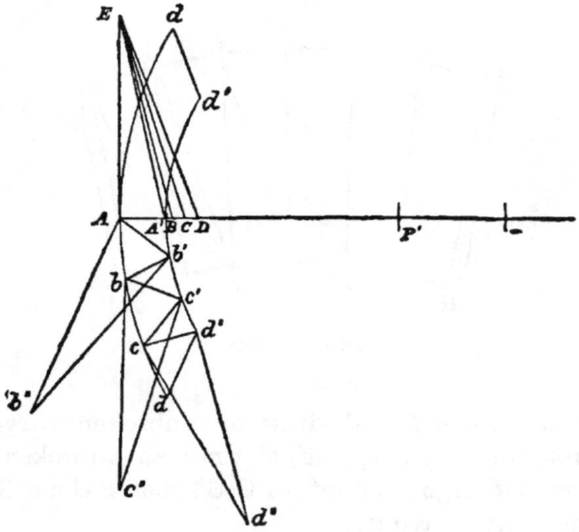

Next draw (Fig. 61) a line A A' equal to A'E (Fig. 60) and with A and A' as centres and radii respectively equal to A b'' and A'b' (Fig. 60) describe arcs right and left of A A', intersecting in b' and b'. With b', right of A', and A as centres, and radii respectively equal to B E and A b (Fig. 60) describe arcs intersecting in b, right of A. With b', left of A', and A as centres, and radii respectively as before, describe arcs intersecting in b left of A. With successively b and b' right and left of A A' as centres, and radii respectively equal to $b\,c''$ and $b'\,c'$ (Fig. 60) describe arcs intersecting in c' and c'. With successively c' and b right and left of A A' as centres, and radii respectively equal to C E and $b\,c$ (Fig. 60) describe arcs intersecting in c and c; and with successively c and c' right

PLATE I. (see p. 169).

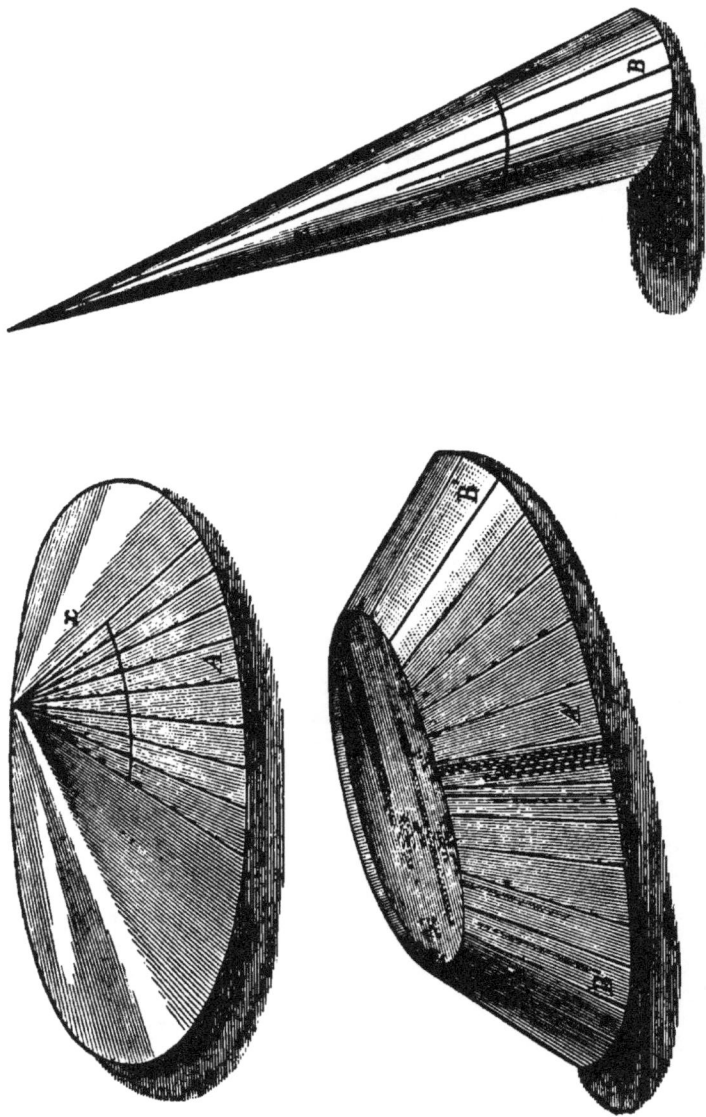

DRAWINGS FROM MODELS IN SOUTH KENSINGTON MUSEUM. (*Part of Exhibit by the Author at the Inventions Exhibition, 1885.*)

and left of A A' as centres, and radii respectively equal to $c\,d''$ and $c'\,d'$ (Fig. 60) describe arcs intersecting in d' and d'. Also with successively d' and c right and left of A A' as centres, and D E and $c\,d$ respectively as radii describe arcs

FIG. 61.

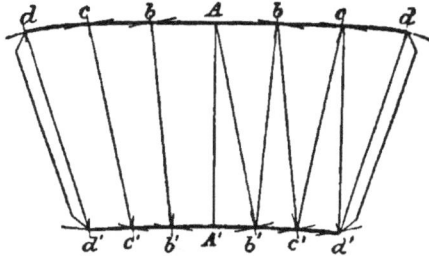

SIDE PATTERN.

intersecting in d and d. Through d, c, b, A, b, c, d, draw an unbroken curved line. Also through d', c', b', A', b', c', d', draw an unbroken curved line. Join $d\,d'$, right and left of A A'; then d A $d\,d'$ A' d' will be the side pattern required.

The remark about lines $f\,f'$, $e\,e'$, &c., in (end pattern) Fig. 59, applies to lines $c\,c'$, $b\,b'$, $b\,c'$, &c., in the present pattern.

PROBLEM XXII.

To draw the pattern for a tapering body with top and base parallel, and having circular top and oblong bottom with semicircular ends (tea-bottle top, for instance), the dimensions of the top and bottom of the body and its height being given.

Four cases will be treated of; three in this problem and one in the problem following (see also § 79, p. 157).

CASE I.—Pattern when the body is to be made up of four pieces.

Draw (Fig. 62) the plan of the body (see Problem VIII., p. 132) preserving of its construction the centres O, O'; the points b in plan of bottom where the extremities of the plan

lines of the sides meet the extremities of the plan semicircles of
the ends; and the points b' in plan of top where the sides and
ends meet in plan. Join $b'b$ at the four corners. The ends
$b\,B\,b\,b'\,B'\,b'$, and $b\,E\,b\,b'\,E'\,b'$ of the body (see g, p. 129), are
portions of frusta of oblique cones. Making the body in four
pieces it will be best that the seams shall correspond with
the lines $A\,b'$, $B\,B'$, $D\,b'$, and $E\,E'$, then one pattern only,
consisting of a half-end with a half-side pattern attached,
will be required.

FIG. 62.

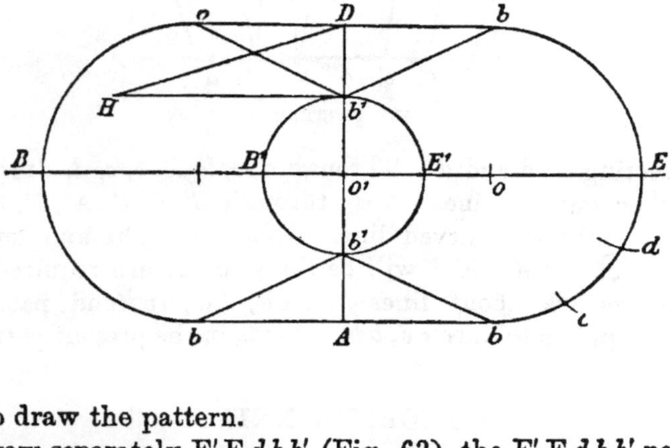

To draw the pattern.

Draw separately $E'\,E\,d\,b\,b'$ (Fig. 63), the $E'\,E\,d\,b\,b'$ portion
of Fig. 62, thus. Draw any line $X\,X$ and with any point O
(to correspond with O, Fig. 62) in it as centre and O E (Fig. 62)
as radius describe an arc (here a quadrant) $E\,d\,b$, equal
to $E\,d\,b$ of Fig. 62. Make $E\,O'$ equal to $E\,O'$ (Fig. 62), and
with O' as centre and $O'\,E'$ (Fig. 62) as radius describe an
arc (here a quadrant) $E'\,b'$ equal to $E'\,b'$ of Fig. 62. Joining
$b\,b'$ completes the portion of Fig. 62 required. Now divide
$E\,b$ into any number of equal parts, here three, in the points
d and c. From E' and O' draw $E'\,E''$, $O'\,O''$ perpendicular to
$X\,X$ and each equal to the given height of the body. Join
$E\,E''$, $O\,O''$; produce them to intersect in P (§ 80, p. 158); from
P let fall $P\,P'$ perpendicular to $X\,X$, and join $E''\,O''$. With P'

as centre and P' d, P' c, and P' b successively as radii describe arcs cutting X X in D, C, and B. Join these points to P by lines cutting O" E" in D', C', and B'

FIG. 63.

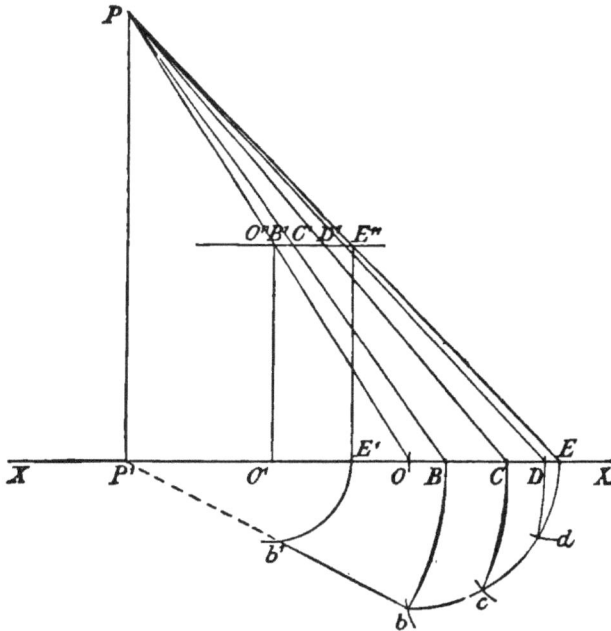

Next draw (Fig. 64) a line P E equal to P E (Fig. 63), and with P as centre, and P D, P C, and P B (Fig. 63) successively as radii describe arcs d, c, and b. With E as centre and radius equal to E d (Fig. 63) describe an arc cutting arc d in D. With same radius and D as centre describe an arc cutting arc c in C, and with C as centre and same radius describe an arc cutting arc b in B. Join the points D, C and B to P. With P as centre and P E" (Fig. 63) as radius describe an arc e' cutting P E in E'. With same centre and P D' (Fig. 63) as radius describe arc d' cutting P D in D'. Similarly

with same centre and P C' and P B' (Fig. 63) successively as
radii find points C' and B'.　　Through *b'* (Fig. 62) draw *b'* H
perpendicular to *b'* D and equal to the given height, and
join D H, then D H will be the true length of the line of
which *b'* D is the plan, that is, will be the length of a slant
of the body at the middle of the side, where one of the seams

Fig. 64.

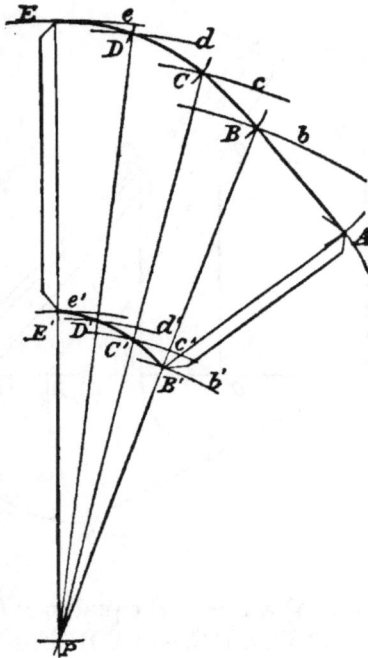

will come.　With B' and B (Fig. 64) as centres and radii
respectively equal to D H and *b* A (Fig. 62) describe arcs
intersecting in A.　Through the points E, D, C, and B draw
an unbroken curved line.　Also through the points E', D', C',
and B' draw an unbroken curved line.　Join B A, A B'; then
E C A B' E' will be the pattern required.

CASE II.—Pattern when the body is to be made up of two pieces.

We will suppose the seams are to correspond with the lines B B' and E E'. It is evident that here we need but one pattern only, which will combine a side of the body and two half-ends, in fact will be double that of Fig. 64.

First draw B B' B' B (Fig. 65) a half-end pattern exactly as the half-end pattern E E' B' B in Fig. 64 is drawn, and with B' as centre and B' B as radius and B as centre and *b b*

FIG. 65.

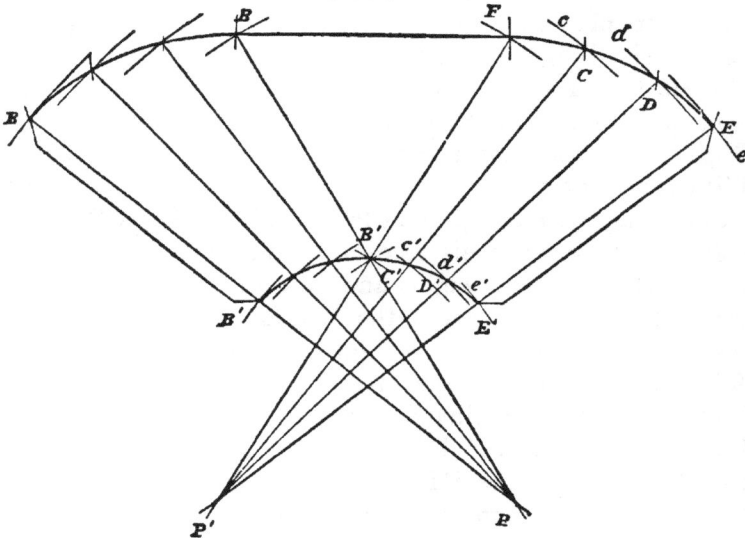

(Fig. 62) as radius, describe arcs intersecting in F. Join B F, F B'; and produce F B' indefinitely. Make F P' equal to P B (Fig. 63), and with P' as centre and P C, P D and P E (Fig. 63) successively as radii describe arcs *c*, *d*, and *e*. With F as centre and *b c* (Fig. 63) as radius describe an arc cutting arc *c* in C, and with C as centre and same radius describe an arc cutting arc *d* in D. With same radius and D as centre describe an arc cutting arc *e* in E. Join C P', D P', E P'. Now with P' as centre and P C' (Fig. 63)

as radius describe an arc c' cutting P' C in C', and with same
centre and P D', P E''' (Fig. 63) successively as radii describe
arcs d' and e' respectively cutting P' D and P' E in D' and E'.
Through F, C, D, and E draw an unbroken curved line.
Also through B', C', D', and E' draw an unbroken curved line.
Then B B F E E' B' B' will be the complete pattern required.

CASE III.—Pattern when the body is to be made up of one piece.

In this case we will put the seam to correspond with B B'
(Fig. 62). We now need an end pattern (the end b E b b' E' b'
in plan), with right and left a side pattern attached (b A b b',
b D b b' in plan), and joined to each of these, a half-end
pattern (b b' B' B, b b' B' B in plan).

First draw Fig. 63; then draw (Fig. 66) P E equal to P E
(Fig. 63) and with P as centre and P D, P C and P B
(Fig. 63) successively as radii describe arcs d d, c c, and b b.
With E as centre and radius equal to E d (Fig. 63) describe
arcs cutting arc d d right and left of P E in D and D. With
points, D, D, successively as centres and same radius describe
arcs cutting arc c c right and left of P E in C and C;
and with same radius and the last found points as centres
describe arcs cutting arc b b right and left of P E in B and B.
Join all the points found to P. With P as centre and
P E''' (Fig. 63) as radius describe an arc cutting P E in E'.
With same centre and P D' (Fig. 63) as radius describe an
arc d' d' cutting lines P D right and left of P E in D' and
D'. With same centre and P C' (Fig. 63) as radius describe
an arc c' c' cutting lines P C right and left of P E in
C' and C'. Similarly by arc b' b' find points B' and B'.
Through B, C, D, E, D, C, B, draw an unbroken curved line.
Also through B', C', D', E', D', C', B', draw an unbroken
curved line. This gives us B E B B B' E' B' a complete end
pattern. Now with B' on the right-hand side of the
end pattern as centre and B' B as radius, and B as centre and
b b (Fig. 62) as radius describe arcs intersecting in F. Join
B F, F B'; produce F B' indefinitely, and to F B' attach the
half-end pattern F E E' B' in precisely the same manner that

F E E' B' the half-end pattern in Fig. 65 is attached to the side pattern B F B'. By a repetition of the foregoing

FIG. 66.

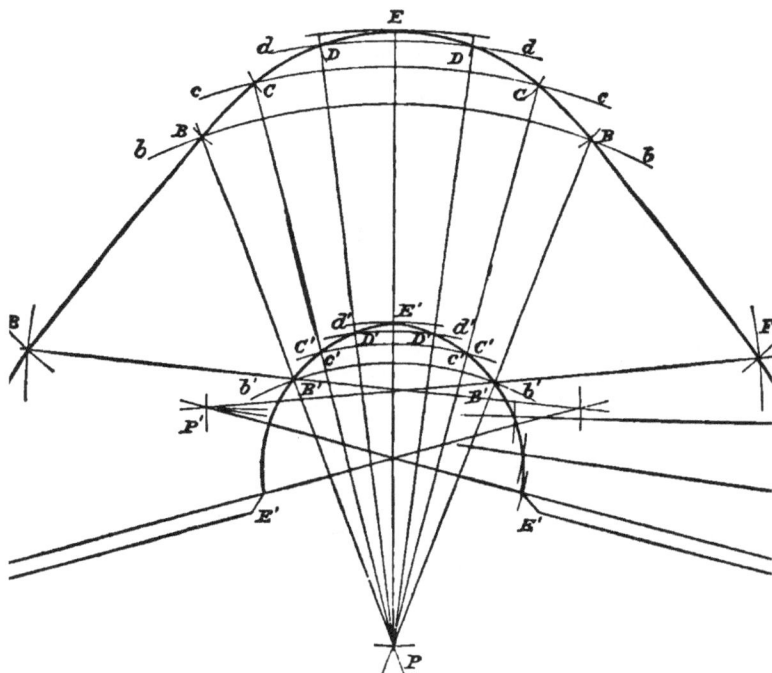

construction on the left of the end pattern B E B B' E' B' we can attach B B E E' B' and complete E E E E' E' E' the pattern required.

PROBLEM XXIII.

To draw, **without long radii,** *the pattern for a tapering body with top and base parallel, and having circular top and oblong bottom with semicircular ends. The dimensions of the top and bottom of the body and its height being given.*

This problem is a fourth case of the preceding, and is exceedingly useful where the work is so large that it is inconvenient to draw the whole of the plan, and to use long radii.

To draw the pattern (with the body in four pieces, as in Case I. of preceding problem).

(81.) Draw (Fig. 67) E' E b A b' one quarter of the plan of the body. Divide the quadrants E b, E' b', each into the same number of equal parts, here three, in the points d, c, d', c';

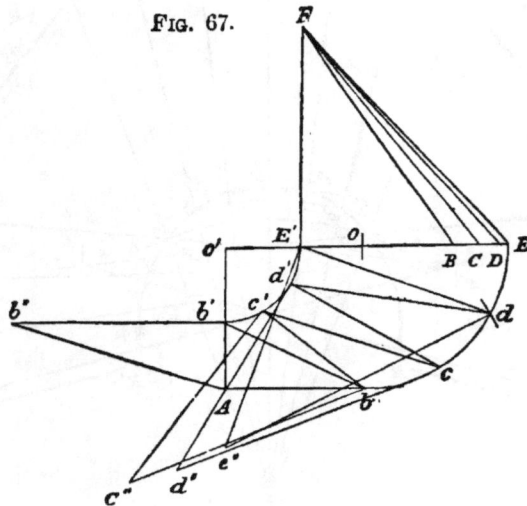

Fig. 67.

join d d', c c'. Through E' draw E' F perpendicular to E' E and equal to the given height of the body. From E' along E' E mark off E' D, E C, and E' B respectively equal to d d', c c', and b b'; and join E F, D F, C F, and B F; then E F, D F, C F, and B F will be the true lengths of E E', d d', c c', and b b' respectively. Next join d E', and draw E' e'' perpendicular to it and equal to the given height, and join d e''; then d e'' may be taken as the true length of d E'. Also join c d' and b c'; through d' and c' draw d' d'' and c' c'' perpendicular to c d' and b c' respectively, and each equal to the given height, and join c d'' and b c''; then c d'' and b c'' may be taken as the true lengths of d' c and c' b respectively. Through b' draw

$b'\,b''$ perpendicular to b' A and equal to the given height, and join A b'', then A b'' will be the true length of b' A.

Next draw (Fig. 68) E E' equal to E F (Fig. 67), and with E' and E as centres and radii respectively equal to $d\,e''$ and E d (Fig. 67) describe arcs intersecting in d, and with d and E' as centre and radii respectively equal to D F and E' d' (Fig. 67) describe arcs intersecting in d'. With d' and d as centres and radii respectively equal to $c\,d''$ and $d\,c$ (Fig. 67) describe arcs intersecting in c, and with c and d' as centres

Fig. 68.

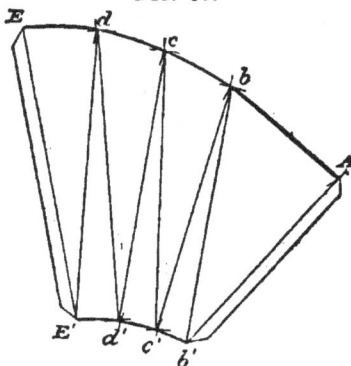

and radii respectively equal to C F and $d'\,c'$ (Fig. 67) describe arcs intersecting in c'. With c' and c as centres and radii respectively equal to $b\,c''$ and $c\,b$ (Fig. 67) d scribe arcs intersecting in b. Similarly with b and c' as centres and radii respectively equal to B F and $c'\,b'$ (Fig. 67) describe arcs intersecting in b'. With b' and b as centres and radii respectively equal to b'' A and b A (Fig. 67) describe arcs intersecting in A. Through E, d, c, b, draw an unbroken curved line. Also through E', d', c', b', draw an unbroken curved line. Join b' A, b A; then E c A b' c' E' is the pattern required.

(82.) The lines $d\,d'$, $c\,c'$, d E', &c., are drawn in Fig. 68 simply to show the position that the lines which correspond to them in Fig. 67 ($d\,d'$, $c\,c'$, d E', &c.) take upon the developed surface of the tapering body.

PROBLEM XXIV.

*To draw the pattern for a tapering body with top and base
parallel, and having an oval bottom and circular top (oval
canister top, for instance).　The height and dimensions of the
top and bottom of the body being given.*

Again four cases will be treated of; three in this problem
and one in the problem following (see also § 79, p. 157).

Draw (Fig. 69) the plan of the body (see Problem IX.,
p. 132), preserving of its construction the centres O, O', P, P,
and the four points (*d*) where the end and side curves of the

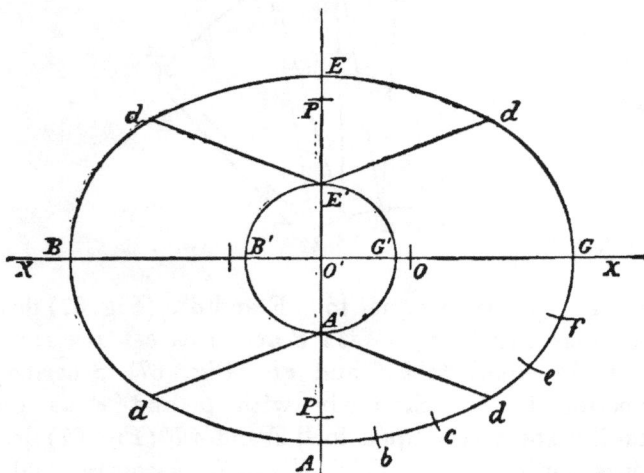

Fig 69.

plan of the bottom meet one another, also the points A' E' where
the axis A E cuts the circular top.　Join *d* A' (two places) and
d E' (two places).　From the plan we know (see *g*, p. 129)
that *d* G *d* A' G' E', *d* B *d* A' B' E', the ends of the body, are like
portions of the frustum of an oblique cone; we also know
that *d* A *d* A', *d* E *d* E', the sides of the body, are like portions
of the frustum of an oblique cone.

(83.) It is evident that in this problem the arcs d G d, E' G' A' and d B d, E' B' A' are, neither pair, proportional (§ 67, p. 124). We have hitherto in Problems XVIII., XX., and XXII. been dealing with proportional arcs. The working will therefore differ, though but slightly, from that of problems mentioned.

In Plate II. (p. 203) is a representation of the tapering body for which patterns are required, also of two oblique cones (x and Z). The oblique cones show to what portions of their surfaces the several portions of the tapering body correspond. Thus the sides, B', of the body correspond to the B portion of cone Z, and the ends, A', correspond to the portion A of cone x. The correspondence will be more fully recognised as we proceed with the problem.

CASE I.—Pattern when the body is to be made up of four pieces.

We will suppose the seams are to correspond with the plan lines G G', B B', A A', E E', of ends and sides, as in Problem XXII. Then one pattern only, consisting of a half-end pattern, with, attached, a half-side pattern will be required.

To draw the pattern.

Draw (Fig. 70) separately G' G f d A' the G' G f d A' portion of Fig. 69, thus. Draw any line X X and with any point O in it (corresponding to O, Fig. 69) as centre and O G (Fig. 69) as radius describe an arc G d, equal to G d of Fig. 69. Make G O' equal to G O' (Fig. 69), and with O' as centre and O' G' (Fig. 69) as radius describe an arc G' A' equal to G' A' of Fig. 69. Joining d A' completes the portion of Fig. 69 required Now divide G d into any number of equal parts, here three, in the points f and e. From G' and O' draw G' G″, O' O″ perpendicular to X X and each equal to the given height of the body. Join G G″, O O″; produce them to intersect in S (§ 80, p. 158); from S let fall S S' perpendicular to X X, and join O″ G″. Now join d to S', by a line cutting arc G' A' in d', then (§ 68, p. 124, and d, p. 126)

o

d' and d are corresponding points and (§ 67, p. 124) the arcs G d, G' d' are proportional; G G' d' d is the plan of a portion of an oblique-cone frustum, lying between the same generating lines, and d d' A' is plan of a portion of the same frustum outside the generating line S' d. With S' as centre

FIG. 70.

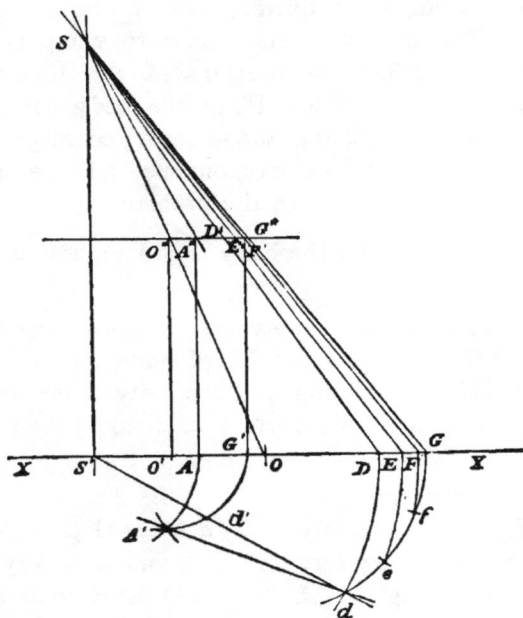

and S'f, S' e, and S' d successively as radii describe arcs cutting X X in F, E, and D; join these points to S by lines cutting O" G" in F', E', and D'. With S' as centre and radius the distance between S' and A' describe an arc A' A cutting X X in A; from A draw A A" perpendicular to X X and cutting O" G" in A", and join A" S.

Next draw any line X X (Fig. 71) and with any point P in it (corresponding to P, Fig. 69) as centre and radius P A (the radius of arc d A d, Fig. 69) describe an arc A d equal to A d of Fig. 69. Make A A' and A O' respectively equal to

A A' and A O' (Fig. 69). Divide A d into any number of
equal parts, here three, in the points b and c. From A and
O' draw A' A", O' O" perpendicular to X X and each equal to
the given height of the body. Join A A", P O"; produce

Fig. 71.

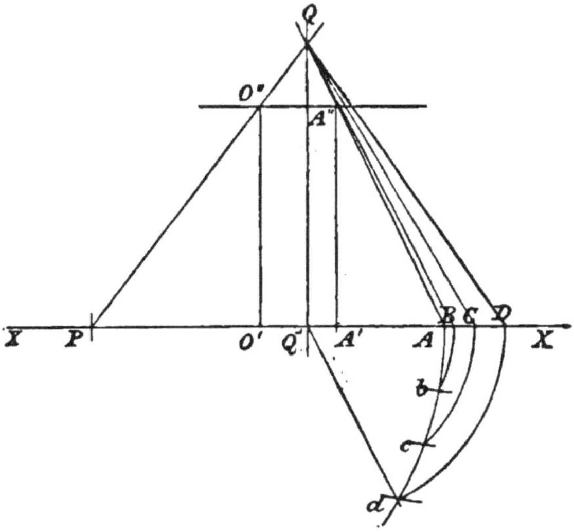

them to intersect in Q (§ 80, p. 158); from Q let fall Q Q' per-
pendicular to X X, and join O" A". With Q' as centre and
Q' b, Q' c, and Q' d successively as radii describe arcs cutting
X X in B, C, and D, and join these points to Q.

Next draw (Fig. 72) a line S G equal to S G (Fig. 70) and
with S as centre and S F, S E, and S D (Fig. 70) successively
as radii describe arcs f, e, and d. With G as centre and
radius equal to G f (Fig. 70) describe an arc cutting arc f in
F. With same radius and F as centre describe an arc cutting
arc e in E, and with E as centre and same radius describe an
arc cutting arc d in D. Join the points F, E, and D to S.
With S as centre and S G" (Fig. 70) as radius describe an
arc g' cutting S G in G'. With same centre and S F' (Fig. 70)

o 2

as radius describe arc f' cutting S F in F". With same centre and S E' (Fig. 70) as radius describe an arc e' cutting S E in E', and with same centre and S D' (Fig. 70) as radius describe arc d' cutting S D in D'. With same centre and S A" (Fig. 70) as radius describe arc a', and with D' as centre and radius d' A' (Fig. 70) describe an arc intersecting arc a' in A'. Make D Q equal to D Q (Fig. 71) and with Q as centre and Q C,

FIG. 72.

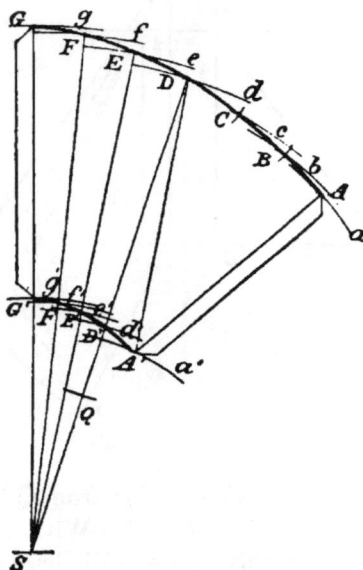

Q B, and Q A (Fig. 71) successively as radii describe arcs $c, b,$ and a. With D as centre and $d c$ (Fig. 71) as radii describe an arc cutting arc c in C, and with C as centre and same radius describe an arc cutting arc b in B. Similarly with same radius and B as centre find point A. Join A A'. Through the points G, F, E, D, C, B, A draw an unbroken curved line. Also through the points G', F', E', D', A' draw an unbroken curved line. Then G D A A' E' G' is the pattern required.

Case II.—Pattern when the body is to be made up of two pieces.

We will suppose the seams are to correspond with the lines G G' and B B'. It is evident that here we need but one pattern only, which will combine a side of the body and two half-ends, in fact will be double that of Fig. 72.

First draw B D A' B' (Fig. 73) a half-end pattern exactly as the half-end pattern G D A' G' in Fig. 72 is drawn, and make D Q equal to D Q (Fig. 71). With Q as centre and

FIG. 73.

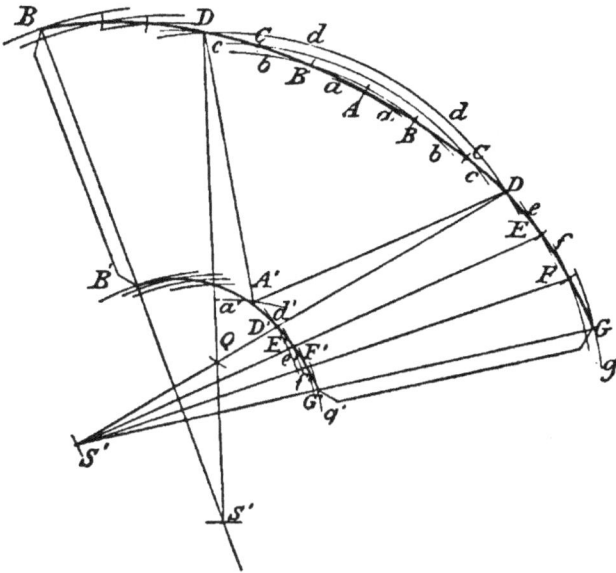

Q C, Q B, and Q A (Fig. 71) successively as radii describe arcs dd, cc, bb, and aa. With D as centre and radius equal to dc (Fig. 71) describe an arc cutting arc cc in C, and with same radius and C as centre describe arc cutting arc bb in B. With B as centre and same radius describe an arc cutting arc a in A, and with same radius and A as centre describe an arc cutting arc bb in B. Similarly with same radius and points B and C, to the right of A, successively as centres find points C and D. Join D Q and produce it indefinitely ; make D S' equal to S D (Fig. 70). With S' as centre and S E, S F, and S G (Fig. 70) successively as radii describe arcs e, f, and g, and with D as centre and $d\ e$ (Fig. 70) as radius describe an arc cutting arc e in E. With same radius and E as centre describe an arc cutting arc f in F. Similarly with F as centre and same radius find point G. Join the points E, F, and G to S'. With S' as centre and S D' (Fig. 70) describe arc d' cutting S' D in D'. With same centre and S E' (Fig. 70) as radius describe arc e' cutting S' E in E'. Similarly with same centre and S F' and S G'' (Fig. 70) successively as radii find points F' and G'. Through the points D, C, B, A, B, C, D, E, F, G, draw an unbroken curved line. Also through points A', D', E', F', G' draw an unbroken curved line. Then B A G G' A B' is the complete pattern required.

CASE III.—Pattern when the body is to be made up of one piece.

In this case we will put the seam to correspond with G G' (Fig. 69). We now need an end pattern (the end d B d A' B' E' in plan), with right and left a side pattern attached (d A d A', d E d E' in plan), and joined to each of these a half-end pattern (d A' G' G, d E' G' G in plan).

First draw Figs. 70 and 71 ; then draw (Fig. 74) S G equal to S G (Fig. 70), and with S as centre and S F, S E, and S D (Fig. 70) successively as radii describe arcs ff, ee, and dd. With G as centre and radius G f describe arcs cutting arc ff right and left of S G in F and F. With points F, F, successively as centres and same radius describe arcs cutting arc ee right and left of S G in E and E, and with same radius, and the last found points as centres describe arcs cutting arc dd right and left of S G in D and D. Join all the points found to S. With S as centre and S G'' (Fig. 70) as radius describe an arc cutting S G in G'. With same centre and S F' (Fig. 70) as radius describe an arc $f'f'$ cutting lines S F right and left of S G in F'' and F''. With same centre and S E' (Fig. 70) describe an arc *e' e'* cutting lines S E right and left of S G in E' and E', and with same centre and S D' (Fig. 70) as radius describe arc *d' d'* cutting lines S D right and left of S G in D' and D'. With S as centre and D' right and left of S G as centres and radii respectively equal to S A'' and *d'* A' (Fig. 70) describe arcs intersecting in A' and E'. Through D, E, F, G, F, E, D, draw an unbroken curved line. Also through E', D', E', F', G', F', E', D', A', draw

Fig. 74.

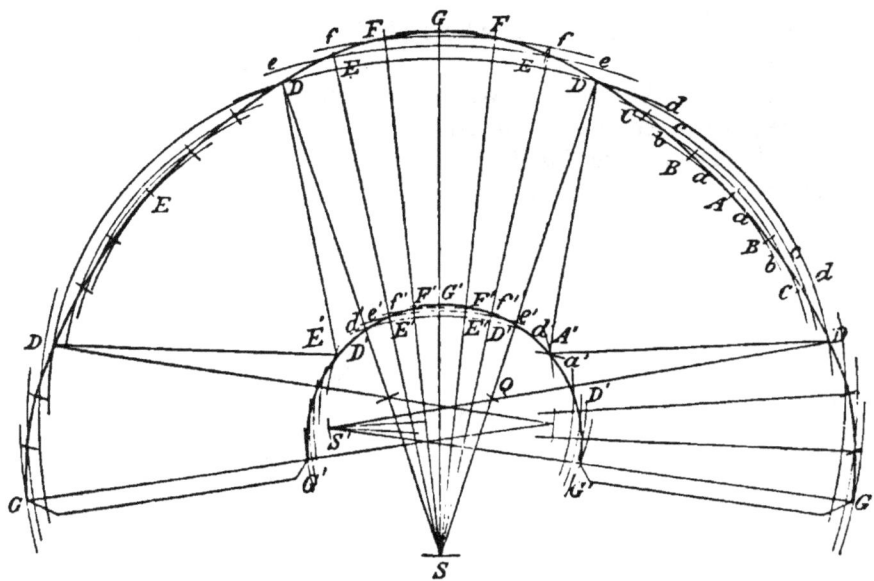

an unbroken curved line, and join D E', D A'. This gives us D G D A' G' E' a complete end pattern. Now attach the side pattern D A D A' and the half-end pattern D G G' A' to the right and left of the complete end pattern we started with, in precisely the same manner that the side pattern D A D A' and half-end pattern D G G' A' in Fig. 73 is attached to D A', which corresponds to D A' in Fig. 74. This will complete G E G A G G' G' G' the pattern required.

PROBLEM XXV.

To draw, **without long radii,** *the pattern for a tapering body with top and base parallel, and having an oval bottom and circular top. The height and dimensions of the top and base of the body being given.*

This problem is a fourth case of the preceding, and is exceedingly useful where the work is so large that it is inconvenient to draw the whole of the plan, and to use long radii.

To draw the pattern (with the body in four pieces, as in Case I. of preceding problem).

(84.) Draw (Fig. 75) E *d b* A A' *d'* E' one quarter of the plan of the body. Join *c* (the point where the end and side curves

FIG. 75.

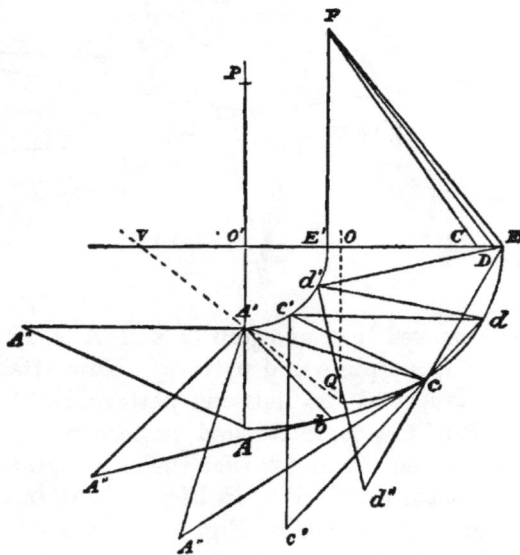

of the plan of the bottom meet) to A', the extremity of the quadrant E A'. We must now get at corresponding points in

the arcs E c, E' A'. To do this, as the arcs are not proportional (§ 67), we must find the plan of the apex of the
oblique cone of a portion of which E c A' E' is the plan. It
is in the finding of these points that our present working
differs from the working of Problems XIX., XXI., XXIII., and
XXVII., where the corresponding points d d', c c' (Figs. 50, 58,
60, and 82) are found. With radius O E produce arc E c indefinitely, and through O draw O Q perpendicular to E' E and
cutting E c produced in Q. Then E Q is a quadrant, and E Q,
E' A' (each a quadrant) are proportional. In Q and A' therefore
we have corresponding points (§ 68), as well as in E and E',
which are points on the longest generating line (b, p. 126) ; and
the intersection of O O' produced, of which E E' is part, and
Q A' joined and produced will give us V, the required plan
of the apex. Next divide E c into any number of equal parts,
here two, in point d. Join d V, c V cutting E' A' in d' and c',
respectively (the lines from d and c are not carried to V in
the fig.), then d and d', c and c' are corresponding points.

Next through E' draw E' F perpendicular to E' E and
equal to the given height; from E' along E' E mark off
E' D, E' C, respectively equal to d d' and c c', and join E F,
D F, and C F; then E F, D F and C F will be the true
lengths of E E', d d', and c c'. Join E d' and d c'; through d'
and c' draw d' d'' and c' c'' perpendicular to d' E and c' d respectively, and each equal to the given height, and join
E d'', d c''; then E d'' and d c'' may be taken respectively as
the true lengths of d' E and c' d. Now divide arc A c into
any number of equal parts, here two, in the point b, and join
b A'; through A' draw three lines A' A''; one perpendicular
to A' c, the second perpendicular to A' b, and the third perpendicular to A' A, and each equal to the given height, and
join c A'', b A'', and A A''; then c A'', b A'', and A A'' may be
taken as the true lengths of A' c, A' b, and A' A respectively.

Next draw (Fig. 76) E E' equal to E F (Fig. 75), and with
E and E' as centres and radii respectively equal to E d'' and
E' d' (Fig. 75) describe arcs intersecting in d', and with d' and
E as centres and radii respectfully equal to D F and E d

(Fig. 75) describe arcs intersecting in d. With d and d' as centres and radii respectively equal to $d\,c''$ and $d'\,c'$ (Fig. 75) describe arcs intersecting in c', and with c' and d as centres and radii respectively equal to CF and $d\,c$ (Fig. 75) describe arcs intersecting in c. With c and c' as centres and radii respectively equal to $c\,A''$ and $c'\,A'$ (Fig. 75) describe

FIG. 76.

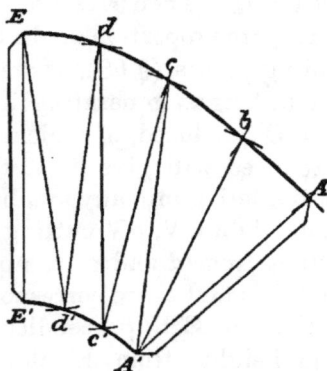

arcs intersecting in A', and with A' and c as centres and radii $b\,A''$ and $c\,b$ (Fig. 75) describe arcs intersecting in b. Similarly with A' and b as centres and radii respectively equal to $A\,A''$ and $b\,A$ (Fig. 75) describe arcs intersecting in A. Join $A\,A'$. Through E, d, c, b, A draw an unbroken curved line. Also through E', d', c', A' draw an unbroken curved line. Then $E\,c\,A\,A'\,E'$ is the pattern required.

The lines $d\,d'$, $c\,c'$, $E\,d'$, $d\,c'$, &c., are not needed for the working, they are drawn for the reason stated in § 82, end of Problem XXIII.

(85.) If V is inaccessible, corresponding points c, c', d, d' can thus be found. From the point E' along the arc E' A' set off an arc proportional to the arc E c in the following manner. Join $O\,c$ (line not shown in fig.) and through O' draw $O'\,c'$ (also not shown in fig.) parallel to $O\,c$ and cutting arc E' A' in c'; then arcs E' c' and E c will be proportional. (The student must particularly notice this

PLATE II. (see p. 193).

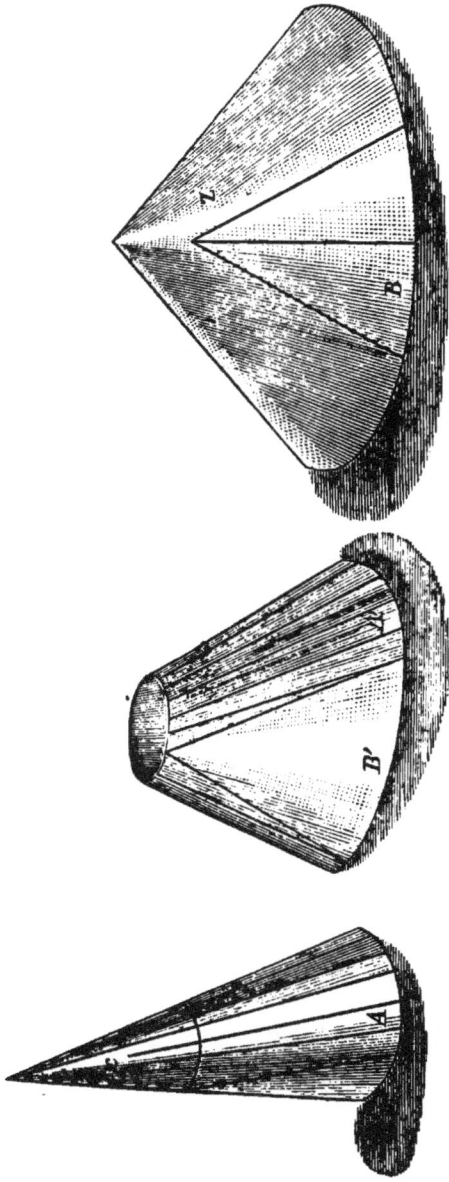

DRAWINGS FROM MODELS IN SOUTH KENSINGTON MUSEUM. (*Part of Exhibit by the Author at the Inventions Exhibition, 1885.*)

method of drawing proportional arcs. It is outside the scope
of the book to prove the method.) Now divide arcs $E' c'$,
$E c$ each into the same number of equal parts, here two, in
the points d and d'; then d, d' and c, c' are corresponding
points.

PROBLEM XXVI.

*To draw the pattern for a tapering body with top and base
parallel, and having oblong bottom with round (quadrant)
corners, and circular top. The dimensions of the top and
base of the body and its height being given.*

Again four cases will be treated of, three in this problem,
and one in the problem following   so § 79, p. 157).

Draw (Fig. 77) the plan of the body (see Problem X.,
p. 133) preserving of its construction the centres O O' and the

FIG. 77.

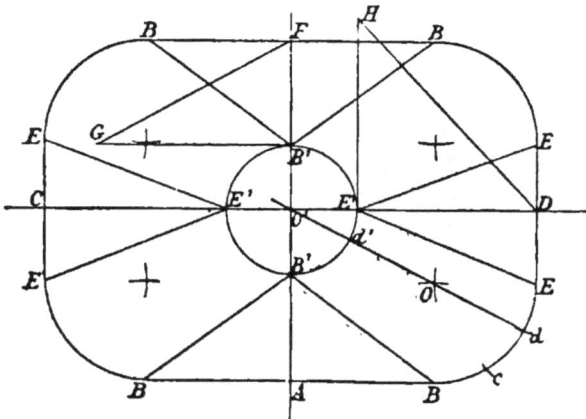

points E' B' where the flat sides and flat ends meet the circle
of the top. Join B B', E E' each in four places. From the
plan we know (see *g*, p. 129) that the round corners of the
body are portions of frusta of oblique cones.

(86.) Looking at the plan, we can at once see that what we have to deal with differs somewhat from what has as yet been before us. Hitherto a line passing through the centres of the plan arcs bisected the arcs, and the cone development was consequently identical each side of a central line. In Fig. 77, however, the line drawn through O O' does not bisect the plan arcs E B, E' B'. This affects the working but little, as will be seen.

In Plate III. (p. 213) the tapering body is represented; also an oblique cone Z, the A portion of which corresponds to the A' portion of the body, and the development of the former is the development of the latter.

CASE I.—Pattern when the body is to be made up of four pieces.

We will suppose the seams to correspond with the plan lines C E', D E', F B', A B', of ends and sides, as in Problems XXII. and XXIV. just preceding. Then one pattern, comprising a half-end, a complete corner, and a half-side, will be the pattern required.

To draw the pattern.

Draw separately (Fig. 78) an E E', B' B portion of Fig. 77, thus. Draw an indefinite line S' d (Fig. 78), and with any point O (corresponding to O, Fig. 77) in it as centre and O B (Fig. 77) as radius describe an arc B E. Join O O' (Fig. 77) and produce it cutting arc B E in d; make d B and d E (Fig. 78) equal respectively to d B and d E (Fig. 77). Now (Fig. 78) make d O' equal to d O' (Fig. 77), and with O' as centre and O' B' (Fig. 77) as radius describe an arc B' E'. Make d' B' and d' E' equal respectively to d' B' and d' E' (Fig. 77). Joining E E', B B' completes the portion of Fig. 77 required. Now divide (Fig. 77) B E into any number of parts. It is convenient to take d as one of the division points, and to make d c equal to d E; leaving c B without further division, thus making the division of B E into three portions not all equal. In actual practice the dimensions of the work will suggest the number of parts expedient. Now (Fig. 78) make d c equal

to dc (Fig. 77), then BE (Fig. 78) will be divided correspondingly to BE (Fig. 77). Draw XX parallel to S'd;
and at d and O draw dD, OQ perpendicular to S'd; and
meeting XX in D and Q; also through d' and O' draw d'D',

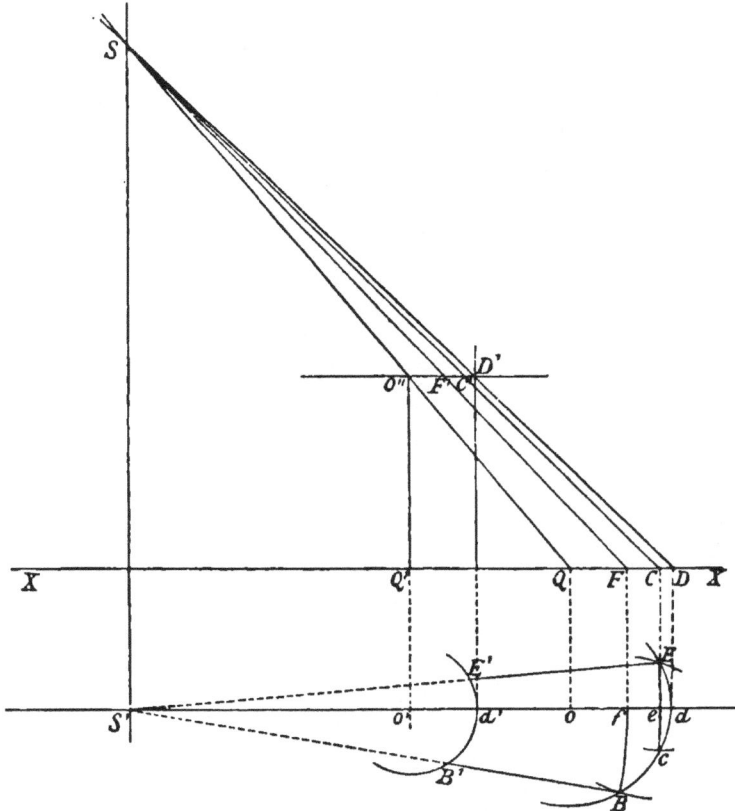

O'O'' perpendicular to S'd; the line O'O'' cutting XX in Q'.
Make Q'O'' equal to the given height of the body, and draw
O''D' parallel to XX. Join DD', QO''; produce them to
intersect in S (§ 80, p. 158); and from S let fall a perpendicular to S'd, cutting S'd in S'. With S' as centre and S'E,

or S′ c (which is equal to S E) and S′ B successively as radii
describe arcs cutting S′ d in e and f. Draw e C, f F perpendicular to X X and cutting it in C and F; also join C S, F S,
cutting O″ D′ in C′ and F′.

Next draw S D (Fig. 79) equal to S D (Fig. 78), and with
S as centre and S C, S F (Fig. 78) successively as radii

FIG. 79.

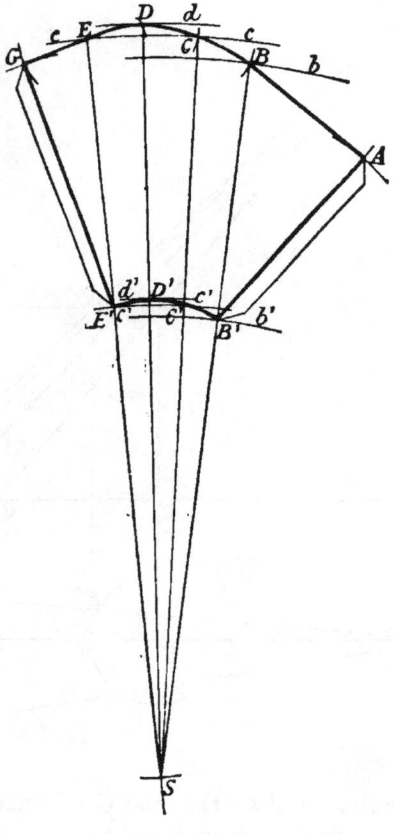

describe arcs c and b. With D as centre and radius equal
to d E (Fig. 78) describe arcs cutting arc c c in E and C.
With C as centre and radius c B (Fig. 78) describe an arc

cutting arc *b* in B. Join E, C, and B to S. Make S D' equal to
S D' (Fig. 78) and with S as centre and S C' (Fig. 78) as radius
describe an arc *c' c'* cutting S E and S C in E' and C' respectively. With same centre and S F' (Fig. 78) as radius describe
an arc *b'* cutting S B in B'. Through E, D, C, and B draw an
unbroken curved line. Also through E', D', C', and B' draw
an unbroken curved line ; this will complete the pattern of
the round corner. To attach the half-end and half-side patterns
to E E' and B B' respectively, the true lengths of E' D and
B' A (Fig. 77) must be found. Draw (Fig. 77) E' H perpendicular to E' D and equal to the given height of the body ;
join D H, then D H is the true length of E' D. The lines
B' E and B' A being equal, their true lengths are equal, we
will therefore for convenience find the true length of B' A
in that of B' F. Draw B' G perpendicular to B' F and equal
to the given height, join F G, then F G is the true length
required. Now with E' (Fig. 79) as centre and D H
(Fig. 77) as radius, and E as centre and radius E D (Fig. 77)
describe arcs intersecting in G. Join E G, G E' ; this
attaches to E' E the half-end pattern. With B' (Fig. 79)
as centre and F G (Fig. 77) as radius, and B as centre and
radius B A (Fig. 77) describe arcs intersecting in A. Join
B A, A B' ; this attaches to B' B the half-side pattern. Then
A B' E' G is the complete pattern required.

CASE II.—Pattern when the body is to be made up of two
pieces.

Here it will be best that the seams shall correspond with
the lines A B', F B', that is with the middle of each side.
The required pattern will then be double that of Case I.

Draw (Fig. 80) E B B' E', the corner pattern, in exactly
the same manner that E B B' E' (Fig. 79) is drawn. With
E' as centre and E' E as radius, and E as centre and E E
(Fig. 77) as radius describe arcs intersecting in F. Join
E F, F E'. Produce F E' indefinitely and make F S' equal to
E S. Using S' as centre, the round corner F G B' E' can be
drawn as was E B B' E. With B' as centre and F G (Fig. 77)

P

as radius, and B as centre and B A (Fig. 77) as radius describe arcs intersecting in A.   Similarly with B' and G as

FIG. 80.

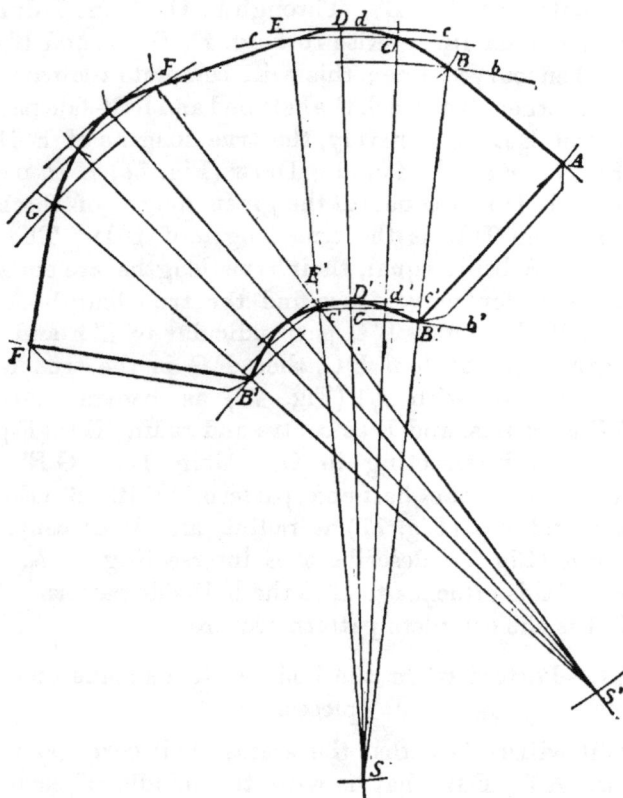

centres and same radii respectively describe arcs intersecting
in F.   Join B A, A B', G F, F B', then A D F F B' E' B' will
be the pattern required.

CASE III.---Pattern when the body is to be made up of one
piece.

We will suppose the seam to correspond with C E' the
middle of one end.   Draw G F D B B' E' B' (Fig. 81) in the

same manner that G F D B B' E' B' (Fig. 80) is drawn. With
B' as centre and B' B as radius, and B as centre and B B
(Fig. 77) as radius describe arcs intersecting in A.   Join
B A, A B'; produce A B' indefinitely.   Make A P equal to
S F (Fig. 78).   With P as centre and S C and S D succes-

FIG. 81.

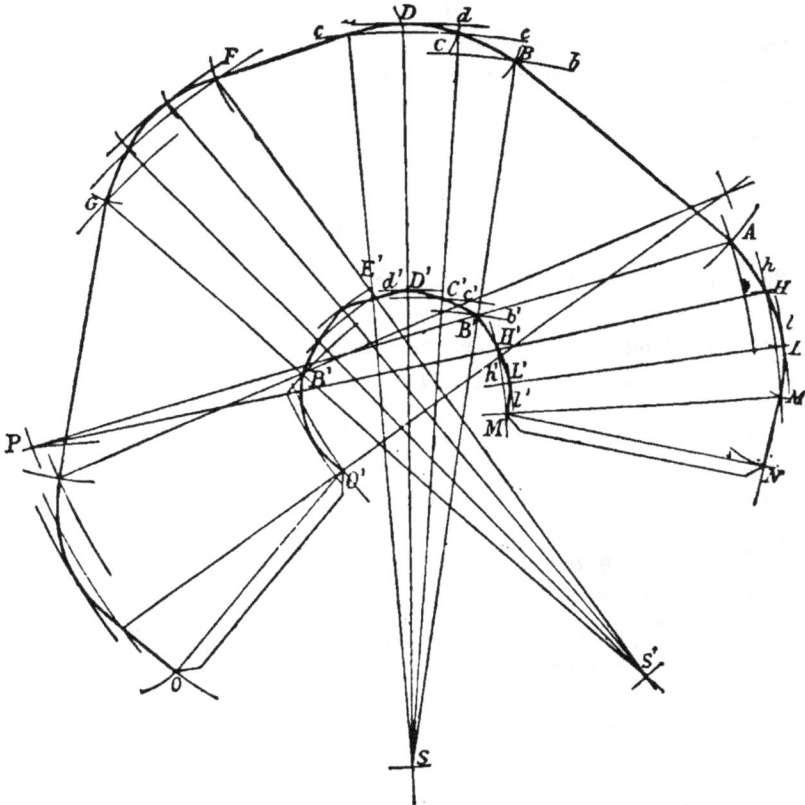

sively as radii describe arcs *h* and *l*, and with A as centre
and B *c* (Fig. 78) as radius describe an arc cutting arc *h* in
H.   With H as centre and *c d* (Fig. 78) as radius describe an
arc cutting arc *l* in L ; and with L as centre and same radius

describe an arc cutting arc h in M. Join H, L, and M to P. With P as centre and S C' (Fig. 78) as radius describe arc h' cutting P H and P M in H' and M' respectively. With same centre and S D' (Fig. 78) as radius describe arc l' cutting P L in L'. Through the points A, H, L, M draw an unbroken curved line. Also through the points B', H', L', M' draw an unbroken curved line. With M' as centre and D H (Fig. 77) as radius, and M as centre and E C (Fig. 77) as radius describe arcs intersecting in N. Join M N, N M'. Repeating the working to the left of B' G, the G O O' B' portion of the pattern can be drawn which completes the pattern required.

PROBLEM XXVII.

To draw, without long radii, *the pattern for a tapering body with top and base parallel, and having oblong bottom with round quadrant corners, and circular top. The dimensions of the top and base of the body and its height being given.*

This problem is a fourth case of the preceding, and is exceedingly useful where the work is so large that it is inconvenient to draw the whole of the plan, and to use long radii.

To draw the pattern (with the body in four pieces as in Case I. of preceding problem).

Draw (Fig. 82) E c b A b' c' d', one quarter of the plan of the body. Divide the arc (quadrant) d b into any number of equal parts, here two, in the point c, and the arc d' b' into the same number of equal parts in the point c'; and join c c'. Through d' draw d' F perpendicular to d' E and equal to the given height of the body. From d' along d' E mark off d' B, equal to b b', and d' D equal to c c' and d d' (which two lines have happened to come in this particular fig. so nearly equal that we may take them as equal), and join B F, E F, and D F; then B F and E F will be the true lengths of b b' and E d' respectively, and D F may be taken as the true

PLATE III. (see p. 206).

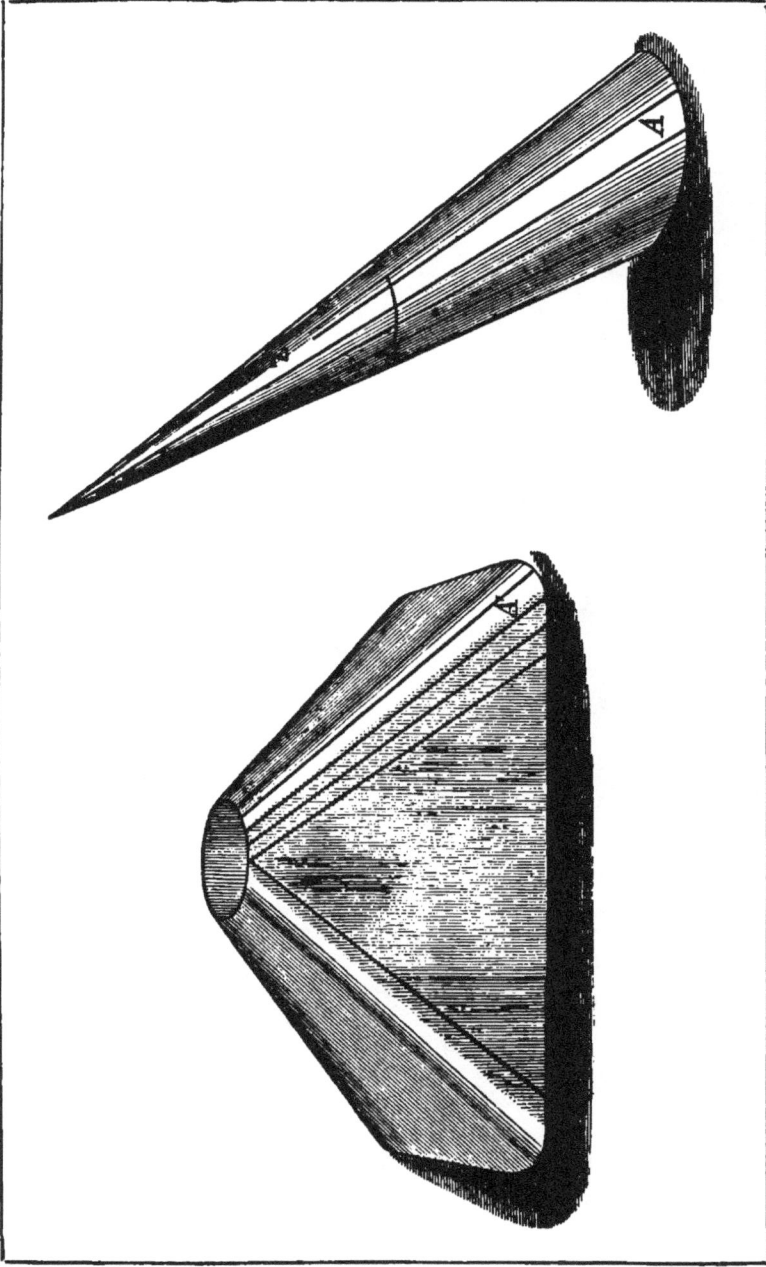

DRAWINGS FROM MODELS IN SOUTH KENSINGTON MUSEUM. (*Part of Exhibit by the Author at the Inventions Exhibition, 1885.*)

length of both $c\,c'$ and $d\,d'$. Next join $c'\,d$, $b'\,c$; draw $c'\,c''$, $b'\,b''$ perpendicular to $c'\,d$, $b'\,c$ respectively, and each equal to the given height; and join $d\,c''$ and $c\,b''$; then $d\,c''$, $c\,b''$ may

Fig. 82.

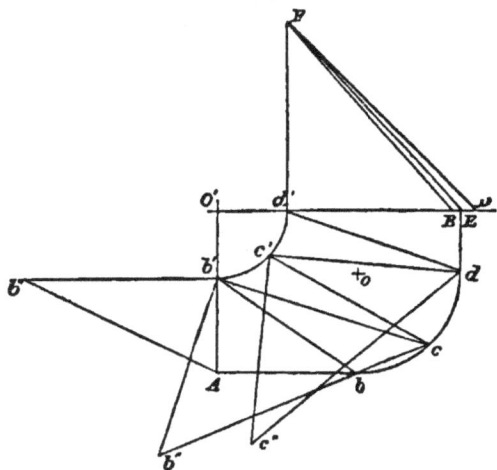

be taken as the true lengths of $c'\,d$ and $b'\,c$ respectively. Draw $b'\,b''$ perpendicular to $b'\,A$ and equal to the given height, and join $A\,b''$, then $A\,b''$ is the true length of $b'\,A$.

Next draw (Fig. 83) $d\,d'$ equal to $D\,F$ (Fig. 82), and with d and d' as centres and radii respectively equal to $d\,c''$ and $d'\,c'$ (Fig. 82), describe arcs intersecting in c'. With c' and d as centres, and radii respectively equal to $D\,F$ and $d\,c$ (Fig. 82) describe arcs intersecting in c; and with c and c' as centres and radii respectively equal $c\,b''$ and $c'\,b'$ (Fig. 82) describe arcs intersecting in b'; also with b' and c as centres and radii respectively equal to $B\,F$ and $c\,b$ (Fig. 82) describe arcs intersecting in b. With d' and d as centres and radii respectively equal to $E\,F$ and $d\,E$ (Fig. 82) describe arcs intersecting in E; and with b' and b as centres and radii respectively equal to $b''\,A$ and $b\,A$ (Fig. 82) describe arcs intersecting in A. Through d, c, b, draw an unbroken curved line. Also through d', c', b' draw an unbroken curved line.

FIG. 83.

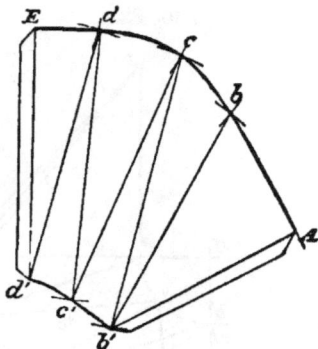

Join d E, d' E, b A, and b' A; then E c A $b'd'$ is the pattern required.

The lines $c c'$, $b b'$, &c., are not needed for the working, they are drawn for the reason stated in § 82, end of Problem XXIII.

PROBLEM XXVIII.

To draw the pattern for an Oxford hip-bath, the like dimensions to those for Problem XI. being given.

It is only necessary to treat of two cases, one in this problem, and one in the problem following (see also § 79, p. 157).

Draw (Fig. 84) the plan of the body (see Problem XI., p. 134), preserving of its construction the centres O, P', P, Q', Q; the points D, D' (two sets) in which the arcs, in plan, of the back and sides meet each other; and the points h, g' (two sets) in which the plan arcs of the sides and front meet each other. Join $h g'$ (two places) as shown in the fig. Examining the plan of the bath we see (d, p. 55) that the back of it,

D A D D′ A′ D′, is a portion of a right cone; that the sides
D D′ g′ h are (g, p. 12ɔ) each of them a portion of an oblique
cone; and that the portion h g′ g′ h is also a portion of an

FIG. 84.

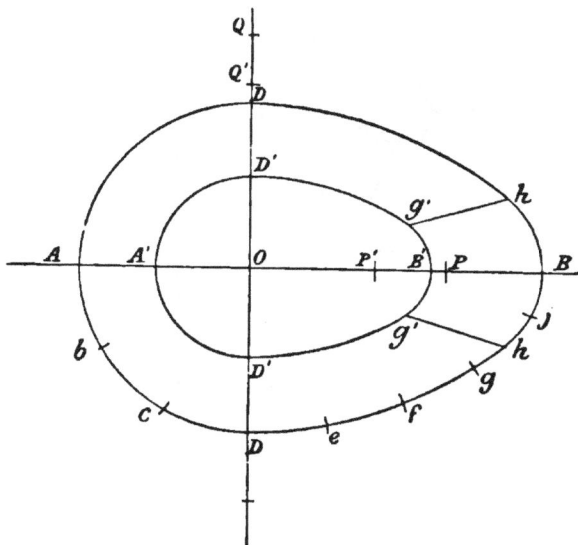

oblique cone. Similarly as in Problem **XXIV.** (§ 83, p. 193)
the arcs D h, D′ g′ and B h, B′ g′ are, neither pair, proportional.

In Plate IV. (p. 227) is a representation of an Oxford hip-
bath, also of a right cone x, and two oblique cones Y and Z.
The cones show to what portions of their surfaces the several
portions of the bath correspond. Thus the back, A′, of the
bath corresponds to the A portion of right cone x; the half-
fronts, C′, of the bath correspond each of them to the C
portion of oblique cone Y, and the sides, B′, of the bath
correspond each of them to the B portion of oblique cone Z.

Patterns when the body is to be made up of three
pieces.

We will put the seams to correspond with the lines D D′

(two places), and B B'. Clearly, only two patterns will be required, one for the back of the bath, and the other for a complete side and a half-front.

To draw the pattern for the back.

Draw E A A' O' O (Fig. 85) the elevation of the back (see

FIG. 85.

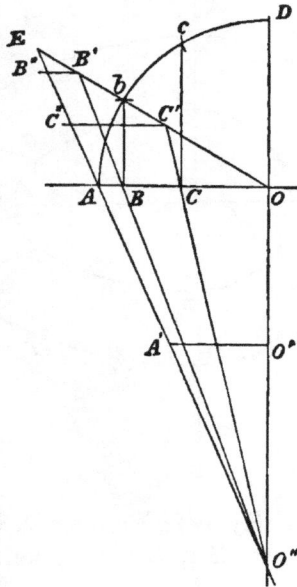

Problem XI. and Fig. 24a, p. 135) and produce A A', O O' to intersect in O". With O as centre and O A as radius describe a quadrant A D (corresponding with A D, Fig. 84), and divide it into any number of equal parts, here three, in the points b and c. Draw c C, b B perpendicular each of them to A O and cutting it in points B and C. Join O" C, and produce it to cut O E in C'; and through C' draw C' C" parallel to A O and cutting O' E in C". Join O" B and produce it to cut O E in B'; and through B' draw B' B" parallel to A O and cutting O" E in B". Then A C" is the true length of C C', and A B" the true length of B B'.

Next draw (Fig. 86) O″ A equal to O″ A (Fig. 85), and with O″ as centre and radius O″ A describe an arc D A D, and with same centre and O″ A′ (Fig. 85) as radius describe an

FIG. 86.

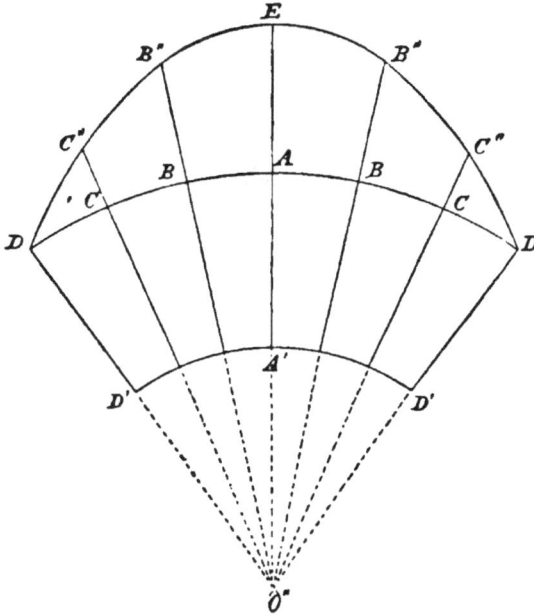

arc D′ A′ D′. Mark off, right and left of A, A B, B C, and C D, each equal to A b (Fig. 85), one of the equal parts into which the quadrant A D is divided. Join D O″, C O″, B O″ right and left of A O″, and produce O″ A, O″ B, O″ C indefinitely. Make A E equal to A E (Fig. 85); B B″ equal to A B″ (Fig. 85); and C C″ equal to A C″ (Fig. 85); and through D, C″, B″, E, B″, C″, D, draw an unbroken curved line. Then D E D D′ A′ D′ is the required pattern for back of bath.

To draw the pattern for a side and a half-front.

Draw separately D′ D *f h g′* (Fig. 87), the D′ D *f h g′* portion
of Fig. 84, thus. Draw any line X X and with any point Q
(to correspond with Q, Fig. 84) in it as centre and (same

Fig. 87.

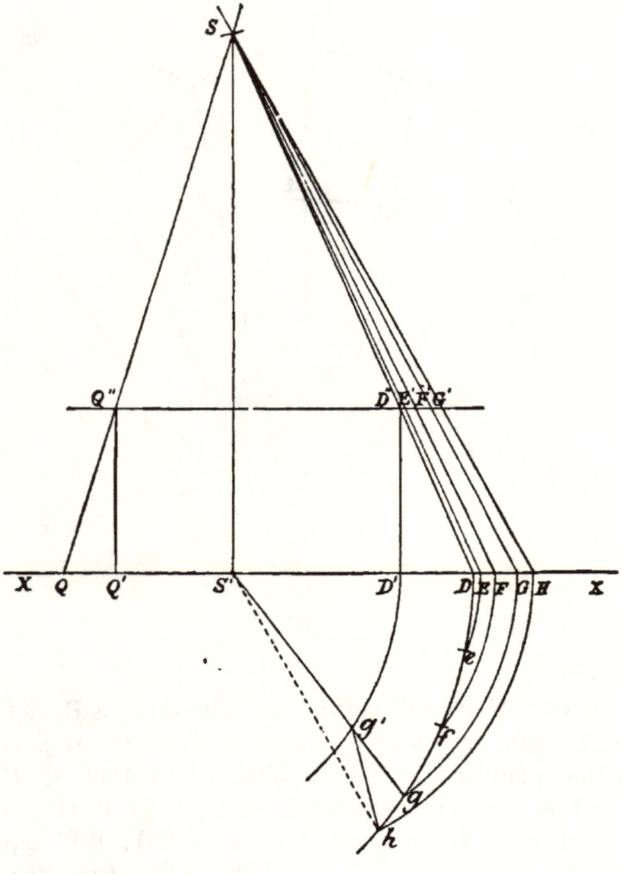

fig.) Q D (the D on the further side of A B) as radius describe
an arc D *h* equal to D *h* of Fig. 84. Make D Q′ equal to D Q′
(Fig. 84), and with Q′ as centre and Q′ D′ (the further D′,
Fig. 84) as radius describe an arc D′ *g′* equal to D′ *g′* of Fig. 84.

Joining hg' completes the portion of Fig. 84 required. Now from D' and Q' draw D' D", Q' Q" perpendicular to X X and each equal to the given height of the D B D portion of the bath. Join D D", Q Q"; produce them to intersect in S (§ 80, p. 158); and from S let fall S S' perpendicular to X X. Join S' g', and produce it to cut arc D h in g, then g and g' will be corresponding points. Divide D g into any number of equal parts, here three, in the points e and f. Join Q" D". With S' as centre and S' e, S' f, and S' g successively as radii describe arcs cutting X X in E, F, and G; join these points to S by lines cutting Q" D", produced, in E', F', and G'. With S' as centre and S' h as radius describe an arc cutting X X in H, and join H S.

Next draw separately B' B h g' (Fig. 88) one of the B' B h g' portions of Fig. 84, thus. Draw any line X X and

FIG. 88.

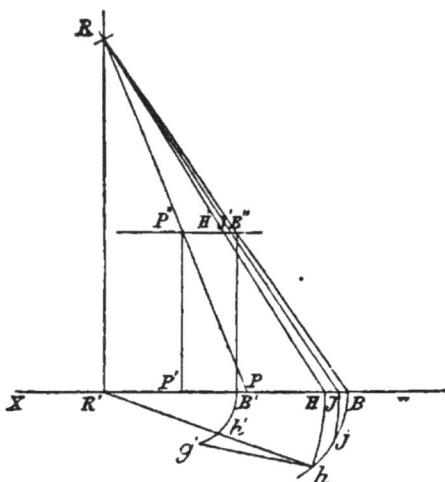

with any point P (to correspond with P, Fig. 84) in it as centre and P B (Fig. 84) as radius describe an arc B h equal to B h of Fig. 84. Make B P' equal to B P' (Fig. 84), and with P' as centre and P' B' (Fig. 84) as radius describe an arc

B' g' equal to B' g' of Fig. 84. Joining h g' completes the portion of Fig. 84 required. Now from B' and P' draw B' B",
P' P" perpendicular to X X and each equal to the given
height of the D B D portion of the bath. Join B B", P P";
produce them to intersect in R (§ 80, p. 158); and from R
let fall R R' perpendicular to X X. Join R' h, cutting arc
B' g' in h', then h and h' will be corresponding points.
Divide B h into any number of equal parts, here two, in the
point j. Join P" B". With R' as centre and R' j and R' h
successively as radii describe arcs cutting X X in J and H;
join these points to R by lines cutting P" B" in J' and H'.

Next draw (Fig. 89) a line D S equal to D S (Fig. 87).

FIG. 89.

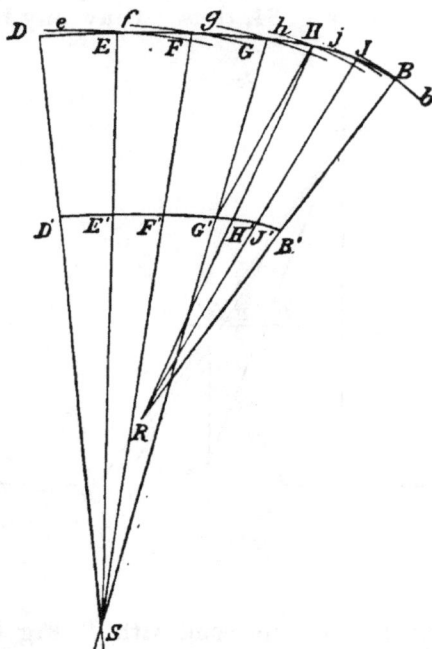

With S as centre and S E, S F, S G, and S H (Fig. 87) successively as radii describe arcs e, f, g, and h. With D as centre

and radius equal to D *e* (Fig. 87) describe an arc cutting arc *e* in E, and with same radius and E as centre describe an arc cutting arc *f* in F. With F as centre and same radius describe an arc cutting arc *g* in G, and with G as centre and radius *g h* (Fig. 87) describe an arc cutting arc *h* in H. Join the points E, F, and G (not H) to S. Make S D' equal to S D" (Fig. 87); and make S E', S F', and S G' respectively equal to S E', S F', and S G' (Fig. 87). Through the points D, E, F, G, H draw an unbroken curved line. Also through points D', E', F', G' draw an unbroken curved line, and join H G'. This completes the side pattern, to which we have now to attach, at H G', a half-front pattern. With H and G' as centres and radii respectively equal to H H' and *g' h'* (Fig. 88) describe arcs intersecting in H'. Join H H'; produce it indefinitely and make H R equal to H R (Fig. 88). With R as centre and R J, R B successively as radii describe arcs *j* and *b*. With H as centre and *h j* (Fig. 88) as radius describe an arc cutting arc *j* in J, and with same radius and J as centre describe an arc cutting arc *b* in B. Join the points J and B to R, and make R J' and R B' respectively equal to R J' and R B"(Fig. 88). Through H, J, and B draw an unbroken curved line. Also through H', J', and B' draw an unbroken curved line. Then D F H B B'G'D' is the complete pattern required.

PROBLEM XXIX.

To draw, **without long radii**, *the pattern for an Oxford hip-bath; given dimensions as before.*

This problem is a second case of the preceding.

Patterns when the body is to be made up of three pieces with seams as in the preceding problem.

To draw the pattern for the back.

Draw E A A' O' O (Fig. 90) the elevation of the back as in Fig. 85, and produce O' O. With O as centre and O A as radius describe a quadrant A D (corresponding with A D, Fig. 84), and divide it into any number of equal parts, here three, in the points b and c. Join b O, c O, and with O as centre and radius O'A' describe a quadrant D'A" (corresponding with D'A', Fig. 84), and cutting lines O c, O b in c' and b' respectively. Then D D' A" A will be the plan of that portion of the back of the bath of which O'A'O A is the elevation. Through b and c draw b B and c C perpendicular to O A; and through b' and c' draw b' B' and c' C' perpendicular to O'A'. (Here part of b' B' happens to coincide with

Fig. 90.

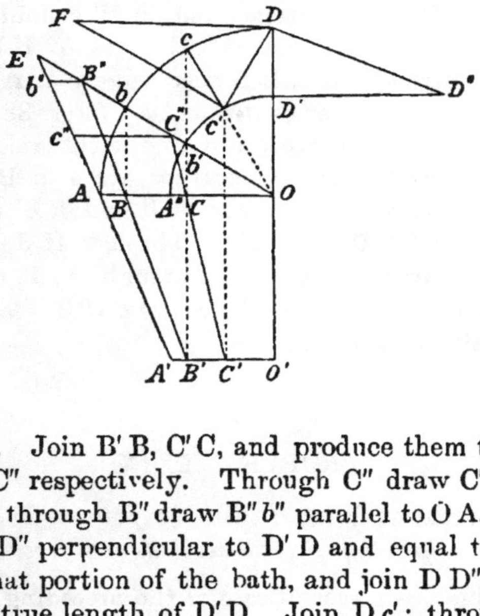

part of c C). Join B' B, C' C, and produce them to meet O E in B" and C" respectively. Through C" draw C" c'' parallel to O A, and through B" draw B" b'' parallel to O A. Through D' draw D' D" perpendicular to D' D and equal to the given height of that portion of the bath, and join D D", then D D" will be the true length of D' D. Join D c'; through c' draw c' F perpendicular to c' D and equal to the given height, and join D F, then D F may be taken as the true length of D c'.

In drawing the pattern, we will first set out that for the O A O'A' portion of the back, and then attach to it the pattern for the O E A portion. It is evident that the O A O'A' portion is half a right cone frustum, and therefore its pattern can be drawn (see Problem VIII., p. 41), thus. Draw (Fig. 91) a line D D' (the line D D' left of E A') equal to D D" (Fig. 90). With D and D' as centres and radii respectively equal to D F and D' c' (Fig. 90) describe arcs intersecting in c'. With D' and D as centres and radii respectively equal to D F and D c (Fig. 90) describe arcs intersecting in c. To find points b and b' proceed as just explained and with the same radii, but c and c' as centres instead of D and D'.

FIG. 91.

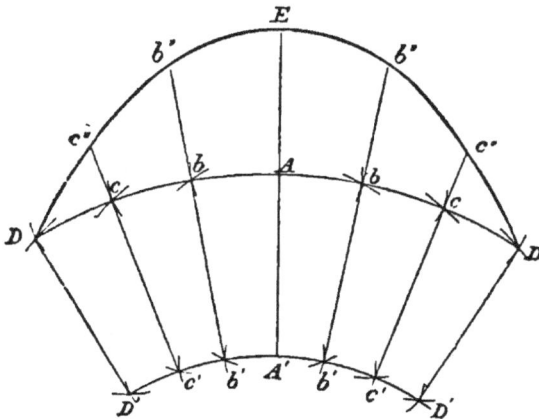

Similarly to find A and A', and the points b and b', &c., on the right-hand side of A A'. Through the points D', c', b', A', b', c', D' draw an unbroken curved line. Also through D, c, b, A, b, c, D draw an unbroken curved line, and join D D'. This completes the pattern for the O A O' A' portion of back of bath, to which we have now to attach, at D A D, the pattern for the O E A portion. Join A' A; produce it indefinitely and make A E equal to A E (Fig. 90). Next join, right and left of A A', b' b, c' c, and produce them indefinitely; make b b", right and left of A A', equal to A b" (Fig. 90). Also make c c", right and left of A A', equal to A c" (Fig. 90), and

through the points D, c'', b'', E, b'', c'', D draw an unbroken
curved line; then D E D D' A' D' is the complete pattern
required.

To draw the side and half-front pattern.

Draw separately (Fig. 92) the B h D D' g' B' portion of
Fig. 84, that is, the plan, of a side and half-front of the bath.
Join $h g'$ (as was done in Fig. 84). We now have to get at
corresponding points in the arcs D h, D' g', and also in the arcs
h B, g' B. We do this by the method given in § 85, p. 202,

FIG. 92.

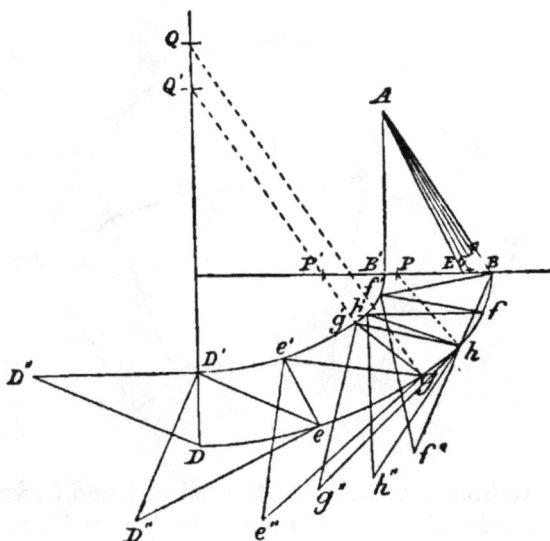

thus. Join Q' g' and through Q draw Q g parallel to Q' g' and
cutting arc D h in g; then the arcs D g and D' g' will be the
proportional. Now divide arc D g into any number of equal
parts, here two only to avoid confusion, in the point e, and
divide the arc D' g' into two equal parts in the point e'; then
e, e' and g, g' are corresponding points. Next join P h and
through P' draw P' h' parallel to P h and cutting arc B' g' in h'.
Divide arcs B h and B' h' each into the same number of equal

PLATE IV. (see p. 217).

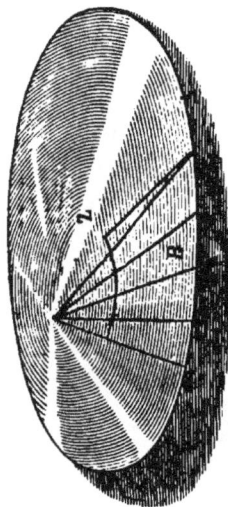

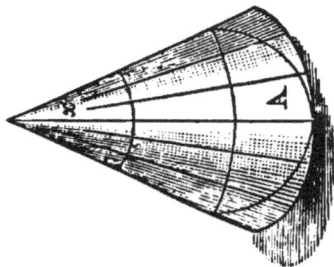

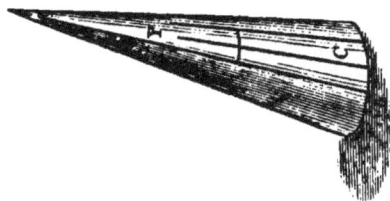

DRAWINGS FROM MODELS IN SOUTH KENSINGTON MUSEUM. (*Part of Exhibit by the Author at the Inventions Exhibition, 1885.*)

Q 2

parts, here two, in the points f and f' respectively; then f, f'
and h, h' are corresponding points. Join $e\,e'$, $g\,g'$, $f\,f'$, and $h\,h'$.
Through B' draw B' A perpendicular to B' B and equal to the
given height of that portion of the bath; and from B' along
B' B mark off B' F, B' H, B' G, and B' E respectively equal to
$f\,f'$, $h\,h'$, $g\,g'$, and $e\,e'$. Join B A, F A, H A, G A and E A;
then B A, F A, II A, G A, and E A will be respectively the
true lengths of B B', $f\,f'$, $h\,h'$, $g\,g'$, and $e\,e'$. Join f' B, $h'f$, $e'g$,
and D' e; through f' draw $f'f''$ perpendicular to f'B, and
equal to the given height, and join Bf'', then Bf'' may be
taken as the true length of f' B. Through h', g', e', and D'
draw $h'h''$, $g'g''$, $e'e''$, and D'D'', perpendicular to $h'f$, $g'h$, $e'g$,
and D' e respectively, and each equal to the given height;
also draw D' D'' perpendicular to D D' and equal to the given
height; and join fh'', hg'' ge'', eD'', and D D''; then fh'',
hg'', ge'', eD'', and D D'' may be taken as the true lengths
of $h'f$, $g'h$, $e'g$, D' e, and D' D respectively.

FIG. 93.

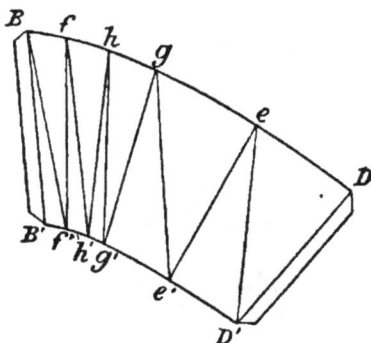

Next draw (Fig. 93) B B' equal to B A (Fig. 92), and with
B and B' as centres and radii respectively equal to Bf'' and
B'f' (Fig. 92) describe arcs intersecting in f, and with f and
B as centres and radii respectively equal to F A and Bf
(Fig. 92) describe arcs intersecting in f. With f and f' as
centres and radii respectively equal to fh'' and $f'h'$ (Fig. 92)

describe arcs intersecting in h', and with h' and f as centres
and radii respectively equal to H A and fh (Fig. 92) describe
arcs intersecting in h.   With h and h' as centres and radii
respectively equal to $h\,g''$ and $h'\,g'$ (Fig. 92) describe arcs intersecting in g', and with g' and h as centres and radii respectively
equal to G A and hg (Fig. 92) describe arcs intersecting in
g.   With g and g' as centres and radii respectively equal to
$g\,e''$ and $g'\,e'$ (Fig. 92) describe arcs intersecting in e', and
with e' and g as centres and radii respectively to E A and $g\,e$
(Fig. 92) describe arcs intersecting in e.   With e and e' as
centres and radii respectively equal to $e\,D''$ and $e'\,D'$ (Fig 92)
describe arcs intersecting in D', and similarly with D' and e
as centres and radii respectively equal to D D″ and $e\,D$
(Fig. 92) describe arcs intersecting in D.   Join D D'.
Through B, f, h, g, e, D draw an unbroken curved line.
Also through B', f', h', g', e', D' draw an unbroken curved
line.   Then B g D D' g' B' is the pattern required.

PROBLEM XXX.

*To draw the pattern for an oblong taper bath, the like dimensions
to those for Problem XIII. being given.*

Again, it is only necessary to treat of two cases—one in this
problem, and one in the problem following (see also § 79,
p. 157).

Draw (Fig. 94) the plan of the body (see Problem XIII.,
p. 140), preserving of its construction the centres O, O'; and
the points b, b', a, a', s, s', h, h', A, A', B, B', in which the
straight lines and arcs meet each other.   Join b, b', a, a', s, s',
h, h', A A' (two places), and B B' (two places) as shown in
the fig.   Examining the plan we see (d, p. 55) that each
round corner A A' B B' of the toe is the same portion of a
right cone frustum ; and each of the round corners $a\,a'\,b'\,b$,
$s\,s'\,h'\,h$, of the head are the same portion (g, p. 129) of an
oblique cone frustum.   As we proceed it will be seen that

the construction of the pattern for the round corners of the head of bath is exactly the same as that for the round corners of the body in Problem XXVI. (see also § 86, p. 206).

In Plate V. (p. 237) is a representation of an oblong

FIG. 94.

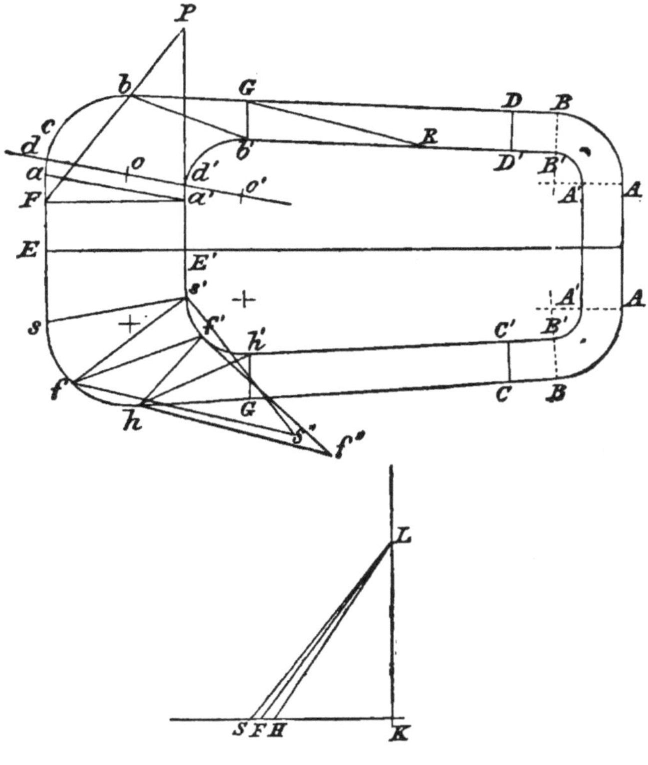

taper bath, also of an oblique cone Z, the A portion of which corresponds to the A' portion of the body, and the development of the former is the development of the latter.

Patterns when the body is to be made up of four pieces.

We will put the seams to correspond with the lines G h', G b', D D', and C C'. The patterns required will be three,

one for the head of the bath, one for the toe, and one for
the sides.   The pattern for the toe can be readily drawn by
Problem XXVII., p. 90.   Likewise the pattern for the sides.
The pattern for the head is drawn as follows.

FIG. 95.

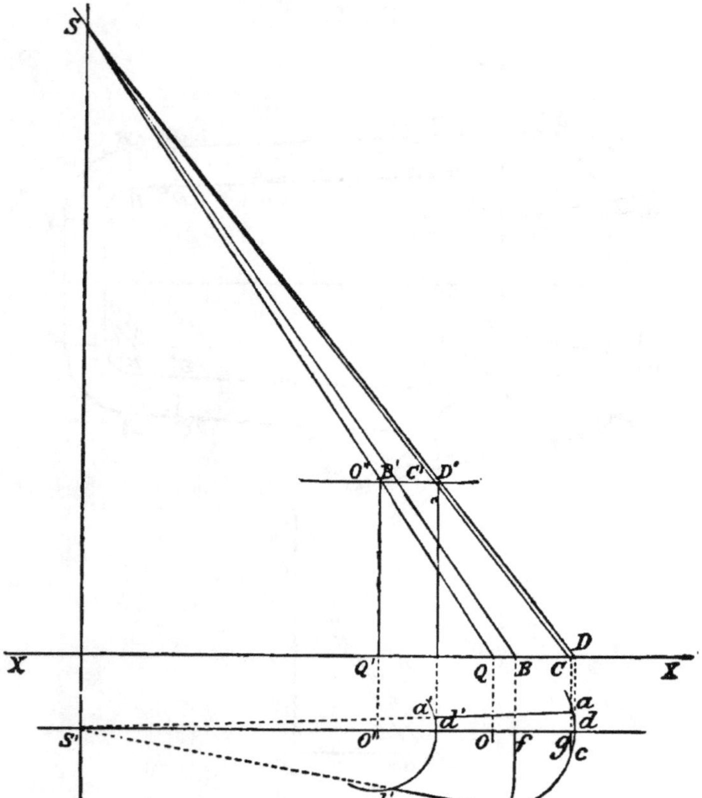

Draw separately (Fig. 95) a head-corner portion of
Fig. 94, say $a\,a'\ b'\,b$, thus.   Draw an indefinite line S'd
(Fig. 95), and with any point O (corresponding to O,
Fig. 94) in it as centre and Oa (Fig. 94) as radius describe
an arc $b\,a$.   Join OO' (Fig. 94) and produce it to cut arc

$b\,a$ in d; make $d\,b$ and $d\,a$ (Fig. 95) equal respectively to $d\,b$ and $d\,a$ (Fig. 94). Now (Fig. 95) make $d\,O'$ equal to $d\,O'$ (Fig. 94), and with O' as centre and $O'\,a'$ (Fig. 94) as radius describe an arc $b'\,a'$. Make $d'\,b'$ and $d'\,a'$ equal respectively to $d'\,b'$ and $d'\,a'$ (Fig. 94). Joining $b\,b'$, $a\,a'$ completes the portion of Fig. 94 required. Now divide (Fig. 94) $b\,a$ into any number of parts. It is convenient to take d as one of the division points, and to make $d\,c$ equal to $d\,a$; leaving $c\,b$ without further division, thus making the division of $b\,a$ into three portions, not all equal. In actual practice the dimensions of the work will suggest the number of parts necessary. Here $b\,c$ is left without further division in order to make clear the correspondence of this problem to Problem XXVI., p. 205. Now (Fig. 95) make $d\,c$ equal to $d\,c$ (Fig. 94), and then $b\,a$ will be divided correspondingly to $b\,a$ (Fig. 94). Draw $X\,X$ parallel to $S'\,d$; and at d and O draw $d\,D$, $O\,Q$ perpendicular to $S'\,d$; and meeting $X\,X$ in D and Q; also through d' and O' draw $d'\,D'$, $O'\,O''$ perpendicular to $S'\,d$; the line $O'\,O''$ cutting $X\,X$ in Q'. Make $Q'\,O''$ equal to the given height of the bath, and draw $O''\,D'$ parallel to $X\,X$, and cutting $d'\,D'$ in D'. Join $D\,D'$, $Q\,O''$; produce them to intersect in S (§ 80, p. 158); and from S let fall a perpendicular to $S'\,d$, cutting $S'\,d'$ in S'. With S' as centre and $S'\,a$, or $S'\,c$ (which is equal to $S'\,a$) and $S'\,b$ successively as radii describe arcs cutting $S'\,d$ in g and f. Draw $g\,C$, $f\,B$ perpendicular to $X\,X$ and cutting it in C and B; and join $C\,S$, $B\,S$, cutting $O''\,D'$ in C' and B'.

Next draw $S\,D$ (Fig. 96) equal to $S\,D$ (Fig. 95) and with S as centre and $S\,C$, $S\,B$ (Fig. 95) successively as radii describe arcs c and b. With D as centre and radius equal to $d\,a$ (Fig. 95) describe arcs cutting arc c in A and C. With C as centre and radius $c\,b$ (Fig. 95) describe an arc cutting arc b in B. Join A, C, and B to S. Make $S\,D'$ equal $S\,D'$ (Fig. 95) and with S as centre and $S\,C'$ (Fig. 95) describe an arc (not shown in the fig.) cutting $S\,A$ and $S\,C$ in A and C respectively; make $S\,B'$ equal to $S\,B'$ (Fig. 95).

Through A, D, C, and B draw an unbroken curved line.
Also through A', D', C', and B' draw an unbroken curved
line; this will complete the pattern for a head-corner. To
attach the patterns for the flat portions of the head to A A'

FIG. 96.

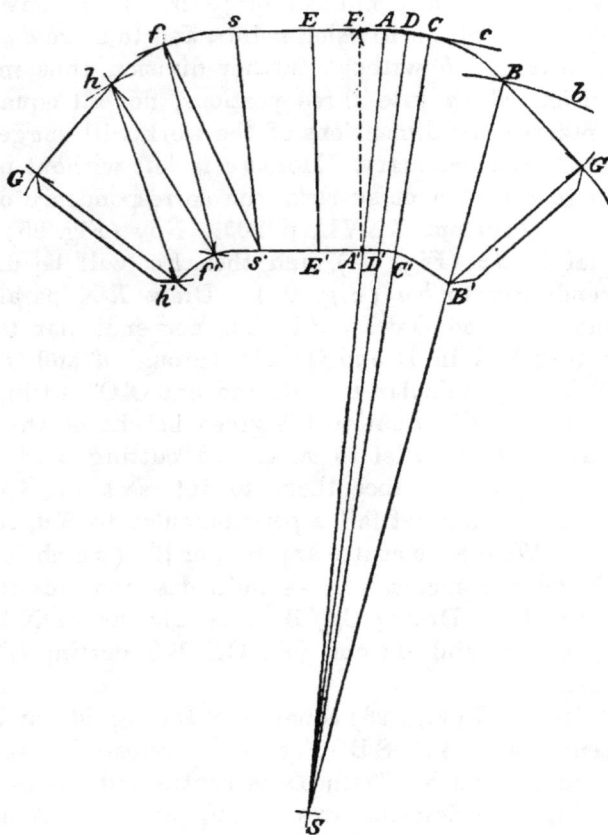

and B B' respectively; draw through a' (Fig. 94) a line a' F
perpendicular to a' s', and through b' draw b' G perpendicular
to b' D'.

Now draw (Fig. 94) a' P perpendicular to a' F and equal

to the given height of the bath, and join F P, then F P is the true length of F a'.

Next draw b' R perpendicular to b' G (b' R will of course coincide with the line b' B') and equal to the given height; join G R, then G R is the true length of G b'. Now with B' (Fig. 96) as centre and G R (Fig. 94) as radius, and B as centre and radius b G (Fig. 94), describe arcs intersecting in G. Join B G, B' G. With A' (Fig. 96) as centre and F P (Fig. 94) as radius, and A as centre and radius a F (Fig. 94) describe arcs intersecting in F. Join A F and produce it indefinitely, and make A s equal to $a s$ (Fig. 94); through A' draw A's' parallel to A s and equal to $a' s'$ (Fig. 94); and join $s s'$. The pattern for the portion, seen in plan in Fig. 94, G b a s s' a' b' of the head of the bath is now completed. It is needless to work out in detail the addition of the portion (Fig. 96) $s h$ G $h'$$f'$$s'$ of the pattern, by which we complete the head pattern G E G h' E' B'. The extra lines in this latter portion of the pattern appertain to the next problem.

PROBLEM XXXI.

To draw, **without long radii,** *the pattern for an oblong taper bath ; given dimensions as in Problem XXX.*

This problem is a second case of the preceding.

Patterns when the body is to be made up of four pieces with seams as in preceding problem.

Again, the patterns required will be three; one for the head of the bath, one for the toe, and one for the sides. The latter pattern needs no description. The pattern for the toe can be readily drawn by Problem XXVIII., p. 94. The pattern for the head can be drawn as follows.

Draw half the plan of the bath, as the lower half of Fig. 94, and divide the arcs $s h$, s' h', each into any number of equal parts, here two, in respectively the points f and f', and join $f f'$. Draw (Fig. 94a) any two lines K S, K L perpendicular to each other, and make K L equal to the given

heignt of the bath.   From K along K S mark off K H equal
to hh', KF equal to ff', and K S equal to ss'; and join L H,
L F, L S; then L H, L F, and L S are the true lengths of
hh', ff', and ss' respectively.   Next join (Fig. 94) $f'h$, $s'f$;
draw $f'f''$, $s's''$, perpendicular to $f'h$, $s'f$ respectively, and
each equal to the given height; and join hf'', fs''; then hf'',
fs'' may be taken respectively as the true lengths of $f'h$ and
$s'f$.   The true length of h'G may be found along h' B' as
was that of bG in Problem XXX. along b'B'; it will of
course be equal to G R, and we shall speak of it as G R.

    Next draw (see Fig. 96, left-hand portion) ss' equal to
L S (Fig. 94a), and with s' and s as centres and radii
respectively equal to fs'' and sf (Fig. 94) describe arcs
intersecting in f. With f and s' as centres and radii
respectively equal to L F (Fig. 94a) and $s'f'$ (Fig. 94)
describe arcs intersecting in f'; and with f' and f as centres
and radii respectively equal to hf'' and fh (Fig. 94) describe
arcs intersecting in h; also with h and f' as centres and
radii respectively equal to L H (Fig. 94a) and $f'h'$ (Fig. 94)
describe arcs intersecting in h'.   With h' and h as centres
and radii respectively equal to G R and hG (Fig. 94)
describe arcs intersecting in G.   Through s, f, h, draw an
unbroken curved line.   Also through s', f', h' draw an unbroken curved line; and join hG, Gh'.   Then $shGh'f's'$ is
the pattern for the portion of the head of the bath represented in plan in Fig. 94 by the same lettering.   It is unnecessary to pursue the pattern further.

PLATE V. (see p. 231).

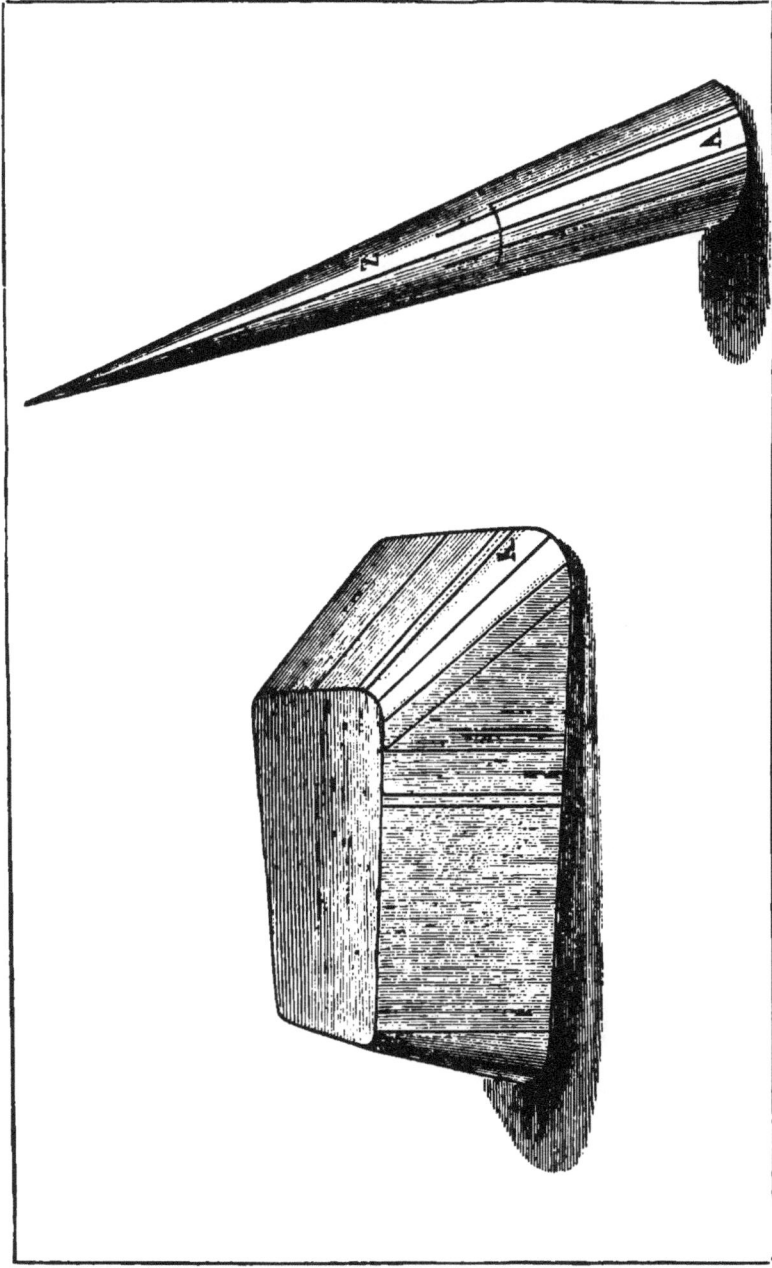

DRAWINGS FROM MODELS IN SOUTH KENSINGTON MUSEUM. (*Part of Exhibit by the Author at the Inventions Exhibition, 1885.*)

Book III.

CLASS III.

Patterns for Miscellaneous Articles.

The book we now reach is made up of miscellaneous problems, all of which are of considerable practical importance, as well as are also typical cases.

PROBLEM I.

To draw a pattern for the elbow formed by two equal circular pipes (cylinders of equal diameters) which meet at any angle.

First draw (Fig. 1) a side elevation and a part-plan of the elbow, as follows. Draw any two lines G G', O G' at an angle to each other equal to the given angle of the elbow; through any point G in G G' draw G A perpendicular to G G' and equal to the diameter of the pipes; and through A draw an indefinite line A A' parallel to G G'. Then from any point O in G' O draw O N perpendicular to G' O and equal to the diameter of the pipe; and through N draw an indefinite line N A parallel to G' O and meeting A A' in A'. Join A' G' (the line A' G' produced always bisects the angle G G' O); then O N A G is a side elevation of the elbow, and the line A' G' represents the joint or 'mitre' of the pipes. On G A describe a semicircle G d A (this will be the part-plan of the elbow); which divide into any number, here six, of equal parts in the points b, c, d, e, f. From each of these points draw lines parallel to A A', namely, the lines b B', c C', d D', e E', and f F', cutting G A in the points B, C, D, E, and F respectively.

FIG. 1.

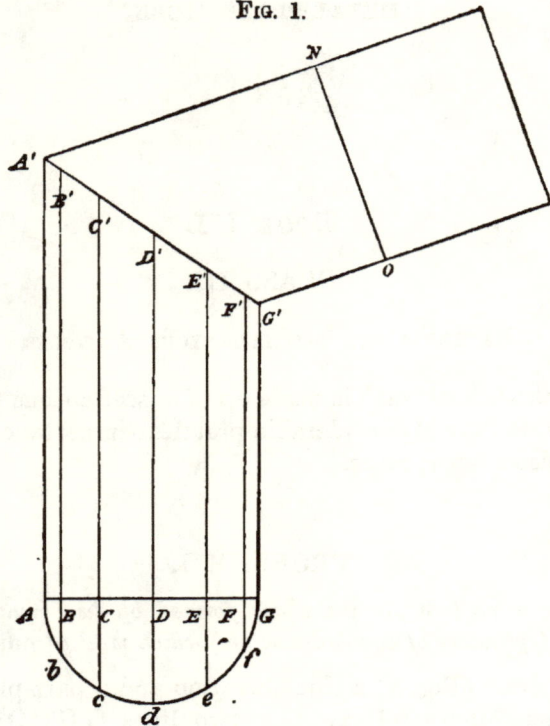

FIG. 2.

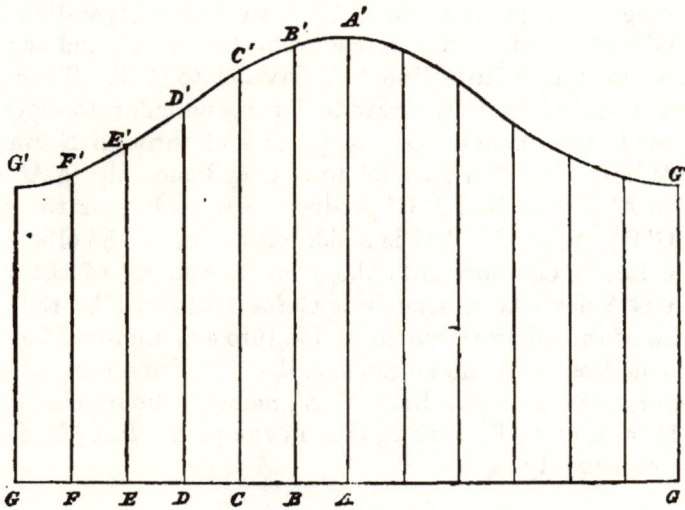

To draw the pattern, with the longitudinal seams of the pipes to correspond with the lines G′ G, G′ O. Draw (Fig. 2) an indefinite line G G, and, from a point G in it set off along it distances G F, F E, E D, D C, C B, and B A, equal respectively to Gf, $f e$, $e d$, $d c$, $c b$, and b A of Fig. 1. For small work the chord distances round the semicircle give satisfactory results; of large work we shall speak immediately. Through G, F, E, D, C, B, and A draw indefinite lines G G′, F F″, E E′, D D′, C C′, B B′, and A A′ perpendicular to G G; and make G G′ equal to G G′ (Fig. 1), F F″ equal to F F″ (Fig. 1), and E E′, D D′, C C′, B B′, and A A′ equal respectively to E E′, D D′, C C′, B B′, and A A′ (Fig. 1). From G′ through F″, E′, D′, C′, B′, and A′ draw an unbroken curved line; this completes one-half the pattern required. The half to the right of A A′ can be drawn in similar manner, setting off along A G the same above distances, but in reverse order and starting from A.

(87.) The line G G (Fig. 2) must of course be equal to the circumference of the pipe. Therefore if a somewhat close accuracy is required, as in large work, it will be found best to make G G equal to the circumference of the pipe (Problem XI., p. 12); then to divide G G into twice as many equal parts as the semicircle A d G (Fig. 1) is divided into; then through each division point to draw lines perpendicular to G G and proceed as above explained.

(88.) By making a start in Fig. 2 with G G′ for an outer line of the pattern, we ensure that the longitudinal seams of the pipes shall, as we desired, correspond with G G′ and G′ O of Fig. 1. If the seams are to correspond with the lines A′ A and A′ N of this fig., we should letter our commencing indefinite line, not G G, but A A, and should start with, for an outer line of the pattern, a perpendicular A A′ equal to A A′ of Fig. 1. The pattern we should then get would be like to that obtained by dividing Fig. 2 into halves along A A′, turning each portion half-way round, and making the lines G G′ coincide.

R

PROBLEM II.

To draw the pattern for the T*-piece formed by two equal or unequal circular pipes (cylinders of equal or unequal diameter), which meet at right angles.*

First draw (Fig. 3) a side elevation and a part-plan of the two circular pieces of pipe, which we will suppose unequal, thus. Draw two indefinite lines Z d, and K J, intersecting each other at right angles in O. Make O Z equal to the

FIG. 3.

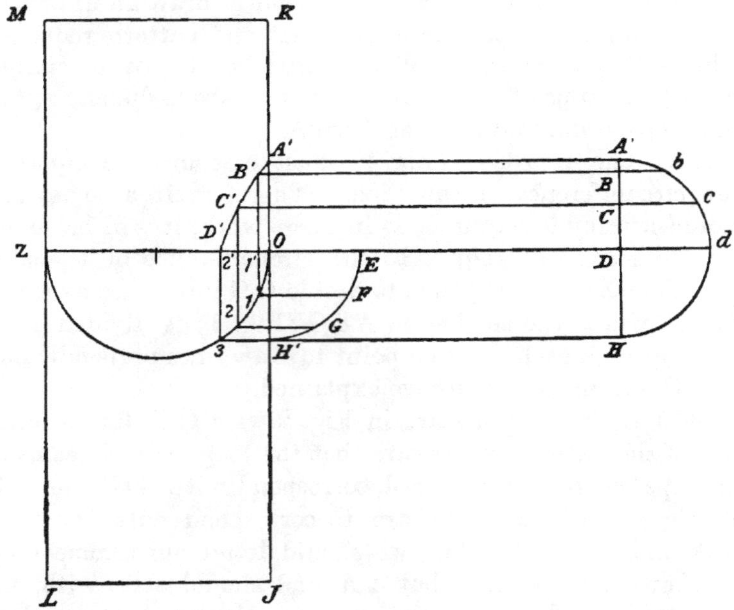

diameter of the larger pipe, and through Z draw an indefinite M L parallel to K J. Make O A' and O H' each equal to half the diameter of the smaller pipe, and through A' and H' draw indefinite lines A' A and H' H each perpendicular to K J. In A' A take any point A, on H' H set off H' H equal to A' A, and join A H cutting O d in D; then A' A H H' will represent, in elevation, a piece of the smaller pipe. Next in A' K take any point K, and through K draw K M perpendicular to K J and meeting M L in M; also in H' J take any point J, and through J draw J L perpendicular to K J and meeting M L in L; then M K J L will represent, in elevation, a piece of the larger pipe, and M K A' A H H' J L a side elevation (except the curve of junction) of the ⊤-piece. With D as centre and radius D A, that is, half the diameter of the smaller pipe, describe a semicircle A d H; divide the quadrant A d of it into any number of equal parts, here three, in the points b and c; and through b and c draw indefinite lines b B' and c C' parallel to A' A. Now on Z O describe a semicircle Z 3 O (this will be a part-plan of the large pipe), and with O as centre and radius D A describe a quadrant H' E (this may be regarded as a part-plan of the smaller pipe) which divide, exactly as quadrant A d was divided in the points F and G; through F, G, and H' draw lines F 1, G 2, and H' 3 parallel to Z d and cutting the semicircle Z 3 O in points 1, 2, and 3. Through point 1 draw a line 1 B' parallel to K J and meeting b B' in B'; through 2 draw 2 C' parallel to K J and meeting c C' in C'; and through 3 draw 3 D' parallel to K J and meeting d D' in D' From D' to A' through the points C' and B' draw an unbroken curved line, then A' C' D' is the elevation of one-half of the curve of junction of the two pipes. In practice it is only necessary to draw the A' O D A (Fig. 3) portion of the elevation of the smaller pipe. The other half elevation H' O D H of it is drawn here simply to make the full side-elevation of the ⊤-piece clearer.

To get at the whole ⊤-piece, it is evident that we require two patterns, one for the smaller piece of pipe, up to its junction with the larger, and one for the larger with the hole in it that the smaller pipe fits to.

To draw the pattern for the larger pipe, with the longitudinal seam to correspond with the line M L.

First set out apart from the pipe itself the shape for the

R 2

hole in it. Draw (Fig. 4) two indefinite lines Z O' and A' A'
intersecting at right angles in O; from O, on Z O', right and
left of A' A' set off distances O 1', 1' 2', and 2' D' equal respectively to O 1, 1 2, and 2 3 (Fig. 3), that is, to the actual
distances on the round curve of the pipe at Z O that the lines
1 B', 2 C', and 3 D' are apart. Through points 1' and 2'

FIG. 4.

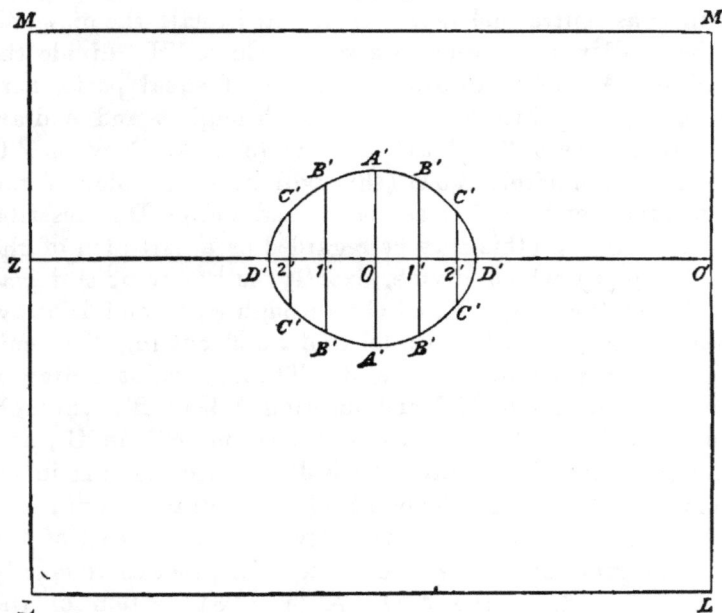

right and left of A' A' draw B' B' and C' C' perpendicular
to Z O'. Make 1' B', above and below Z O', and right and
left of A' A', equal to 1' B' (Fig. 3); and make 2' C', above and
below Z O', and right and left of A' A', equal to 2' C' (Fig. 3)
and through all the points as above found, namely D', C',
B', A', B', C', D', C', B', &c., draw an unbroken curved line;
then D' A' D' A' D' will be the shape of the hole required.

To complete the pattern for the M K J L (Fig. 3) piece
of the larger pipe, make O Z and O O' each equal to half

its circumference (Problem XI., p. 12); and through Z and O' draw indefinite lines M L perpendicular to Z O'. Make Z M of left-hand line M L, and O'M of right-hand line M L each equal to Z M (Fig. 3); similarly make Z L and O'L each equal to Z L (Fig. 3). Then M L L M will complete the pattern required.

(89.) We have shown how to mark out by itself the hole the larger pipe, because in cases where the pipe is already made up, it is convenient to be able to mark out the shape of the hole apart from the pipe, on, say, a thin piece of sheet metal, which shape can then be cut out and used as a template; being applied to the pipe and bent to it, and the shape of the hole marked on it from the template. Even when the pipe is not made up, it is useful when the pipe is large to be able to mark out the hole quite apart from the pipe itself.

To draw the pattern for the smaller piece of pipe, the longitudinal seam to correspond with the line A' A (Fig. 3).

FIG. 5.

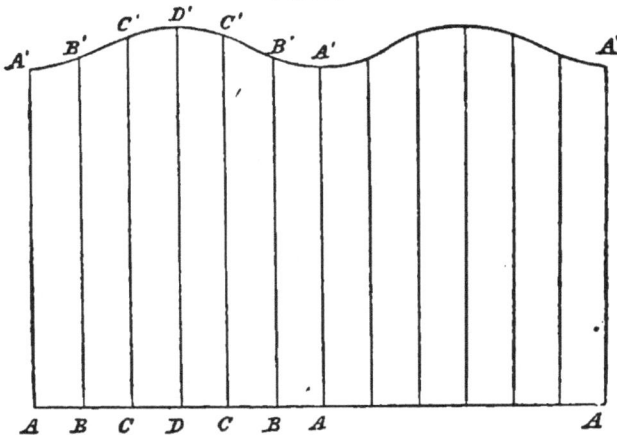

Draw (Fig. 5) an indefinite line A A. In it take any point D, and from D, right and left, set off distances D C,

C B, B A, equal respectively to dc, cb, and ba (Fig. 3), that is, equal to one another; and from the point D and each of the points C, B, and A draw lines perpendicular to A A. Make D D' equal to D D' (Fig. 3), and the lines C C', B B', A A', right and left of D D', equal respectively to C C', B B', and A A' (Fig. 3). From either point A' to A' on the other side of D D' draw through B', C', D', C', and B' an unbroken curved line; then A A' A' A will be half the required pattern. The other half can be similarly drawn. If a somewhat close accuracy in the length of A A (Fig. 5) is required proceed as in Problem I. (§ 87, p. 241). See also § 88, p. 241.

PROBLEM III.

To draw the pattern for the ' slanting T *' formed by two equal or unequal circular pipes (cylinders of equal or unequal diameter) which meet otherwise than at right angles.*

We will suppose the pipes unequal. First draw (Fig. 6) a side elevation and a part-plan of the pipes, thus. Draw two indefinite lines K O, and A" A' at that angle to one another that the pipes are to be, the angle O A" A', say. In A" A' take any point A'; through it draw A' G' perpendicular to A" A' and equal to the diameter of the smaller pipe, and through G' draw a line parallel to A" A' and meeting K O in G"; then A" A' G' G" will represent, in elevation, a piece of the smaller pipe. Next in K O take any point O, and through it draw an indefinite line L G perpendicular to K O. Make O L equal to the diameter of the larger pipe; through L draw an indefinite line L M parallel to K O; in L M take any point M, and through M draw M K perpendicular to M L. Then M K O L will represent, in elevation, a piece of the larger pipe; and M K G" G' A' A" O L a side elevation (except the curve of junction) of the slanting T. On A' G' describe a semicircle A' d' G'; and divide it into any number of equal parts, here six, in the points

b', c', d', e', and f'; and through these division points draw indefinite lines b' B'', c' C'', d' D'', e' E'', and f' F'' parallel to G' G'', and cutting A' G' in the points B', C', D', E', and F'.

Fig. 6

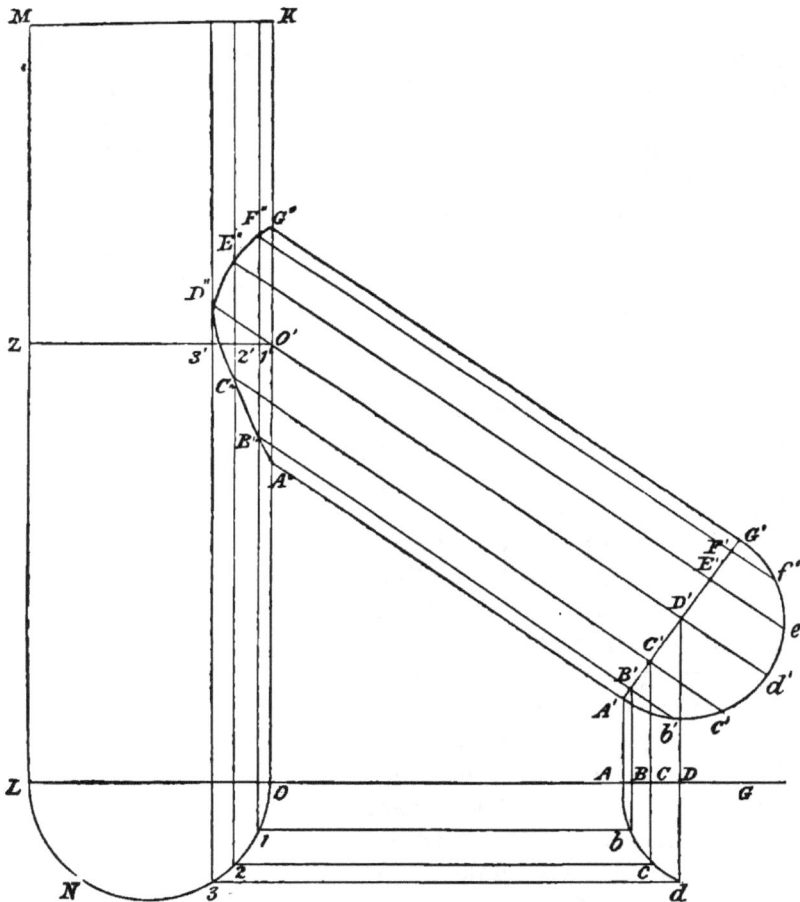

Now from A' let fall a perpendicular to L G meeting L G in A, and through B', C' and D' draw indefinite lines B' b, C' c, and D' d perpendicular to and cutting L G in B, C, and D.

Make B b equal to B' b', C c equal to C' c', and D d equal to D' d';
and from A through b and c to d draw an unbroken curved
line. On L O describe the semicircle L 3 O, and through the
points b, c, d, just found draw lines b 1, c 2, and d 3 parallel
to L G and cutting the semicircle L 3 O in points 1, 2, and 3;
then L N 3 d D is a part-plan of the slanting T. Through
point 1 draw a line 1 F''' parallel to O K and meeting B' B''
and F' F'' in B'' and F'' respectively; through 2 draw 2 E''
parallel to O K and meeting C' C'' and E' E'' in C'' and E''
respectively; and through 3 draw 3 D'' parallel to O K and
meeting D' D'' in D''. From A'' through the points B'', C'',
D'', E'', and F'' to G'' draw an unbroken curved line, then
this curve is the elevation of the curve of junction of the two
pipes. In the straight line A'' G'' take any point O', about
midway between A'' and G'', and through it draw O' Z perpendicular to K O, and cutting lines 1 F'', 2 E'', and 3 D'' in
points 1', 2', and 3'.

It is evident that for slanting T we require two patterns,
one for the smaller piece of pipe up to its junction with
the larger, and one for the larger with the hole in it that
the smaller pipe fits to.

To draw the pattern for the larger pipe, the longitudinal
seam to correspond with the line M L.

First to set out apart from the pipe itself the shape of the
hole in it. Draw (Fig. 7) two indefinite lines Z O and G'' A''
intersecting at right angles in O'; and from O', on Z O, right
and left of G'' A'' set off distances O' 1', 1' 2', and 2' 3' equal
respectively to O 1, 1 2, and 2 3 (Fig. 6), that is, to the actual
distances on the curve of the pipe at Z O that the lines
B'' F'', C'' E'', and 3' D'' are apart. Through points 1', 2', and
3', right and left of G'' A'' draw F'' B'', E'' C'', and D'' 3' perpendicular to Z O. Make O' G'' equal to O G'' (Fig. 6),
O' A'' equal to O' A'' (Fig. 6). Also make 1' F'', 2' E'', 3' D'',
right and left of G'' A'' and above Z O, equal respectively to
1' F'', 2' E'', 3' D'' above Z O' (Fig. 6); and 1' B'', 2' C'' right
and left of G'' A and below Z O, equal respectively to 1' B'',
2' C'', below Z O' (Fig. 6). Through all the points as above

found, namely D″, E″, F″, G″, F″, E″, D″, C″, B″, &c., draw
an unbroken curved line; then D″ G″ D″ A″ will be the
shape of the hole required.

FIG. 7.

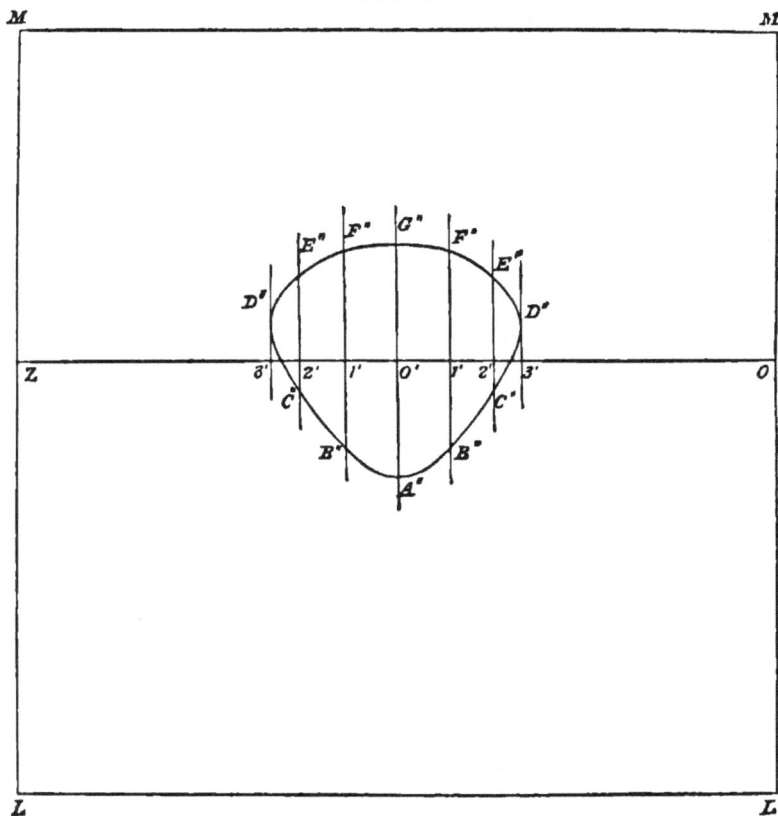

To complete the pattern for the M K O L (Fig. 6) piece of
the larger pipe, make O′ Z and O′ O each equal to half its
circumference (Problem XI., p. 12); and through Z and O
draw indefinite lines M L perpendicular to Z O. Make Z M
of left-hand line M L, and O M of right-hand line M L each
equal to Z M (Fig. 6); similarly make Z L and O L each

equal to Z L (Fig. 6). Then M L L M will complete the
pattern required. In connection with this see § 89, p. 245.

To draw the pattern for the smaller piece of pipe, the
longitudinal seam to correspond with the line A″ A′ (Fig. 6).

Draw (Fig. 8) an indefinite A′ A′. In it take any point
A′, and from A′ set off distances A′ B′, B′ C′, C′ D′, D′ E′,
E′ F′, and F′ G′ equal respectively to A′ b′, b′ c′, c′ d′, d′ e′, e′ f,

Fig. 8.

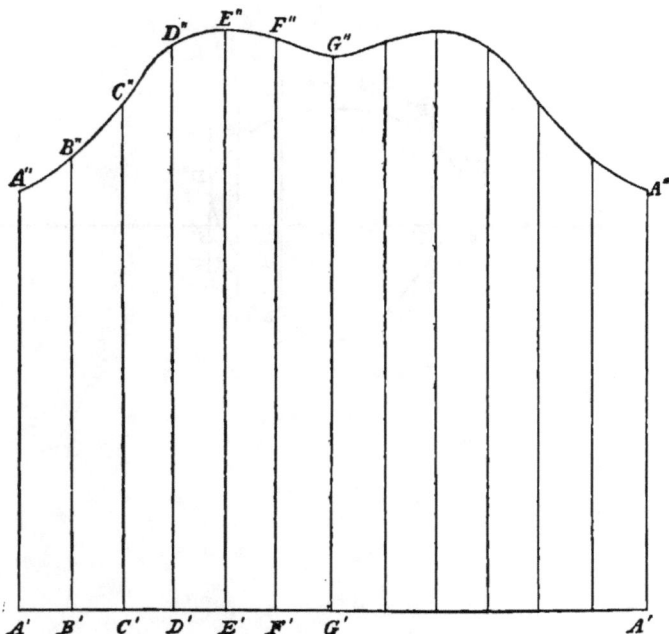

and f′ G′ (Fig. 6), that is, equal to one another. Through
A′, B′, C′, D′, E′, F′, and G′ draw indefinite lines A′ A″,
B′ B″, C′ C″, D′ D″, E′ E″, F′ F″, and G′ G″ perpendicular to
A′ A′; and make A′ A″ equal to A′ A″ (Fig. 6), B′ B″ equal
to B′ B″ (Fig. 6), C′ C″ equal to C′ C″ (Fig. 6), and D′ D″,
E′ E″, F′ F″, and G′ G″ equal respectively to D′ D″, E′ E″,
F′ F″, and G′ G″ (Fig. 6). From A″ through B″, C″, D″, E″,

F''', and G'' draw an unbroken curved line; this completes one-half the pattern required.   The half to the right of G' G'' can be drawn in similar manner, setting off along G' A' the same above distances, but in reverse order and starting from G'.   With this see § 87, p. 241, and § 88, p. 241.

PROBLEM IV.

To draw the pattern for the T *formed by a funnel-shape piece of pipe and a circular piece, the former being ' square' to the latter (part-cone joined ' square' to a cylinder); the diameter of the circular pipe, and the diameters of the ends of the funnel-shape pipe and its length being given.*

By ' being square' we mean that the axes of the pieces of pipe intersect and are at right angles.   The given diameter of the smaller end of the funnel-shape pipe is the diameter in the direction of the length of the circular pipe and that coincides with its surface.

First draw a side elevation and a part-plan of the T, thus. Draw (Fig. 9) any two indefinite lines Z d and K J, intersecting at right angles at O.   Make O Z equal to the diameter of the circular pipe; and through Z draw M L parallel to K J ; make O D equal to the length of the funnel-shape pipe and through D draw a line A H perpendicular to O d.   Now make D A and D H each equal to half the given diameter of the larger end of the funnel, and O A' and O H' each equal to half the given diameter of its smaller end, which small end we will suppose is not let into, but fits against the circular pipe; join A A' and H H', then A' A, H H' will be, in elevation, the main portion of the funnel-shape pipe.   Next in A' K take any point K, and through K draw K M perpendicular to K J and meeting M L in M ; also in H' J take any point J, and through J draw J L perpendicular to K J and meeting M L in L ; then M K J L will represent, in elevation, a piece of the circular pipe : and M K A' A H H' J L a side elevation (except the curve of

junction) of the ⊤. Produce A A' and H H' to intersect Z O
in V. With D as centre and radius D A describe a semicircle A d H and divide the quadrant A d of it into any
number of equal parts, here three in the points b and c;

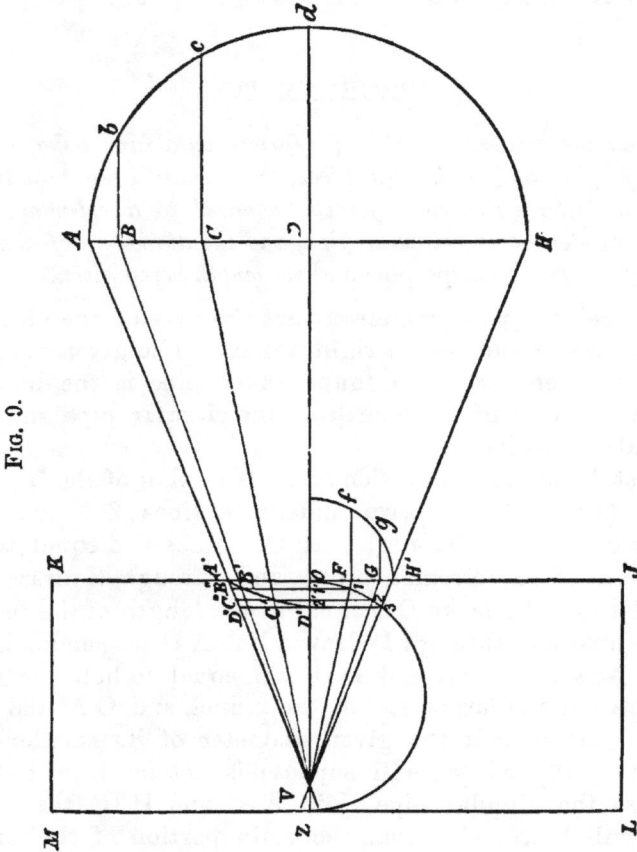

Fig. 9.

through b and c draw b B and c C each perpendicular to A H
and cutting it in B and C, and join B V and C V. Now on
Z O describe a semicircle Z 3 O (this will be a part-plan of
the circular pipe) cutting V H' in point 3. With O as centre
and radius O H' describe a quadrant H' E (this may be
regarded as a part-plan of the funnel-shape pipe) which

divide into the same number of equal parts that the quadrant A d is divided into, in the points f and g. Through f and g draw f F and g G each perpendicular to A' H' and cutting it in F and G ; join F V, G V, cutting the semicircle Z 3 O in points 1 and 2 respectively. Through point 1 draw a line 1 B" parallel to K J, meeting A V in B" and cutting Z O and B V in 1' and B' respectively; through 2 draw 2 C" parallel to K J, meeting A V in C", and cutting Z O and C V in 2' and C' respectively; and through 3 draw 3 D" parallel to K J, cutting Z O in D' and meeting A V in D". From D' through C' and B' to A' draw an unbroken curved line, then A' C' D' is the elevation of one-half the curve of junction of the two pipes. In practice it is only necessary to draw the O A' A D (Fig. 9) portion of the elevation of the smaller pipe. The other half-elevation O H' H D of it is drawn here simply to make the full side elevation of the T clearer.

It is evident that the T requires 2 patterns, one for the circular pipe with the hole in it that the funnel-shape pipe fits to, and one for the funnel-shape itself.

To draw the pattern for the circular pipe; the longitudinal seam to correspond with the line M L.

Proceed in exactly similar manner as explained for the pattern of the corresponding pipe in Problem II., p. 242.

To draw the pattern for the funnel-shape pipe; the longitudinal seam to correspond with the line A A' (Fig. 9).

With V (Fig. 10) as centre and V A (Fig. 9) as radius describe an arc A A, and from any point in it set off along the arc distances A B, B C, C D, D C, B C, and B A each equal to A b (one of the equal parts in which quadrant d A (Fig. 9) is divided). Join A V, B V, C V, D V, C V, B V, and A V ; and make the extreme lines A A' right and left of D V equal to A A' (Fig. 9), also the lines B B", right and left of D V, equal to A B" (Fig. 9), C C" right and left of D V, equal to A C" (Fig. 9) and D D" equal to A D" (Fig. 9). Through points A', B", C", D", C", B", A' draw an unbroken curved line, then A A' A' A will be one-half the pattern required. By continuing to the right, say, the arc A A, and setting off on it the same above equal distances A B, B C, &c., and proceeding in exactly similar manner the other half pattern can be drawn to complete the pattern required.

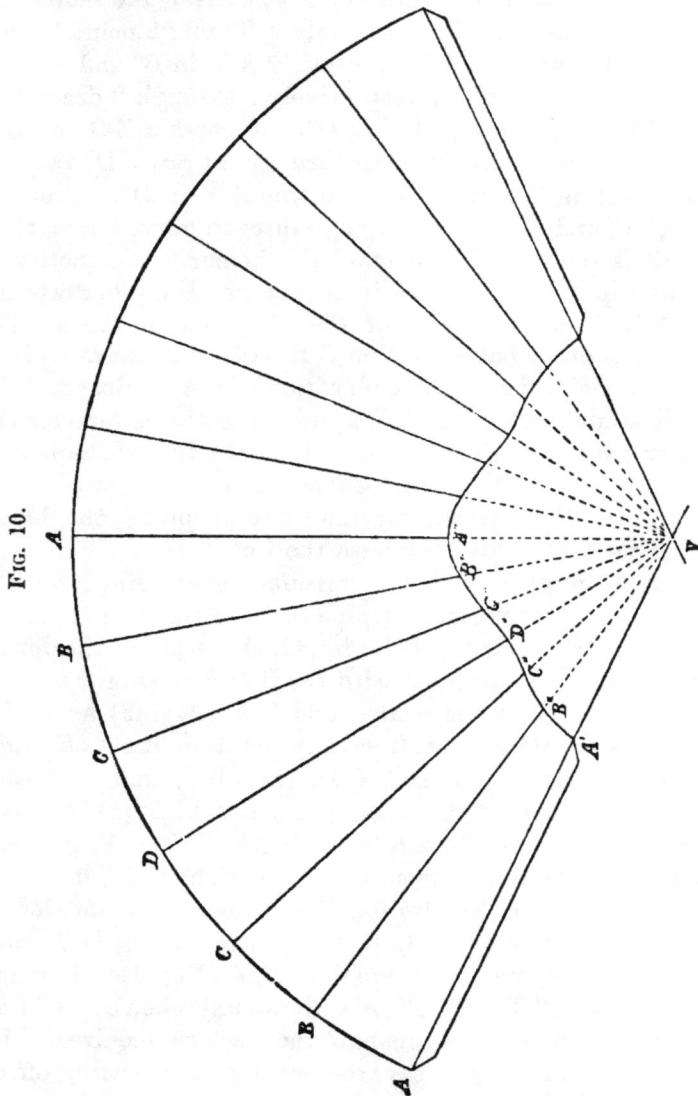

FIG. 10.

PROBLEM V.

To draw the pattern for a tapering piece of pipe joining two circular pipes of unequal diameter and not in the same line with each other.

A representation of such tapering piece of pipe joining two circular pipes will be found in Plate VI. (p. 259); also (except apex portion) of an oblique cone (Z) the frustum B′ D′ D E B of which (the line B′ D′ should be continued to the outside of the cone) corresponds to the tapering piece of pipe.

This tapering piece of pipe being an oblique cone frustum, its pattern can be drawn by Problem II., p. 113, Book II. In connection here see also §§ 58, 59, p. 111, and Figs. 5 and 6, p. 111.

PROBLEM VI.

To draw the pattern for a piece of pipe joining two circular pipes of equal diameters, and not in the same line with each other.

This problem is a special case of the preceding, and the piece of pipe joining the two equal pipes is an oblique cylinder. Being so, we can draw its pattern by Problem V., p. 121, Book II. See also § 61, p. 112, and Figs. 7a and 7b, p. 112.

PROBLEM VII.

To draw the pattern for the ' Y ' formed by two tapering pieces of pipe uniting two equal circular pipes at the arms of the Y to a third larger piece of circular pipe at its stem, the Y being both ways symmetrical.

In Plate VI. (p. 259) is a representation of such ' Y . The frustum B′ D′ D E B of the oblique cone Z in same

Plate (the line B′ D′ being continued to the outside lines of
the cone) corresponds to either of the tapering pieces of pipe
of the Y, except that at their junction C′ D′ a piece (B C D of

FIG. 11.

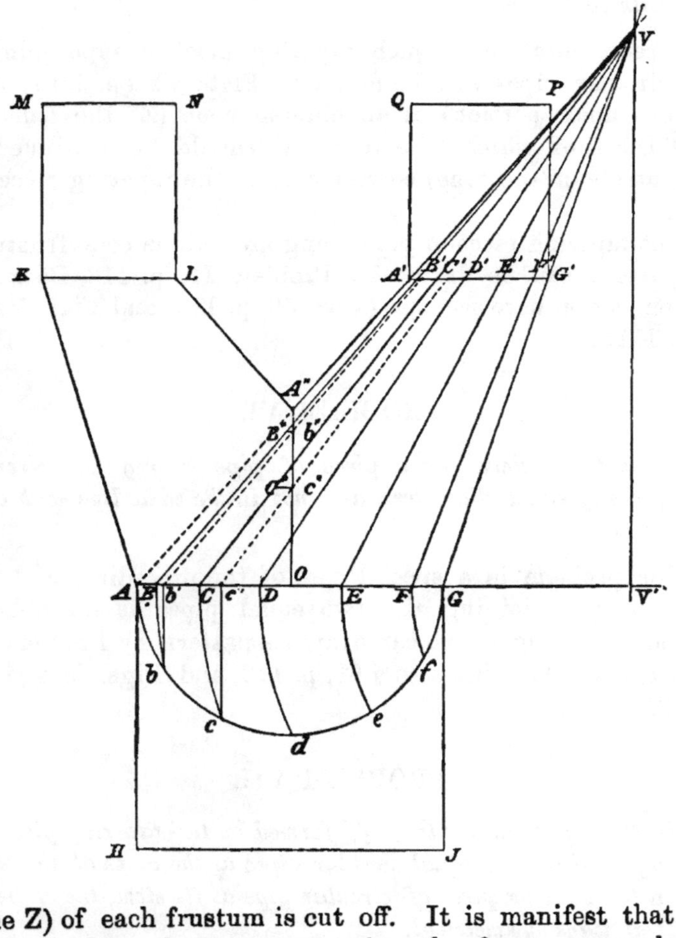

cone Z) of each frustum is cut off. It is manifest that the
base of either frustum, supposing the frustum completed,
would coincide with the top of the stem of the Y.

Let M A H J G P Q A″ N (Fig. 11) represent the Y in

elevation. As it is both ways symmetrical, G' G is the shortest generating line. Produce A' A" to A; then A' A is the longest generating line. Also produce A A', G G', to their intersection in V. On A G describe a semicircle A d G, which divide into any number of equal parts, here six, in the points b, c, d, e, f. Produce A G indefinitely and from V let fall a perpendicular to the produced line, meeting it in V'. With V' as centre and V'f, V'e, V'd, V'c, and V'b respectively as radii, describe arcs meeting A G in F, E, D, C, and B. Join each of these points to V by lines cutting A' G' in F', E', D', C', and B'. From c and b draw cc' and bb' perpendicular to A G and join c' and b' to V, by lines cutting A" O in c'' and b''. Through c'' draw c'' C" parallel to A G and cutting V C in C"; also through b'' draw b'' B" parallel to A G and cutting V B in B".

Now to draw the pattern (Fig. 12), so that the seam shall correspond with G' G (Fig. 11) the shortest generating line. Draw V A (Fig. 12) equal to V A (Fig. 11) and with V (Fig. 12) as centre and radii successively equal to V B, V C, V D, V E, V F, and V G (Fig. 11), describe, respectively, arcs bb, cc, dd, ee, ff, and gg. With A as centre and radius equal to A b (Fig. 11), describe arcs cutting the arc bb right and left of V A in B and B. With these points B and B as centres and radius as before, describe arcs cutting the arc cc right and left of V A in C and C. With same radius and the last-named points as centres, describe arcs cutting dd right and left of V A in D and D. With D and D as centres and same radius, describe arcs cutting ee right and left of V A in E and E, and with E and E as centres and same radius describe arcs cutting ff in F and F. Similarly, with same radius and F and F as centres find points G and G. Join the points B, C, D, E, &c., right and left of V A to V. With V as centre and V A' (Fig. 11) as radius describe an arc cutting V A in A'. With same centre and V B' (Fig. 11) as radius describe an arc $b'b'$ cutting V B right and left of V A in B'. With same centre and V C' (Fig. 11) as radius describe an

s

arc $c' c'$ cutting V C right and left of V A in C′. Similarly, with same centre and V D′, V E′, V F′, and V G′
(Fig. 11) successively as radii describe, respectively, arcs
$d' d'$, $e' e'$, $f' f'$, and $g' g'$, cutting V D, V E, V F, and V G,
right and left of V A respectively in D′, E′, F′, and G′.

FIG. 12.

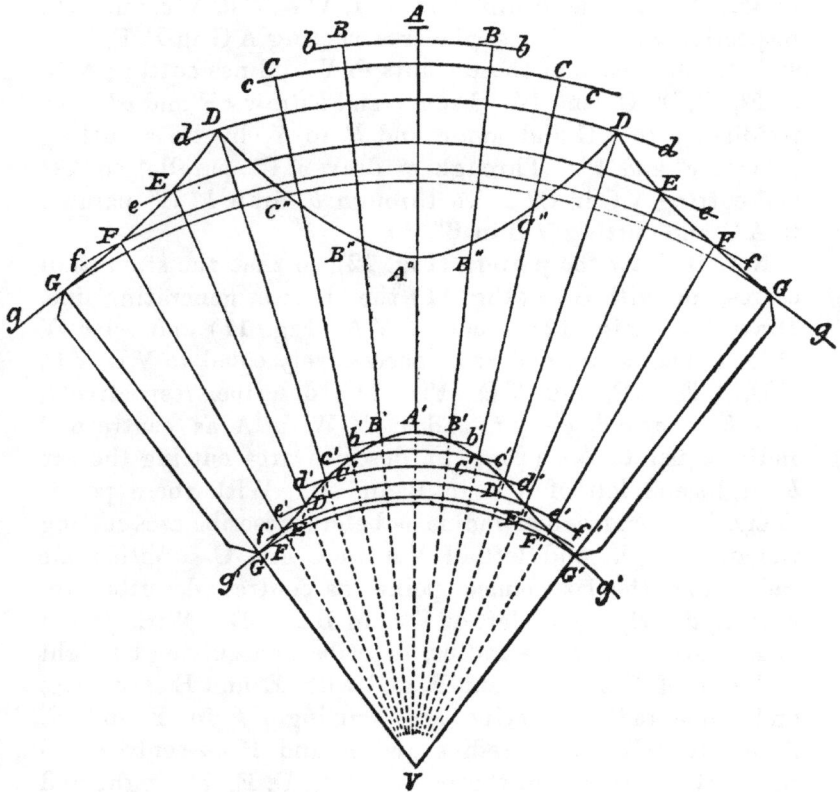

Draw through A′, and the points B′, C′, D′, E′, F′, and G′,
right and left of V A, an unbroken curved line. Also
through the points G, F, E, D, right and left of V A draw
unbroken lines. Now make A A″ equal to A A″ (Fig. 11);

PLATE VI. (see p. 255).

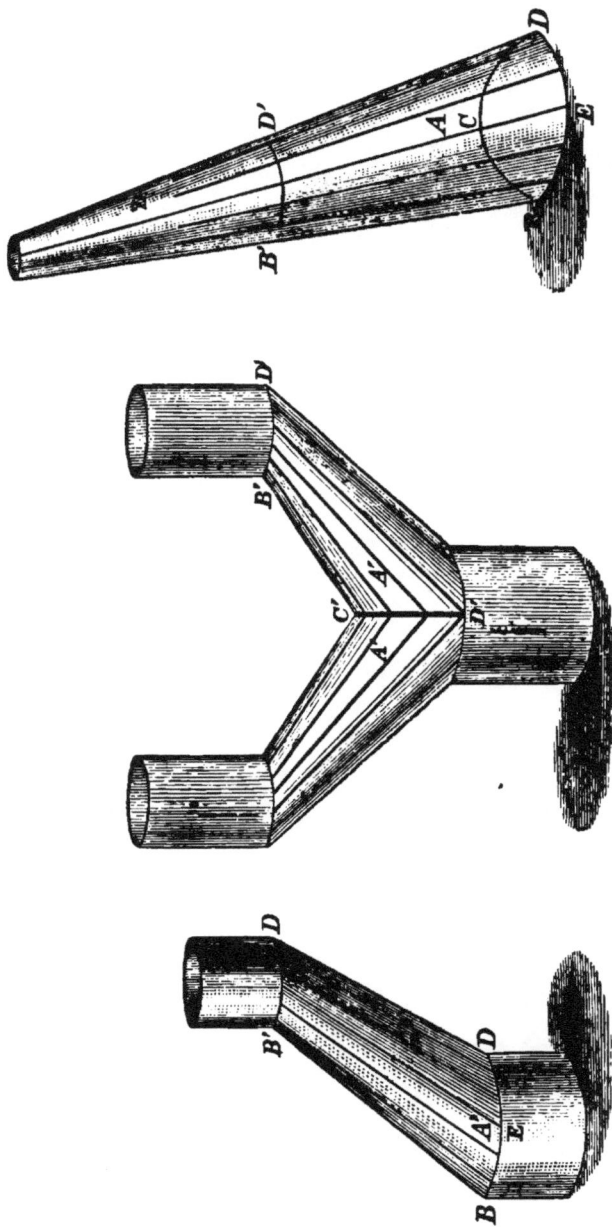

DRAWINGS FROM MODELS IN SOUTH KENSINGTON MUSEUM. (*Part of Exhibit by the Author at the Inventions Exhibition*, 1885.)

B B″ right and left of A V equal to B B″ (Fig. 11); and C C″ right and left of A V equal to CC″ (Fig. 11). Through A″, and the points B″, C″, and D, right and left of V A draw an unbroken curved line. Then G D A″ D G G′ A′ G′ will be the pattern for the pipe A″ O G G′ A′ (Fig. 11) as well as also for the pipe A″ O A K L. The patterns for the circular pipes need no explanation.

(90.) In this problem, the Y being symmetrical, the two tapering pieces of pipe are equal, and the mitre or joint line A″ O, perpendicular to A G, bisects A G, and, produced, bisects also the angle L A″ A′. If the Y were unsymmetrical, the line of junction of the tapering pipes should still be made perpendicular to A G; it will, however, neither bisect it nor the angle L A″ A′, and a distinct pattern for each tapering piece of pipe will be required.

PROBLEM VIII.

To draw the pattern for the Y formed by two cylindrical pieces of pipe of equal diameter uniting two further pieces at the arms of the Y to a third piece at its stem, the Y being both ways symmetrical.

This problem is a special case of the preceding, the oblique cone frusta of that problem now becoming oblique cylinders.

Let M A′ H J G′ P O a′ N (Fig. 13) represent the Y in elevation. Produce A″ a′ to A′. On A′ G′ describe a semi-circle A′ 3 G′, which divide into any number of equal parts, here six, in the points 1, 2, 3, 4, and 5; through each point of division draw lines perpendicular to A′ G′, meeting it in the points B′, C′, D′, E′, and F′, and through these points draw lines B′ B″, C′ C″, D′ D″, &c., parallel to A′ A″. From any point A in A′ A″ draw A G perpendicular to A′ A″, meeting G′ G″ in G, and cutting the lines B′ B″, C′ C″, &c., in the points B, C, D, E, and F. Next make B b equal to

B' 1, C c equal to C' 2, D d equal to D' 3, E e equal to E' 4, and F f equal to F' 5, and draw a curve from A through the points b, c, d, e, and f to G.

FIG. 13.

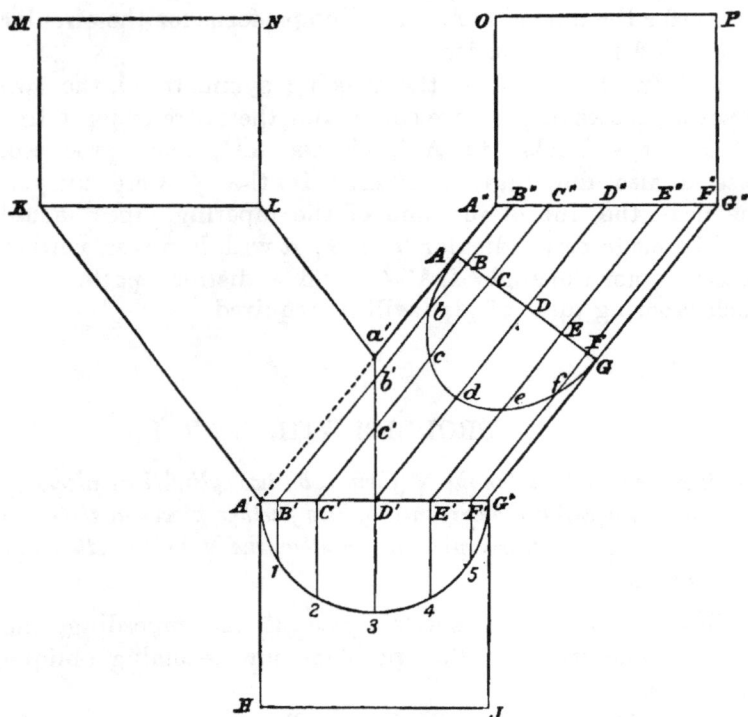

To draw the pattern (Fig. 14), so that the seam shall correspond with G' G" (Fig. 13). Draw any line G G, and at or about its centre draw any line A" a' perpendicular to it and cutting it in A. From A, right and left of it, on the line G G mark distances A B, B C, C D, D E, E F, and F G equal respectively to the distances A b, b c, c d, d e, ef, and f G (Fig. 13). Through the points B, C, D, &c., right and left of A, draw lines parallel to A" a'. Make A a', A A" equal to A a', A A" (Fig. 13) respectively. Similarly make

B b', B B'', C c', C C'', D D', D D'', &c., right and left of A'' a'
equal respectively to B b', B B'', C c', C C'', D D', D D'', &c.
(Fig. 13). Draw an unbroken curved line from A'' through
B'', C'', &c., right and left of A'', to G''. Also draw unbroken
curved lines from D' to G', right and left of a', and an

FIG. 14.

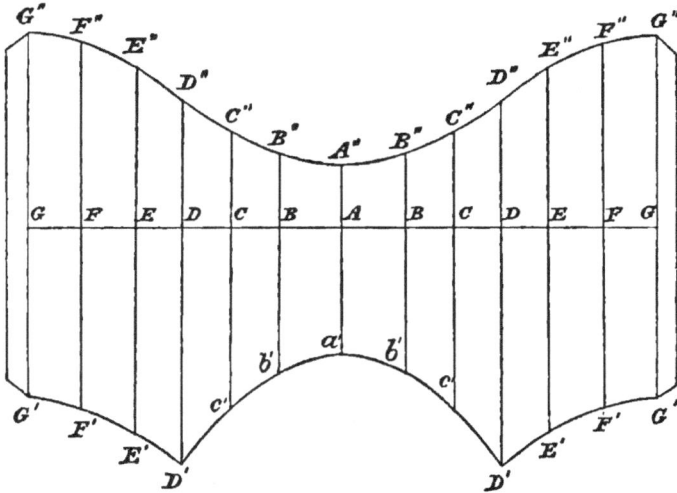

unbroken curved line from D' through c', b', a', b', and c', to
D'. Then G'' A'' G'' G' D' a' D' G' will be the pattern required for either of the portions of oblique cylinders
a' A'' G'' G' D', or a' L K A' D'. The patterns for the circular
pipes need no explanation.

(91.) In this problem, as in the preceding, the Y being
symmetrical the two obliquely cylindrical pipes are equal,
and the mitre or joint line a' D', perpendicular to A' G',
bisects A' G', and, produced, bisects also the angle L a' A''.
If the Y were unsymmetrical, the line of junction of the
pipes should still be made perpendicular to A' G'; it will,
however, neither bisect it nor the angle L a' A'', and a distinct
pattern for each obliquely cylindrical piece of pipe will be
required.

PROBLEM IX.

To draw the patterns for an elbow formed by two equal rectangular pipes (pipes of oblong section) meeting at any angle.

In the case under consideration, we assume that the throat of the elbow is on the broad face of the pipe, which is what happens when pipes of oblong section are fitted to a wall or between a floor and ceiling.

FIG. 15.

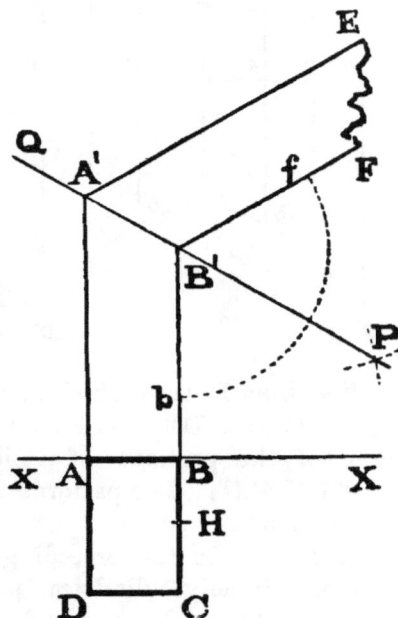

First draw a section of the pipe at right angles to it; also a side elevation of the elbow. Taking any point B (Fig. 15) to represent a definite point in one of the straight lengths of pipe, draw any two lines B·B' and F B' at an angle equal to the given angle of the elbow, and through B draw an indefinite line X X perpendicular to B B'. From B on X X set off B A equal to the narrow face of the pipe, and through B and A draw B C and A D each perpendicular to X X and equal to the broad face of pipe; join D C; then A B C D will be the section of the pipe; it is also part plan (see Problem I. p. 239) of the elbow.

To get the side elevation of the elbow. Through A draw an indefinite line A A′ and find the joint or mitre line B′ A′ as follows:— With B′ as centre and any convenient radius, describe an arc cutting B′ B and B′ F in b and f respectively,

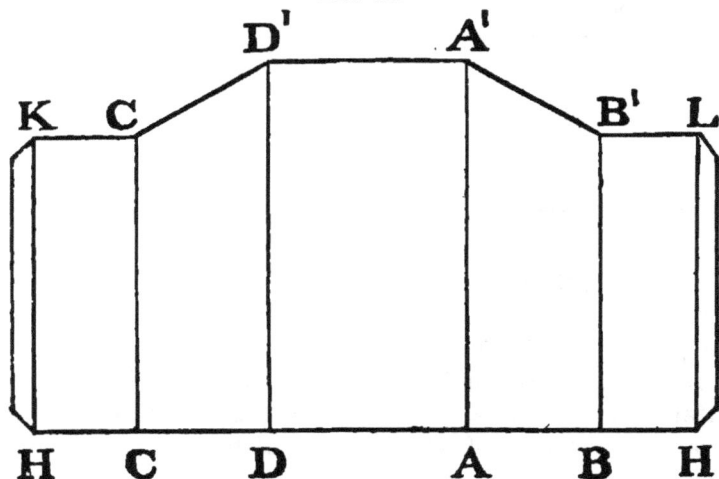

FIG. 16.

and with b and f as centres and any radius greater than half the length of the arc $b\,f$, describe arcs intersecting in P; join P B′ and produce it indefinitely to Q cutting the line A A in A′; then B′ A′ is the mitre line. The line A′ E drawn through A′ parallel to B′ F completes the elevation.

To draw the pattern for the A A′ B′ B piece of pipe, the seam to be in the middle of the broad face, of which the line B B′ is the end elevation. Bisect B C in H, and draw (Fig. 16) an indefinite line H H, and from a point H in H H, set off distances H C, C D, D A, A B, and B H equal respectively to H C, C D, D A, A B, and B H (Fig. 15), and through H, C, D, A, B, and H draw indefinite lines H K, C C', D D', A A', B B' and H L perpendicular to H H; make H K, C C', B B', and H L, each equal to B B' (Fig. 15), and D D', A A' each equal to A A' (Fig. 15). Join K C', C' D', D' A', A' B' and B' L by straight lines, then the figure H K C' D' A' B' L H will be the pattern required. The pattern will also be that of an equal length of the pipe E A' B' F; and the two patterns will be the pattern for the elbow.

The procedure is exactly similar if the throat of the elbow has to be on the narrow face of the pipe.

PROBLEM X.

To draw the patterns for a curved elbow (bend) to join two equal rectangular pieces of pipe.

Let F G H K (Fig. 17) be a side elevation of the bend, and A B C D a section of the pipe at right angles to its length. The pattern for the F G H K faces of the bend will obviously be of the same size and shape as the figure F G H K. The pattern for the throat of the bend from G to H will be a rectangle, two sides of which are equal to D C, and the other two sides equal in length to the 'stretch out' of the curve from G to H. The pattern for the outside of the bend will also be a rectangle, having two of its sides equal to A B and the other two sides equal to the 'stretch out' of the curve from F to K. Thus the necessary patterns for the elbow consist of two rectangles and two pieces of shape F G H K; and the elbow is formed by seaming them at the angles corresponding with the lines F K and G H.

FIG. 17.

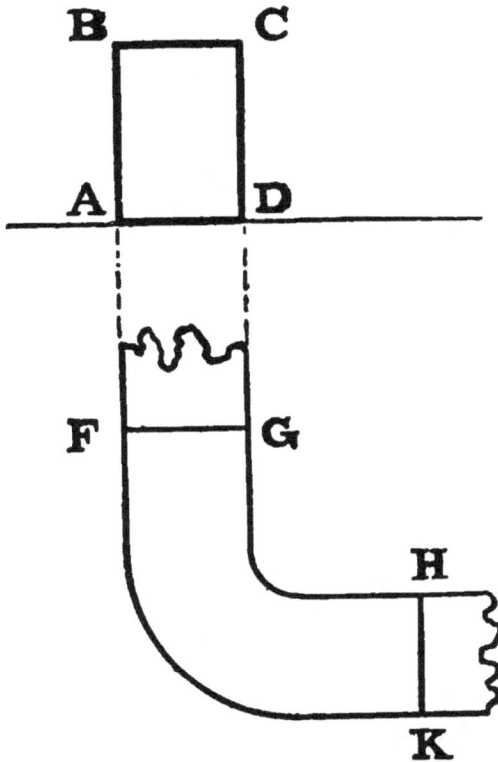

PROBLEM XI.

To draw the patterns for the ⊤-piece formed by two circular pipes,
one of which is an oblique cylinder, the axes of the pipes being
at right angles and intersecting.

Let A′ G′ G″ A″ (Fig. 18), drawn as in Problem V. p. 121
(Fig. 13), be the elevation of the oblique cylinder pipe on
which the ⊤ is to be formed. On A′ G′ describe a semicircle A′ D G′; this will be part plan of the pipe. In A′ A″
take any point A, draw A G perpendicular to A′ A″, and on

it describe, as in the problem referred to, the semi-ellipse A 3 G. Produce A G indefinitely to r, and let G r be the axis of the pipe which, of smaller diameter than the oblique cylinder, is to form a T with it. Make G N' and G M' each equal to half the given diameter of the smaller pipe, and through N' and M' draw indefinite lines N' N and M' M perpendicular to G' G''. In N' N take any point N; on M' M set off M' M equal to N' N and join N M; then N' M' M N will be the elevation of a piece of the smaller pipe. With point R where N M cuts A r as centre and R M as radius,

FIG. 18.

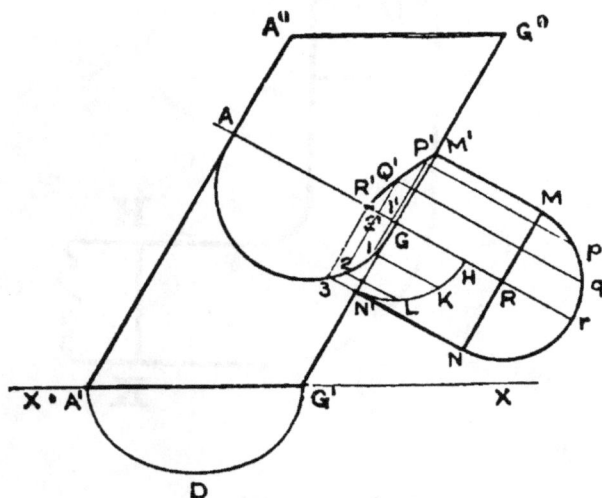

describe a semicircle N r M, which will be part plan of the smaller pipe. Divide the quadrant M r into any number of equal parts, here three, in the points p and q, and through p and q draw indefinite lines p P', q Q' parallel to M M'. With G as centre and radius G N' describe an arc terminating at H in the line A r, which arc, as the lines G N' and A r are perpendicular to each other, will be a quadrant, and may be regarded as part plan of the pipe N' M' M N. Divide this quadrant, exactly as the quadrant M r was divided, in the points K and L, and through K, L, and N' draw lines K 1, L 2, and N' 3 parallel to A R and cutting the semi-ellipse A 3 G in the points 1, 2, and 3. Through point 1 draw 1 P' parallel to G' G'', cutting A r in 1' and meeting p P' in P'; through 2 draw 2 Q' parallel to G' G'', cutting A r in 2' and meeting q Q' in Q'; and through 3 draw 3 R' parallel to G' G'', and meeting A r in R'. From M' to R', through P' and Q' draw an unbroken curved line, then M' P' Q' R' is the elevation of one quarter of the curve of junction of the two pipes.

(92A.) Turning back now to Problem II. p. 242 (which problem the student should read with that in course of solution), and comparing its Fig. 3 with the Fig. just constructed, it will be seen that the *method* of obtaining the required part elevation of the curve of junction is the same in each problem; but that, while in the former problem the required part plan of the larger pipe (Z 3 O, p. 242; and see p. 243) is always a semicircle (sections of a right cylinder cut by planes perpendicular to its axis being circles), in the latter the required part plan of the larger pipe (A 3 G) is a semi-ellipse (sections of an oblique cylinder cut by planes perpendicular to its axis being ellipses, the ellipse varying according to the obliquity of the oblique cylinder). This characteristic difference between the larger pipes in the two instances should be carefully noted, as errors will thus be avoided, and the advantages be secured which arise from correct practice.

Just as with the T-piece of p. 242, so with the T now being dealt with, two patterns are required, one for the oblique cylinder pipe with the hole in it where it receives the smaller pipe, and one for the smaller pipe as it fits to the larger.

The pattern for the larger pipe is set out by Problem V., p. 121. To get out the shape of the hole in it draw (Fig. 19) two indefinite lines A A, G'' G' at right angles to one another, and intersecting in G. (These lines correspond respectively to the lines A A, G'' G' of Fig. 14, p. 122. From G on A A, right and left of G'' G', set off distances

G 1', 1' 2', and 2' R' equal respectively to the curved distances G 1, 1 2, and 2 3 (Fig. 18), that is to the actual distances, as shown on the plan curve A 3 G, that the lines 1 P', 2 Q', and 3 R' are apart from one another. Through points 1' and 2' right and left of G' G'', draw P' P' and Q' Q' perpendicular to A A. Make G M' above and below A A equal to G M' (Fig. 18); make 1' P' above and below A A and right and left of G'' G' equal to 1' P' (Fig. 18); and make 2' Q' above and below A A and right and left of G'' G' equal to 2' Q'

FIG. 19.

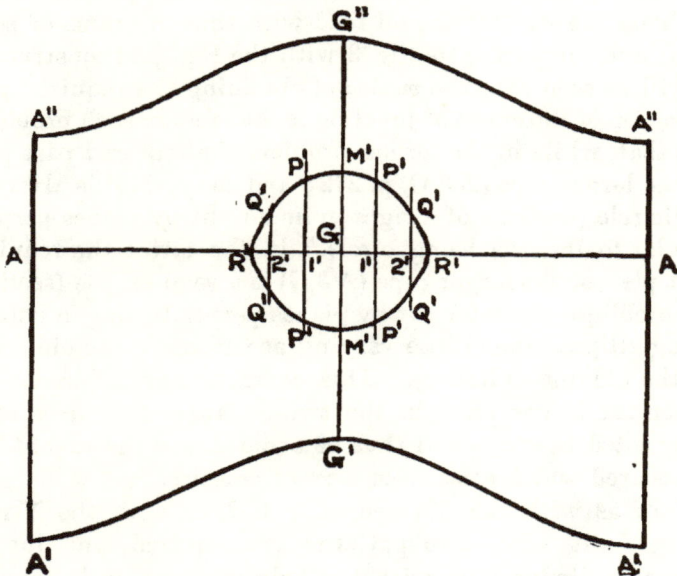

(Fig. 18). Through the points just found, namely R', Q', P', M', P', Q', R', above and below the line A A, draw an unbroken curved line, the shape thus obtained will be the required pattern for the hole in the inclined pipe.

To draw the pattern for the smaller pipe at its junction with the larger, proceed in exactly the same manner as described for the corresponding pattern (shown in Fig. 5 p. 245) of Problem II. already referred to.

SECTIONS: PATTERNS OF SECTIONED CONES.

Workers in sheet metal come across in practice a large number of cases in which a knowledge is required of the sections of solids, especially of the cone and cylinder, and of short direct methods of finding the true shapes of the sections and of drawing the patterns for them. With such knowledge, many problems, said by some men to be very difficult, become extremely simple, and the serious errors arising in attempting their solution without the knowledge, are entirely avoided. We now briefly deal with the sections referred to.

Fig. 20.

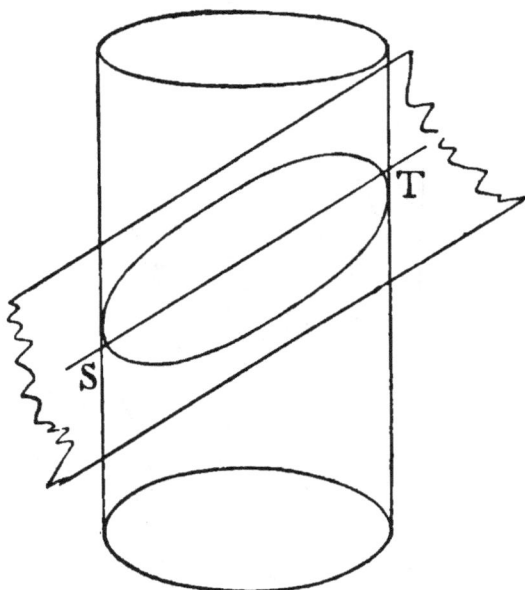

DEFINITIONS.—(93A.) The section of a right cone or cylinder, or of an oblique cone or cylinder, formed by a plane which

Fig. 20a.

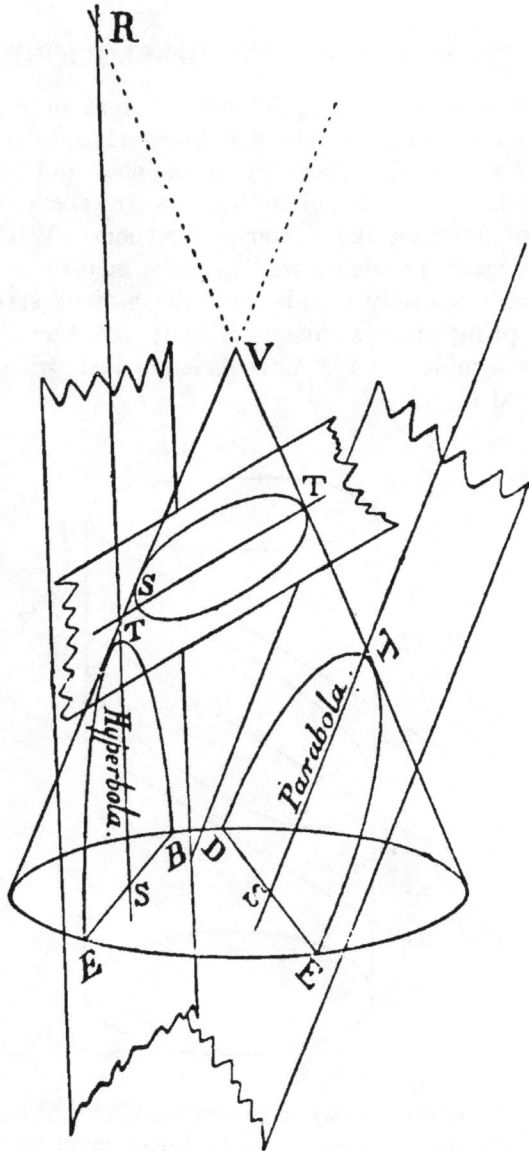

cuts all the generating lines of either solid, but is not parallel to the base of the solid, is an ellipse.

In connection with this definition the student must read §§ 44, 45, 46, and 47 (pp. 105 and 106), noting especially § 47. Further, he will find it convenient to regard a cylinder, whether or not oblique, as an extreme case of frustum of a cone, that is, as being a frustum of which the ends are so nearly equal that they may in practice be regarded as equal. The oblique cylinder has already (§ 61, p. 112) been spoken of as an instance of such frustum.

For examples of elliptic section see Figs. 20 and 20a. Also see Problem XI., p. 267, where the section of the cylinder, A′ A″ G″ G′ is an ellipse; Problem XII., p. 274; and Problem XX., p. 284, where the section of the cone V′ A′ G′, by the plane (T S) of the end of the pipe B T S A, is an ellipse.

(94A.) The section of a right or oblique cone by a plane parallel to one of its generating lines is a parabola (see Fig. 20a and Problem XVI., p. 278).

(95A.) Every other section of a right or oblique cone by a plane not containing the axis or parallel to the base is an hyperbola.

For instances of the hyperbola see Fig. 20a; also see Problem XVIII., p. 281, and Problem XXI., p. 286, where the conical cap is cut by the planes of the sides of the square pipes.

The foregoing definitions are general, and will be useful knowledge to the workman; in the few problems that follow, however, only sections of the right cone and of the cylinder will be dealt with, cases of other sections so seldom occurring in practice, that they cannot be treated here.

PROBLEM XII.

Given a cylinder cut obliquely (§ 93A), to find the dimensions of the cut and draw its ellipse.

Let Fig. 21 represent in elevation a cylinder A A′ B′ B cut by a plane perpendicular to the plane of the paper in the direction S T. The plane of the paper being the plane in which the axis (§ 61, p. 113) of the cylinder lies, the points S and T will be the extreme points of intersection, in the plane of the paper, of the cylinder with the plane of section. The ellipse of the section will then have S T for its major axis or length, and A B, the diameter of the cylinder, for its minor axis or width (see B, p. 17). It may be drawn by either of the methods, as most convenient, given in Chap. II. of Book I., pp. 15–18, or by Problem XXXV., p. 330.

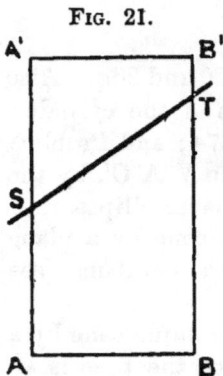

Fɪɢ. 21.

PROBLEM XIII.

To draw the pattern for a cylinder cut elliptically.

This problem has been already solved in Problem I., p. 239, where (Fig. 1, p. 240) A A′ G′ G is a cylinder cut elliptically by a plane A′ G′. The pattern required for the present problem is to be obtained exactly as was the pattern (Fig. 2, p. 240), of A A′ G′ G (Fig. 1, p. 240).

PROBLEM XIV.

Given a cone cut elliptically (§ 93A), to find the dimensions of the cut and draw the ellipse.

Let Fig. 22 represent in elevation a cone V′ A′ G′ cut by a plane perpendicular to the plane of the paper in the direc-

tion S T.   The plane of the paper being the plane in which
the axis of the cone lies; the points S and T will be the
extreme points of intersection, in the plane of the paper, of
the cone with the plane of section.   The ellipse of the section
will then have S T for its length; and the width is found as
follows :—

Bisect S T in E, and through E draw a line A G parallel

FIG. 22.

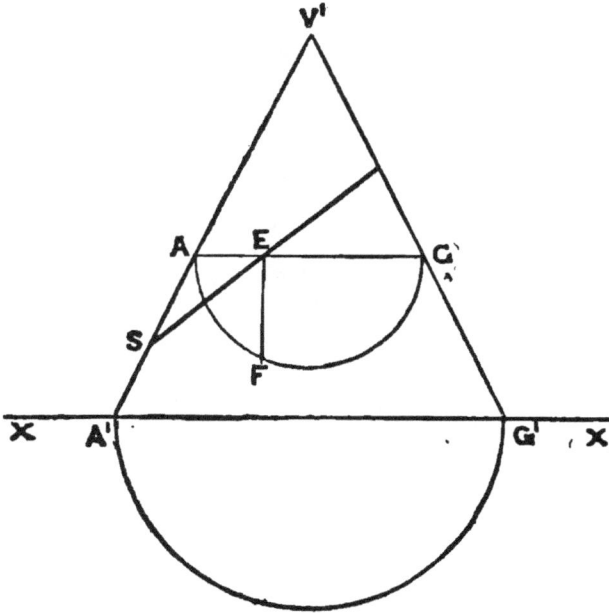

to A′ G′ and cutting V′ A′ and V′ G′ in A and G respectively.
On A G describe a semicircle A F G and from E let fall a
perpendicular E F cutting the semicircle A F G in F; then
E F is half the minor axis or width required.   An ellipse
whose length is S T, and width twice E F, will be the
ellipse required.   For how to draw it see references just
given in Problem XII., p. 274.

PROBLEM XV.

To draw the pattern for a cone cut elliptically.

Let V′ A′ G′ (Fig. 23) represent the cone in elevation, cut elliptically by a plane perpendicular to the plane of the paper in the direction S T. The points S and T will be the extremities (see preceding Problem) of the ellipse of section. The pattern is required of A″S T G′ portion of the cone.

FIG. 23.

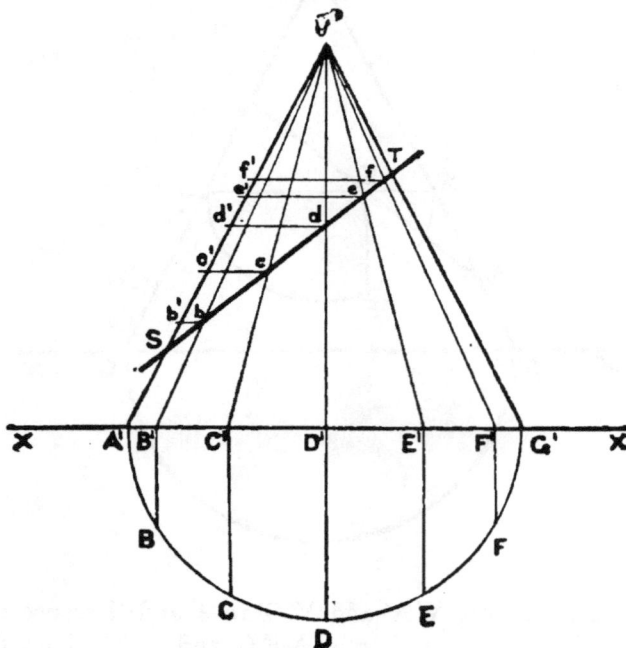

Draw the semicircle A′ D G′, half-plan of the cone and divide it into any number of equal parts, here six, in the points B, C, D, E, and F, and through these points draw lines perpendicular to A′G′ and intersecting it in B′, C′, D′, E′ and F′ respectively. Join B′ V′, C′ V′, D′ V′, E′ V′, and F′ V′ by lines cutting S T in the points b, c, d, e, and f respectively, and through b, c, d, e, and f, draw lines $b\,b'$, $c\,c'$, $d\,d'$, $e\,e'$, and $f\,f'$ parallel to A′ G′, and cutting A′ V′ in the points b', c', d', e', and f' respectively. The lengths A′ b', A′ c', A′ d', A′ e' and A′ f' will be the true lengths respectively of the lines B′ b, C′ c, D′ d, E′ e, and F′ f, these latter lines being elevation lengths only, and not true lengths.

(96A.) In dealing with a cone true lengths on its surface must always be obtained, as shown, and the workman should particularly notice this, because even in books mistakes are made in this important matter.

To draw the pattern; the seam to correspond with the line A′ S. With V′ (Fig. 24) as centre and V′ A′ (Fig. 23)

Fig. 24.

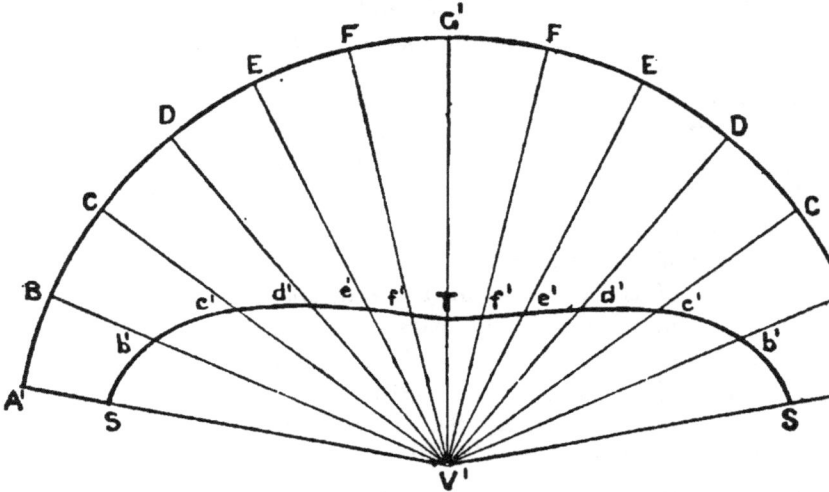

as radius, describe an arc A′ G′ A′, and take any point in it as G′. Join G′ V′, and set off along the arc right and left of G′ V′, distances G′ F, F E, E D, D C, C B, and B A′ equal each to A′ B (Fig. 23), one of the equal parts into which the semi-circle A′ D G′ is divided. Join F V′, E V′, D V′, C V′, B V′, and A′ V′ right and left of G′ V′, and make G′ T equal to G′ T

(Fig. 23); also make Ff', Ee', Dd', Cc', Bb', and A'S right
and left of G'V', equal respectively to A'f', A'e', A'd', A'c',
A'b', and A'S (Fig. 23). Through the points S, b', c', d', e', f',
T, f', e', d', c', b', S, draw an unbroken curved line; then A'G'
A'Sd'Td'S will be the pattern for that portion of the
cone of which the pattern is required, the portion, namely,
A'S T G' (Fig. 23).

It should be noted that V'Sd'Td'S V' (Fig. 24) is pattern
for the V'S T (Fig. 23) portion·of the cone V'A'G' should
the pattern of that portion be needed.

PROBLEM XVI.

*Given a cone cut in parabolic section (§ 94A), to find the
dimensions of the cut and draw the parabola.*

Let Fig. 25 represent in elevation a cone V'A'H' cut by a
plane perpendicular to the plane of the paper in the direction S T; the semicircle A'D H' being half-plan of the cone.
The points S and T will be the extreme points of intersection, in the plane of the paper, of the cone with the plane of
section, and the line S T will be the representation in end
elevation of the parabola of the section. The line S T will
also be the length of the parabola. And as dividing the
curve into two equal parts (see Fig. 20a and Fig. 65, p. 331),
S T is also its "axis." The width at S of the parabola can be
found by drawing through S a line S E perpendicular to A'H'
cutting the semicircle A'D H' in E; then S E is half the
width, at S, of the section of the cone. A parabola having
S T for its length, and twice S E for its width, or "double
ordinate" at S that is, will be the parabola required. It may
be drawn by the method given in Problem XXXVI., p. 331.

FIG. 25.

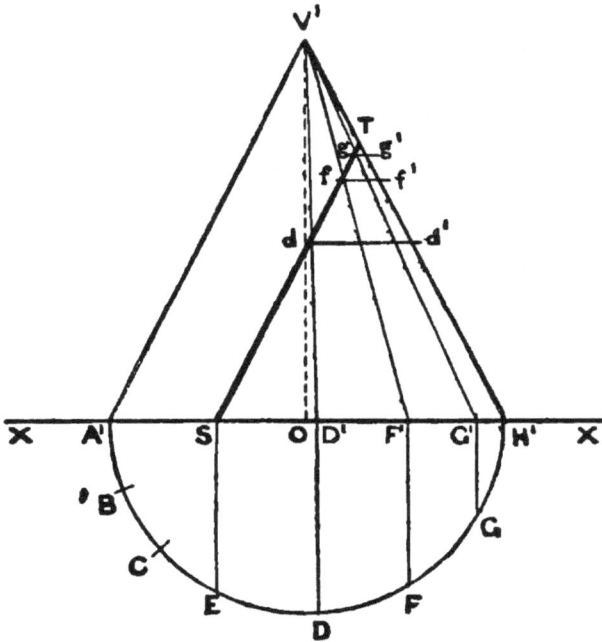

PROBLEM XVII.

To draw the pattern for a cone cut in parabolic section.

Before commencing this Problem the student should again read § 96A, p. 277.

Let V'A'H' (Fig. 25) represent in elevation a cone (axis V'O), cut in parabolic section by a plane perpendicular to the plane of the paper in the direction S T, the semicircle A'D'H' being half-plan of the cone. The points S and T will be the extremities of the length (see preceding Problem), of the parabola. The pattern is required of, say, either the S T H' or A'V'T S portion of the cone. Through S draw S E perpendicular to A'H', cutting the half-plan A'D'H' in E. Divide the E H' portion of A'D'H' into any number of equal parts, here four, in the points G, F and D, and through these points draw lines perpendicular to A' H' and intersecting it in G', F', and D' respectively. Join G' V', F" V', and D'V" by lines intersecting S T in the points g, f, and d, respectively, and through g, f, and d draw lines $g g'$, $f f'$ and $d d'$ parallel to A' H', and cutting H' V' in the points g', f', and d' respectively. The lines H' g', H' f', and H' d' will be the true lengths respectively of the lines G' g, F' f, and D' d, these latter lines being only elevation lengths and not true lengths.

To draw the pattern of the S T H' portion of the cone. With V' (Fig. 26) as centre and V' A' (Fig. 25) as radius,

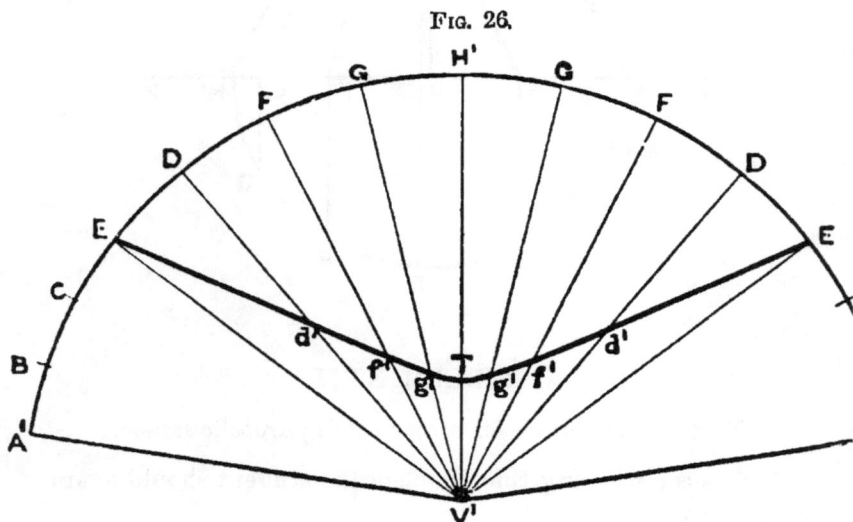

FIG. 26.

describe an arc A' H' A' and take in it any point H'. Join H' V', and set off along the arc right and left of H' V' distances H' G, G F, F D, and D E, equal each to H' G (Fig. 25), one of the equal parts into which the arc H' E is divided. Join G V', F V', and D V', right and left of H' V', and make H' T equal to H' T (Fig. 25); also make G g', F f, and D d', right and left of H' V' equal respectively to H'g', H'f', and H' d' (Fig. 25), and through the points E, d', f', g', T, g', f', d', E draw an unbroken curved line. Then E H′ E d′ T d′ is the pattern required.

The pattern for the A′ V′ T S portion of cone is obtained by dividing the arc E A′ (Fig. 25) into any number of equal parts, here three, in the points C and B, and setting off from E (Fig. 26) (a point obtained in getting the pattern for the S T H′ portion of the cone) along the arc, A′ H′ A′, right and left of H′ V′, distances E C, C B, and B A′, equal each to E C (Fig. 25), one of the equal parts into which the arc E A′ is divided. Then joining A′ V′, right and left of H′ V′, the figure E d′ T d′ E A′ V′ A′ will be the pattern sought. The seam will correspond with A′ V′ (Fig. 25).

PROBLEM XVIII.

Given a cone cut in hyperbolic section (§ 95A), to find the dimensions of the cut and draw the hyperbola.

Let Fig. 27 represent in elevation a cone V′ A′ H′ cut by a plane perpendicular to the plane of the paper in the direction S T; the semicircle A D H′ being part-plan of the cone. The points S and T will be the extreme points of intersection, in the plane of the paper, of the cone with the plane of section, and the line S T will be the representation in end elevation of the hyperbola of the section. The line S T will also be the length of the hyperbola. And as dividing the curve into two equal parts (see Fig. 20a and Fig. 66, p. 332), S T is also its axis. Another line is required, which is called the "major" axis of the hyperbola, and can be found as follows:—Produce A′V′ and S T to intersect in R; then T R is the major axis required. The width at S of the hyperbola is ascertained by drawing a line S E drawn through S perpendicular to A′ H′, and cutting the semicircle A′ D H′ in E; then S E will be half the width at S, of the section of the cone. A hyperbola having T R for its major axis, twice S E for its double ordinate at S, or its width, that is, and T S for its

length, will be the hyperbola required. It may be drawn by
the method given in Problem XXXVII., p. 332.

FIG. 27.

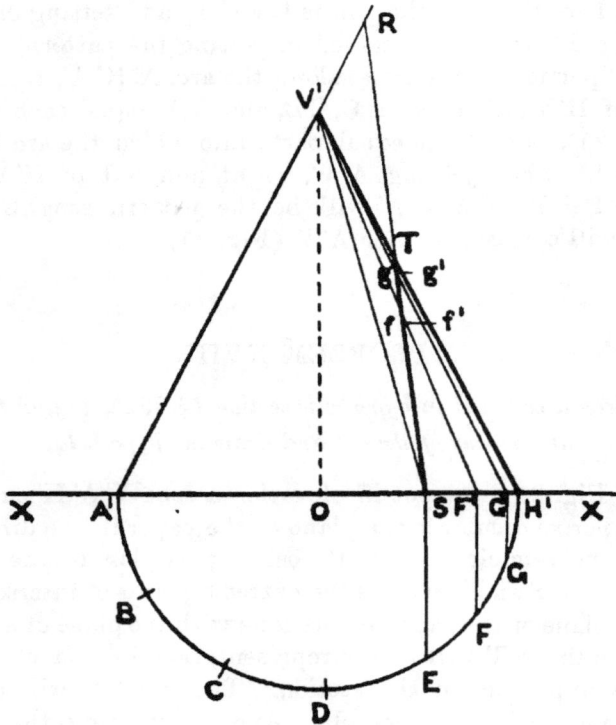

PROBLEM XIX.

To draw the pattern for a cone cut in hyperbolic section.

The student is here again referred to the caution of § 96A,
p. 277.

Let V' A' H' (Fig. 27) represent in elevation a cone (axis
V' O), cut in hyperbolic section by a plane perpendicular to
the plane of the paper in the direction S T, the semicircle
A' D H' being half-plan of the cone. The points S and T will

be the extremities (see preceding Problem) of the length of
the hyperbola of section, and the line S T the representation of the hyperbola in end elevation. The pattern is
required of, say either the S T H', or A' V' T S portion of
the cone. Through S draw S E perpendicular to A' H',
cutting the half plan A' D H' in point E. Divide the E H'
portion of semicircle A' D H' into any convenient number of
equal parts, here three, in the points G and F, and through
these points draw lines perpendicular to A' H', intersecting
it in G' and F' respectively. Join G' V' and F'' V' by lines

FIG. 28.

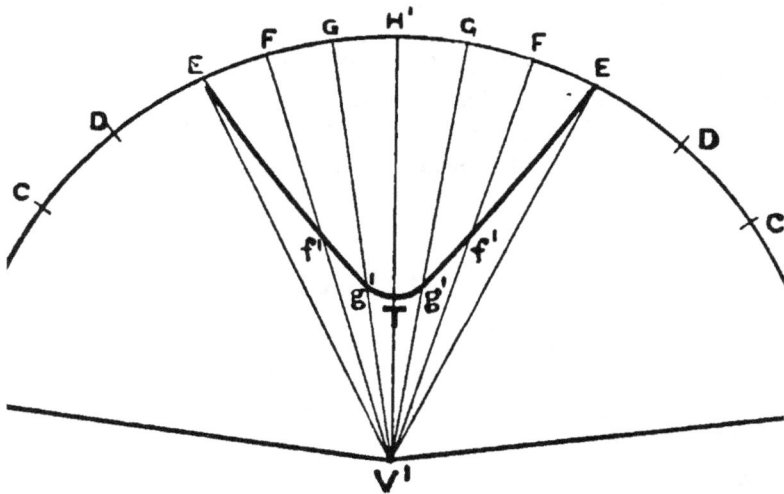

cutting S T in the points g and f respectively ; and through
g and f draw lines gg' and ff' parallel to A' H', and cutting
H' V' in the points g' and f' respectively. The lines H' g'
and H' f' will be the true lengths respectively of the lines
G'g and F'f, these latter lines being only elevation lengths.

To draw the pattern for the S T H' (Fig. 27) portion of
the cone. With V' (Fig. 28) as centre and V' A' (Fig. 27) as
radius, describe an arc A' H' A', and take in it any point H'.
Join H'V' and set off along the arc right and left of H' V' dis-

tances H' G, G F, and F E, equal each to H' G (Fig. 27) one of the equal parts into which the arc H' E is divided. Join G V' and F V', right and left of H' V', and make H' T equal to H' T (Fig. 27); also make G g' and F f' right and left of H' V', equal respectively to H' g' and H' f' (Fig. 27). Through the points E, f', g', T, g', f', E draw an unbroken curved line, then E H' E f' T' f' will be the pattern required.

The pattern for the A' V' T S portion of cone is obtained by dividing the arc E A' (Fig. 27) into any number of equal parts, here four, in the points D, C, and B; and setting off from E (Fig. 28) (a point obtained in getting the pattern of the S T H' portion of the cone), along the arc A' H' A', right and left of H' V', distances E D, D C, C B, and B A', equal each to A' B (Fig. 27), one of the equal parts into which the arc E A' is divided. Then joining A' V' right and left of H' V' the figure V' A' E f' T f' E A' will be the pattern sought. The seam will correspond with V' A' (Fig. 27).

PROBLEM XX.

To draw the patterns for the elbow formed by a circular pipe fitting on to an elliptic section of a conical pipe (part of right cone, § 93A, p. 271); also to determine the size of the circular pipe and the angle at which it must be cut to fill the cone ellipse.

The student should give attention to this Problem, because it is one, in the working of which, notwithstanding its simplicity, mistakes are made in several books. Elevation lengths of lines are taken as true lengths, nothing is said of determining the size of the circular pipe to fit the cone ellipse, and the elevation of the elbow is drawn more or less at random. It is plainly necessary for the formation of a correct elbow that the ellipse of section of the circular pipe must exactly equal the ellipse of the conical pipe, that is, that the diameter of the circular pipe (see Problem XII., p. 274), must be equal to the minor axis of the cone ellipse (see Problem XIV., p. 274), and that the pipe must be cut at such an angle that the major axis of the cone ellipse of section shall equal the major axis of the cone ellipse.

Let A′ T S G′ (Fig. 29) represent a conical pipe or cap, the line S T being an elliptical section of it in side elevation; it is required to fit on a circular pipe to the cone at S T.

First to find the size of the circular pipe, and the angle

FIG. 29.   FIG. 30.

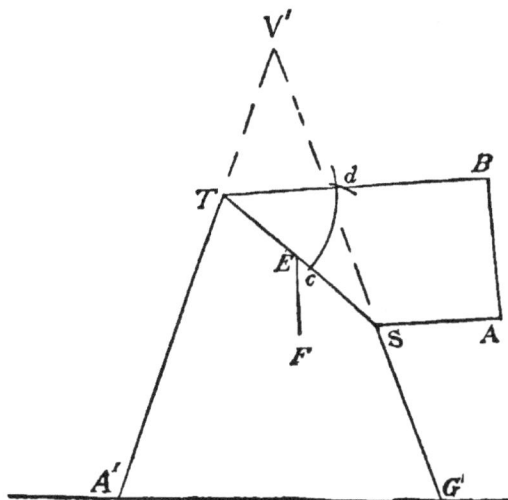 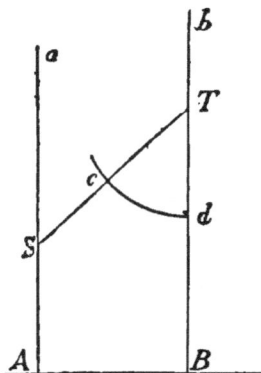

at which it must be cut. Produce A′ T and G′ S, the sides of the cone to meet in V′ and find E F the half of the minor axis of the ellipse S T by Problem XIV., p. 274, then twice E F will be the diameter of the required circular pipe. Now draw (Fig. 30) a line A B equal to this diameter, equal, that is, to twice E F, and through A and B draw A a, B b perpendicular to A B. With any point S on A a as centre and radius equal to S T (Fig. 29) describe an arc cutting B b in T and join S T; then A S T B will represent a cylinder cut at the necessary angle.

To complete the elevation of the conical cap with circular pipe attached (Fig. 29). With T (Fig. 30) as centre and any convenient radius describe an arc cutting T S and T B in *c* and *d* respectively. With T (Fig. 29) as centre and radius equal to T *c* (Fig. 30) describe an arc cutting T S in *c*, and make this arc from *c* equal to the arc *c d* (Fig. 30). The arc *c d* (Fig. 29) is the arc thus equal to *c d* (Fig. 30). Join T *d* and produce it as T B, and draw a line S A from S parallel to the line drawn from T. The required elevation is now completed.

To draw the patterns for the elbow. First, that of the conical pipe A' T S G' (Fig. 29). This being a cone cut elliptically, its pattern can be drawn by Problem XV., p. 276. The pattern for the piece of circular pipe A S T B is that of a cylinder cut obliquely, and can be found by Problem XIII., p. 274.

PROBLEM XXI.

To draw the patterns for a conical cap having a square pipe fitted to it concentrically.

First draw an elevation and a part plan of the arrangement, of which Fig. 31*a* is a perspective drawing. Draw an indefinite line X X (Fig. 31), and from any point V in it as centre, and radius equal to half the given diameter of the larger end of the cap, describe a semicircle A' E K. Also from V on X X set off distances V 4, V *h*, each equal to half the given side of pipe, through points 4 and *h* draw lines 4 1, *h e*, perpendicular to X X, also each equal to half the given side of pipe, and join 1 *e*. Then A' E K with 4 *h e* 1 is the part plan required. Through V draw V O perpendicular to X X, and make it equal to the height from the base of the conical cap to the highest point of the cap, that is, the highest point where the square pipe meets it. Through O draw an indefinite line 4' K' perpendicular to V O, and set off distances O 4' and O K', each equal to half the given side of the square pipe. Now join K K', A' 4', and draw indefinite lines $4'q$, $K'p$ perpendicular to $4'K'$. Produce the lines $A'4'$ and KK' to intersect each other in V'; V' will be the apex of the cone, of which the cap is a part. Join $V1$, and produce it to intersect the semicircle $A'EK$ in D; from D draw DD' perpendicular to XX, and intersecting it in D', and join $D'V'$, intersecting $q4'$ produced in the point $1'$.

Fig. 31.   Fig. 31a.

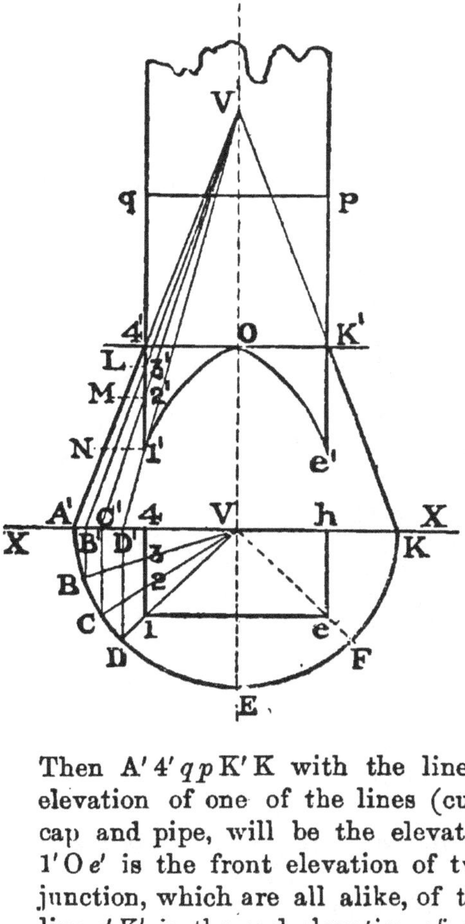

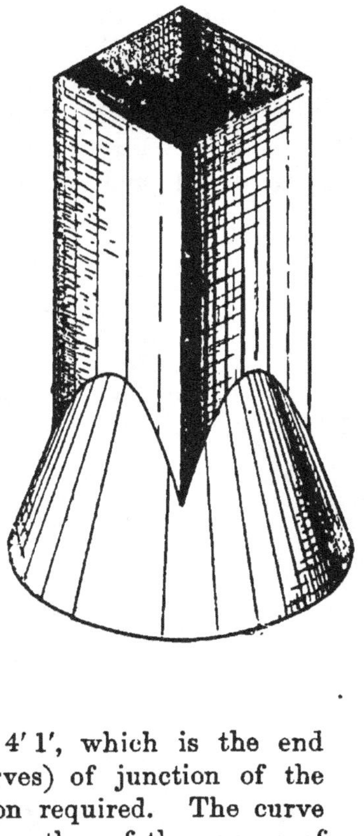

Then $A'4'qpK'K$ with the line $4'1'$, which is the end elevation of one of the lines (curves) of junction of the cap and pipe, will be the elevation required. The curve $1'Oe'$ is the front elevation of two other of the curves of junction, which are all alike, of the cap and pipe, and the line $e'K'$ is the end elevation of another of these junctions.

For these lines no construction is given; the lines indeed
are not required for the problem, and are added to the Fig.
merely to make clear the arrangement of cap and pipe
under consideration. Next divide the arc A' D into any
number of equal parts, here three, in the division points
B and C, and join C V, B V by lines cutting 4 1 in points
2 and 3 respectively; also through C and B draw lines
C C', B B' perpendicular to X X, and intersecting it in the
points C' and B' respectively; join C' V' and B' V' by lines
cutting the line 1' 4' in points 2' and 3' respectively. The
points 2' and 3' will be the elevations of points in the curve
of which the line 1' 4' is the elevation.

To obtain the true lengths of the lines B' 3', C' 2', and
D' 1', draw lines through the points 3', 2', and 1' perpendicular to 4' 1', and cutting A' V' in the points L, M, and N; then

FIG. 32.

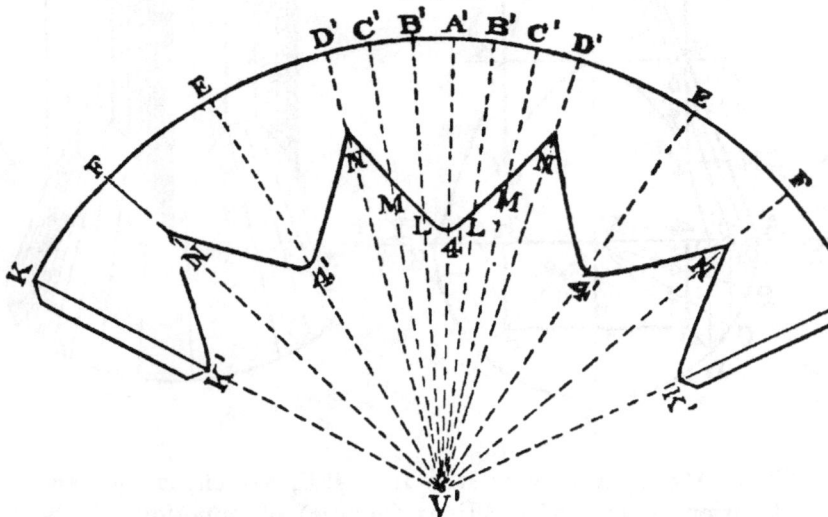

the lengths A' L, A' M, and A' N are respectively the true
lengths required; the lines B' 3', C' 2', D' 1' being elevations
only, and not true lengths (§ 96A, p. 277).

To draw the pattern for the conical cap A' 4' 1' O' e' K' K,

with the seam to correspond with the line K K'. With any
point V' (Fig. 32) as centre and V' A' (Fig. 31) as radius,
describe an arc K A' K. In the arc take any point A' and
join A' V'; then from A' right and left along the arc set off
distances A' B', B' C', and C' D' equal each to A' B (Fig. 31),
one of the equal parts into which the arc A' D is divided.
Now right and left of A' join B' V', C' V', and D' V', and make
A' 4' equal to A' 4' (Fig. 31), the lines B' L, right and left,
equal to A L (Fig. 31), the lines C' M, right and left, equal to
A' M (Fig. 31), and the lines D' N, right and left, equal to
A' N (Fig. 31). Through the points N M L 4' L M N draw an
unbroken curved line, then D' A' D' N 4' N is the pattern for
one-quarter of the conical cap. To complete the pattern of
the cap set off from the points D' along the arc, right and
left of A' V', distances D' E, E F, and F K each equal to
A' D', and draw the further N 4' N curves and the half
curves N K' in exactly the same manner as for the curve
N M L 4' L M N. The figure K A' K K' N 4' N 4' N 4' N K'
will be the complete pattern for the conical cap.

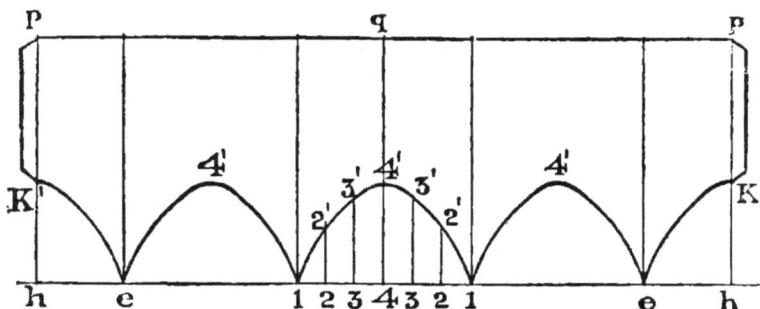

To draw the pattern for the square pipe, with the seam
in the middle of one face, say the face of which e' P (Fig. 31)
is an end elevation. Draw (Fig. 33) an indefinite line h h,
and from any point in it as 4 draw a perpendicular line 4 q;
then set off along h h, right and left of point 4, distances 43

U

32, 21, 1 e and $e h$ respectively, equal to the distances 43, 32, 21, 1 e and $e h$ (Fig. 31), and through the points 4, 3, 2, 1, e and h, right and left of 4, draw lines perpendicular to $h h$. Make 44' equal to 1' 4' (Fig. 31), and 33' and 22', right and left, equal respectively to 1' 3' and 1' 2' (Fig. 31), and through the points 1, 2', 3', 4', 3', 2', 1 draw an unbroken curved line, this will be the curve for one face of the pipe pattern. The curves 1 4' e and half curves e K' can be drawn in exactly similar manner. Assuming the line $q p$ (Fig. 31) to represent the top of the pipe, make 4' q equal to 4' q (Fig. 31), and through q draw $p p$ parallel to $h h$, cutting lines $h p$ in points p and p. Then $p p$ K' e 4' 1 4' 1 4' e K' will be the complete pattern for the square pipe, and the lines to $h h$, from 1 and e, right and left of 4, will be the lines for bending up.

PROBLEM XXII.

To draw the pattern for a conical cap having a rectangular pipe fitted to it concentrically.

This problem differs from the preceding in that all four curves of junction of pipe and cap are not alike, but opposite curves only; that is to say, two pairs of like curves have to be dealt with, and points found by which to draw their patterns.

First draw an elevation and a part plan of the arrangement. Draw an indefinite line X X (Fig. 34), and from any point V in it as centre, describe a semicircle A' G K. Also from V on X X set off distances V 4, V h, each equal to half the given dimension of the narrow side of the pipe, and through the points 4 and h draw lines 4 1, $h e$ perpendicular to X X, each equal to half the given dimension of the wider side of the pipe, and join 1 e. Then A' G K with 4 $h e$ 1 is the part plan required. Through V draw V O perpendicular to X X, and equal in length to the height from the base of the conical cap to the highest point of the cap, that is, the highest point where the oblong pipe meets it, which will be the point 4 of the 14 face of the pipe. Through O draw an indefinite line 4′ K′ perpendicular to V O, and from O set off distances O 4′ and O K′, each equal

FIG. 34.

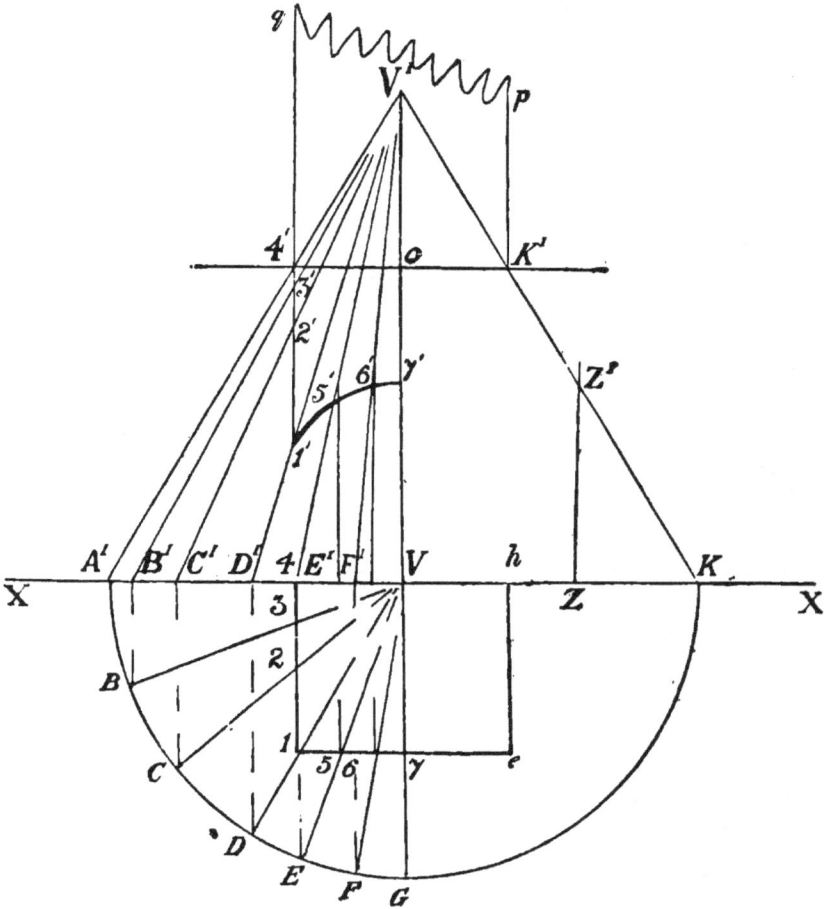

to half the narrow side of the pipe. Join K K′, A′ 4′, and produce the lines to meet in V′ (V′ is the apex of the cone of which the cap is a part); also draw, through 4′ and K′. indefinite lines q 1′, K′p perpendicular to 4′ K′. Join V 1.

and produce it to intersect the semicircle A′ G K in D ; from
D draw D D′ perpendicular to X X and intersecting it in D′,
and join D′ V′ intersecting q 1′ in the point 1′; the line
1′ 4′ will be the side elevation of that one of the curves of
junction of cap and pipe of which 4 1 is the plan, and the
point 1′ will be necessarily a lowest or starting point of the
part elevation of the curve of junction of which 1 e is the
plan. Next divide the arc A′ D into any number of equal
parts, here three, in the division points B and C, and join
C V, B V by lines cutting 4 1 in points 2 and 3 respectively ;
also through C and B, draw lines perpendicular to X X, and
intersecting it in the points C′ and B′ respectively, and join
C′ V′ and B′ V′ by lines cutting the line 1′ 4′ in points 2′ and
3′ respectively.   The points 2′ and 3′ will be the elevations
of points in that curve of the cap and pipe of which 1′ 4′ is
the side elevation, of the points 2 and 3 in the part plan
of the pipe.   To find the elevation of the other pair of curves
of junction; divide the arc D G into any number of equal
parts, here three, in the division points E and F, and join
E V, F V, and G V by lines cutting 1 e in points 5, 6, and
7 respectively ; also through E and F draw lines perpendicular to X X, and intersecting it in the points E′ and F′
respectively.   Part of the line E E′ in this particular figure
happens to coincide with the line 1 4, and the point E with
the point 4.   Join E′ V′, and F′ V′, and through points 5 and
6 draw lines perpendicular to X X, and intersecting lines
E′ V′ and F′ V′ in points 5′ and 6′ respectively ; these points
5′ and 6′ will be the elevations of the points 5 and 6 in the
part plan of the pipe.   The elevation of point 7 will be found
by setting off, on X X, V Z equal to V 7, through Z drawing
a line perpendicular to X X and intersecting K K′ in Z′, and
making V 7′ equal to Z Z′.   Through the points 1′, 5′, 6′, and 7′
draw an unbroken curve; this curve will be the front
elevation of that part-curve of cap and pipe of which the
line 1 7 is the plan, as well as also a representation for
other like portions of the curves of junction.   The necessary
patterns can now be drawn by proceeding exactly as described
for the patterns in the preceding Problem.

PROBLEM XXIII.

To draw the pattern for a body of which the bottom is rectangular and the top circular (base of a tall-boy).

First draw the plan of the body, thus. Draw (Fig. 35) A B C D the rectangle of the bottom of the body. The rectangle is here a square, but the working is the same of whatever nature the rectangle may be. Draw the diagonals of the rectangle, and with their intersection, which will be at O, as centre, and radius equal to half the given diameter of the circular top, describe a circle. Through O draw O 1 perpendicular to B A and cutting the circle in 1'.

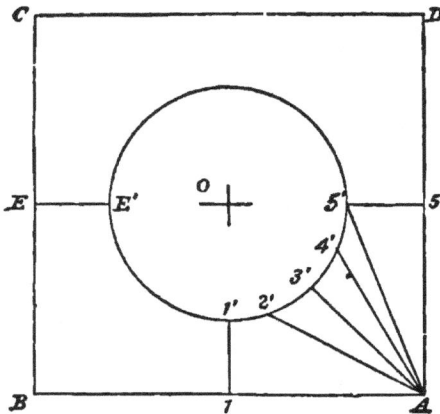

Also through O draw a line perpendicular to A D, meeting C B and D A in points E and 5 respectively, and cutting the circle in E' and 5'. (Neither O 1 nor E 5 are fully drawn in the figure in order to keep this clear of confusing lines.) Divide the arc from 1' to 5' into any number of equal parts, here four, in the points 2', 3', and 4'. Join A 2', A 3', A 4', and A 5'; then 1' 1, 2' A, 3' A, 4' A, 5' A, and 5' 5 are the plans of lines on the body of which we need the true lengths, lengths which we find as follows :

Draw (Fig. 36) any two lines A F, 5' F at right angles to each other and intersecting in F. From F on F A set off F A equal to the height of the body, and from F on F 5' set off F 1' equal to 1' 1 or 5' 5 (Fig. 35), F 2' equal to A 2' (Fig. 35), F 3' equal to A 3' (Fig. 35), F 4' equal to A 4' (Fig. 35) (the point 4' here in Fig. 35 coincides with 2', because of A B C D (Fig. 35) being a square), and F 5' equal to A 5' (Fig. 35). Join the points 5', 2', 3', and 1' to A; then A 1' will be the true length of 1' 1 or 5' 5 (Fig. 35), A 2' will be the true length of A 2' or A 4 (Fig. 35), and A 3' and A 5' will be respectively the true lengths of A 3 and A 5' (Fig. 35).

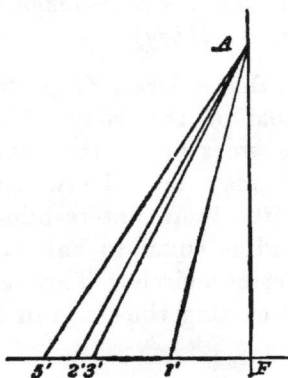

FIG. 36.

FIG. 37.

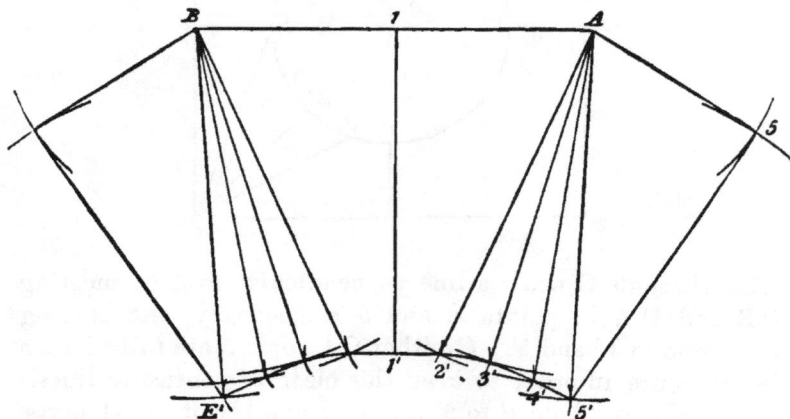

To draw the pattern (body in two pieces; seams to correspond with E E' and 5' 5, Fig. 35). Draw (Fig. 37) any line

A B and from point 1 at or about its centre draw a line 1 1'
perpendicular to A B. Make the line 1 1' equal to A 1'
(Fig. 36). Now make 1 A and 1 B equal each to 1 A or 1 B
(Fig. 35), and with A and 1' as centres and radii respectively
equal to A 2' (Fig. 36) and the arc 1' 2' (Fig. 35) describe
arcs intersecting in 2'. With A and 2' as centres and radii
respectively equal to A 3' (Fig. 36) and 2' 3' (Fig. 35)
describe arcs intersecting in 3'. With A and 3' as centres
and radii respectively equal to A 2' (Fig. 36) and 3' 4'
(Fig. 35) describe arcs intersecting in 4'. And with A and
4' as centres and radii respectively equal to A 5' (Fig. 36)
and 4' 5' (Fig. 35) describe arcs intersecting in 5'. Through
the points 1', 2', 3', 4', and 5' draw an unbroken curved line.
With A and 5' as centres and radii respectively equal to A 5
(Fig. 35) and A 1' (Fig. 36) describe arcs intersecting in
point 5. Join A 5, 5 5'; then 1 A 5 5' 1' is half the required
pattern. The other half can be similarly drawn, and
E B A 5 5' 1' E' will be the complete pattern.

PROBLEM XXIV.

To draw the pattern for a dripping-pan with " well."

The pattern for the bottom can be drawn by the method
given in the preceding problem, the rim and corners being
added by means of Problem XXIII., p. 77.

PROBLEM XXV.

To draw the pattern for a tall-boy base to fit a rectangular opening in the slant of a roof.

Let A B C D (Fig. 38) represent the opening in the roof,
and let Q P X (Fig. 39) be the angle of its pitch or slant.
On P Q from any point E' in it set off a distance equal to the

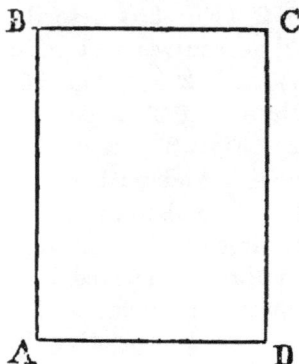

length of one of the sides of the rectangle that are in the
direction of the slant of the roof, here a distance E' G' equal
to A B (Fig. 38). The length of
the tall-boy base has now to be
marked. If this has been given
from where the lowest point portion of it meets the roof, through
E' draw an indefinite line perpendicular to P X, cutting it in A and
on it set off a distance E' a equal
to the given height. If this height
has been given from where the
highest portion of the base meets
the roof, then set off that height,
G' b say, on an indefinite line
through G' perpendicular to P X
and cutting that line in B. Next through a (or b) draw an
indefinite line parallel to P X; bisect E' G' in H', and from
H' draw an indefinite line perpendicular to P X cutting that
line in F, and cutting a b in 1'; from 1' set off 1' 4' and 1' 7'
each equal to half the given diameter of the circular end of
the tall-boy base; and join E' 7' and G' 4'; then E' 7' 4' G' will
be the elevation of the tall-boy base.

FIG. 38.

To draw the complete plan of it, set off from B on the
line through G', a distance B C, and from A on the line
through E' a distance A D, each equal to B C (Fig. 38), and
join D C, cutting the line through H' in H. Then A B C D
will be the plan of the opening in the roof, and consequently
of the bottom of the tall-boy base. The addition of its top
will make the plan complete, and this is obtained by bisecting
F H in O and describing a circle with O as centre and the
distance 1' 4 as radius.

Let the points 1 and 8 be the points in which the circle
cuts the line F H; the line 1 8 is a diameter of the circle.
Bisect A D in E and from E draw E G parallel to A B,
cutting the circle in the points 7 and 4; the line 7 4 will
be another diameter of the circle. Divide each of the
quadrants 1 4 and 1 7 into any number of equal parts, here

three, in the points 3, 2, and 5, 6 respectively, and join
C 3, C 2, C 1, D 1, D 5, and D 6.   The lines H 1, E 7, F 8,

FIG. 39.

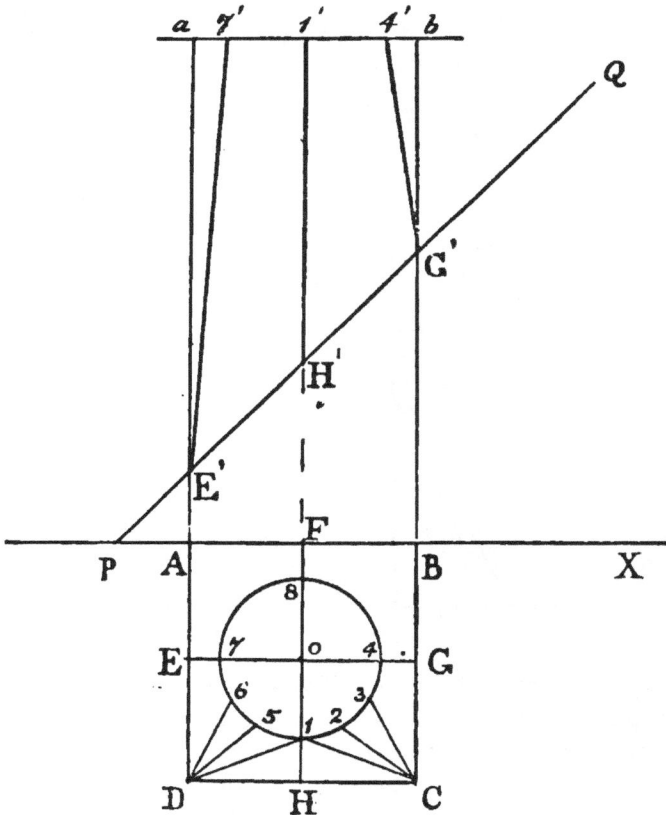

G 4. and the lines meeting the circle from C and D, are plans
of lines on the tall-boy base.   The true lengths of two of
these lines, E 7 and G 4, we already have in E' 7' and G' 4'.
To find the true lengths of C 3, C 2, and C 1, draw (Fig. 40)
any two lines K 1, K I' at right angles to each other and
intersecting in K.   From K on K I' set off K C equal to G' b
(Fig. 39) the height of the points 1, 2, and 3 above the
point G', (G' being the elevation of C); and from K on K 1
set off K 3 equal to C 3 (Fig. 39), K 2 equal to C 2 (Fig. 39),

and K 1 equal to C 1 (Fig. 39), and join C 3, C 2, and C 1 ; these lines will be respectively the true lengths of C 3, C 2, and C 1 (Fig. 39). The true length of H 1 is found by making K H equal to H 1 (Fig. 39), and K I' equal to H' I' (Fig. 39), and joining H I'. This length H I' will be the true length of F 8 (Fig. 39) also. To find the true lengths

FIG. 40.   FIG. 41.

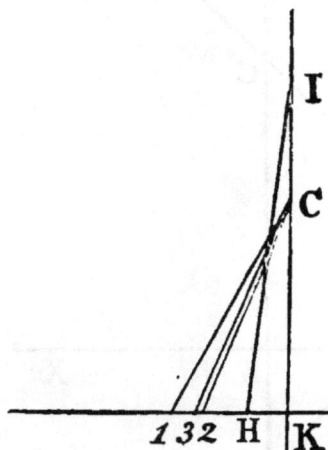
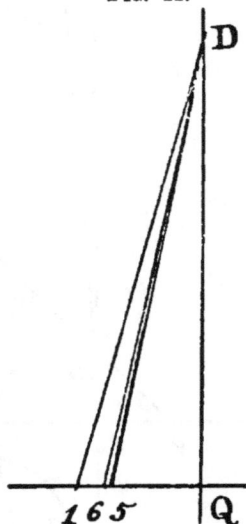

of D 1, D 5, and D 6, draw (Fig. 41) any two Q D, Q 1 at right angles to each other and intersecting in Q ; from Q on Q D set off Q D equal to E' a (Fig. 39), the height of the points 1, 5, and 6 above the point E' (E' being the elevation of D) ; from Q on Q 1 set off Q 1 equal to D 1 (Fig. 39), Q 5 equal to D 5 (Fig. 39), and Q 6 equal to D 6 (Fig. 39), and join D 1, D 5, and D 6 ; these lines will be respectively the true lengths of D 1, D 5, and D 6 (Fig. 39).

CASE I.—To draw the pattern for the tall-boy base, to be made in two pieces ; the seams to correspond with the lines F 8 and H 1 (Fig. 39). Two patterns will be required, one for the F B C H 1 4 8 portion and one for the F A D H 1 7 8 portion.

The pattern for the F B C H 1 4 8 portion. Draw (Fig. 42)

any line B C, and from a point G at or near its centre, draw
a line G 4 perpendicular to it, and make G 4 equal to G' 4'
(Fig. 39). Make G B and G C each equal to G B or G C
(Fig. 39), and with C and 4 as centres and radii respectively
equal to C 3 (Fig. 40) and the arc 4 to 3 (Fig. 39), describe
arcs intersecting in point 3. With C and 3 as centres and

Fig. 42.

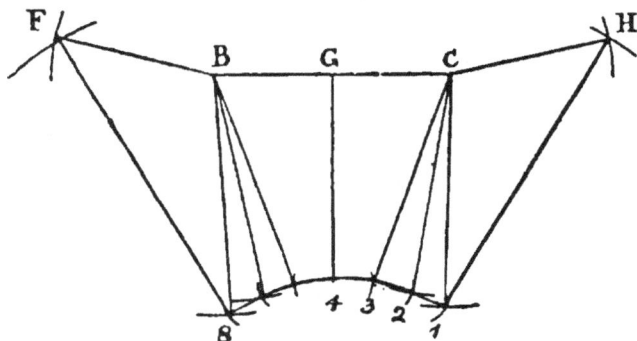

radii respectively equal to C 2 (Fig. 40) and the arc 3 to 2
(Fig. 39), describe arcs intersecting in point 2. And with
C and 2 as centres and radii respectively equal to C 1
(Fig. 40) and the arc 2 to 1 (Fig. 39), describe arcs intersecting in point 1. Through the points 4, 3, 2, and 1 draw
an unbroken curved line. With C and 1 as centres and radii
respectively equal to G' H' (Fig. 39) and I' H (Fig. 40)
describe arcs intersecting in H. Join C H and H 1; then
4 G C H 1 is half the pattern required. The other half can
be drawn in like manner, and F B C H 1 4 8 will be the
complete pattern.

The pattern for the F A D H 1 7 8 portion. Draw (Fig. 43)
any line D A, and from a point E at or near its centre, draw
a line E 7 perpendicular to it, and make E 7 equal to E' 7'
(Fig. 39). Make E D and E A each equal to E D or E A
(Fig. 39), and with D and 7 as centres and radii respectively
equal to D 6 (Fig. 41) and the arc 7 to 6 (Fig. 39), describe
arcs intersecting in point 6. With D and 6 as centres and
radii respectively equal to D 5 (Fig. 41) and the arc 6 to 5

(Fig. 39), describe arcs intersecting in point 5. And with D and 5 as centres and radii respectively equal to D 1 (Fig. 41) and the arc 5 to 1 (Fig. 39) describe arcs intersecting in point 1. Through the points 7, 6, 5, and 1 draw an unbroken curved line. With D and 1 as centres and

FIG. 43.

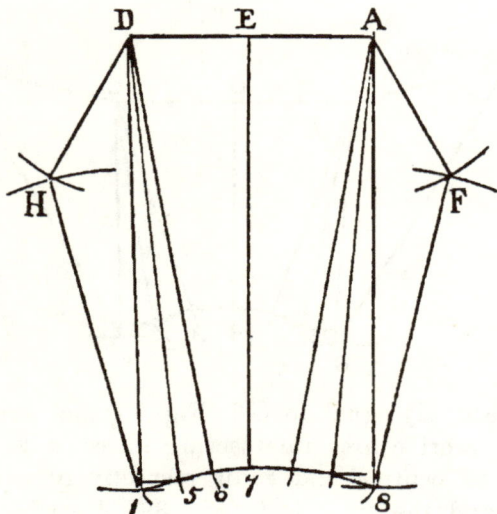

radii respectively equal to E′ H′ (Fig. 39) and I′ H (Fig. 40) describe arcs intersecting in H. Join D H, H 1; then 7 E D H 1 is half the pattern required. The other half can be drawn in like manner, and the F A D H 1 7 8 pattern completed.

CASE II.—To draw the pattern for the tall-boy base to be made in two pieces; the seams to correspond with the lines E 7 and G 4 (Fig. 39). In this case it is evident that only one pattern will be required; the E D C G 4 1 7 and E A B G 4 8 7 portions (Fig. 39) being equal.

The pattern can be obtained by drawing the G C H 1 4 (Fig. 42) part of the pattern as described in the preceding case and joining to it, on the line H 1, the H D E 7 1 portion of the pattern in Fig. 43. This will give the required pattern.

PROBLEM XXVI.

To draw the pattern for a tall-boy base to fit on the ridge of a roof.

Let E′ 7′ 4′ G′ H′ (Fig. 44) represent in elevation a tall-boy base on the ridge of a roof, and A B C D 1 7 8 4 the plan, drawn as described in the last problem. On referring to the preceding problem, it will be seen that the pattern for the tall-boy base in the present problem will be obtained in exactly similar manner to that of the F A D H 1 7 8 portion of Fig. 39, the pattern for which is shown in Fig. 43; the seams will then correspond with the lines F 8, and H 1, and, as the E 7′ 1′ H′, and G′ 4′ 1′ H′ parts are alike, one pattern only will be required for the body to be made in two pieces.

Fig. 44.

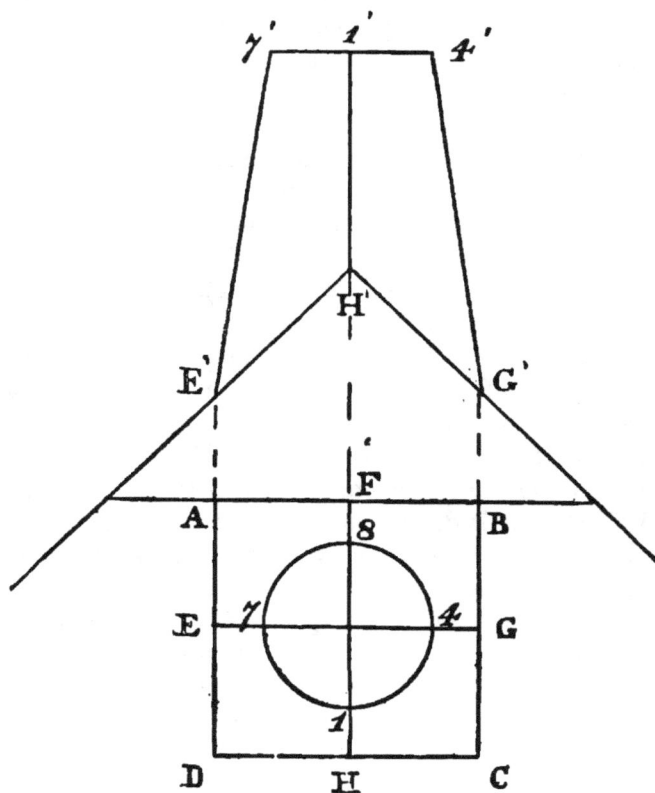

MOULDINGS.

INTRODUCTION.

(97A.) The problems now following treat of mouldings. With these, just as with elbows (p. 239), the essential condition for the formation, with two pieces of moulding, of a joint at any angle, is that the section or end-shape of each piece at the " mitre " (see Problem I., p. 239) or junction shall be the same. This condition should always be borne in mind, and it is fulfilled when, in the plan of the joined pieces on the horizontal plane, the mitre-joint shows as a line bisecting the angle that the two pieces make with one another; for then, as a little consideration will show, there is the same section or cut on each piece.   Thus, if this angle is a right angle, that is, if the two pieces of moulding meet " square," the joint line makes an angle in plan of 45 degrees with the internal and external edges of each of the pieces; if the angle of meeting is of 120 degrees, as with six pieces of moulding forming a regular hexagon in plan, each joint line makes an angle of 60 degrees with the internal and external edges of either adjacent piece, and so on.

(98A.) In dealing for pattern-making purposes with the section or shape of a moulding (§ 99A), it is convenient to draw it on two straight lines at right angles, the extremities falling one on either line, and the shape being arranged in respect of one of the lines just as the moulding itself is disposed to the surface to which it is applied; this line is thus part plan line—or part elevation line—of that surface. Looking to Fig. 45 (Problem XXVII.) the section or shape K I H E D C B A is drawn on the indefinite lines A K', K' K, at right angles, one extremity of the shape standing at a point A in A K', and the other extremity at a point K in K' K.   The plane of the paper being the plane of the section, then, if A K' represent the surface to which the moulding is applied, A K' is the depth of the moulding, and K' K its outstretch or span. If K' K represent that surface, then K′ K is the depth of the moulding and K′ A its span. Thus, that plane of section of a moulding which gives its shape is a plane that contains the lines both of its depth and span.

(99A.) It is necessary to observe that when the section of a moulding or shape of it is spoken of, with no qualifying words, the section is supposed to be on the plane just referred to.

(100A.) The shape K I H E D C B A appears as a line, because metal plate, with which our problems deal, may be regarded as of superficial dimensions only, that is, as having no thickness. In some of the Figs. that follow, with a view of helping the student, the space contained between the line of shape, and the lines of depth and span, is shaded, as in representing a solid moulding ; the student will however remember that only the shaped line is the moulding.

The solution of problems is now easy.

PROBLEM XXVII.

Given the shape and length of a piece of moulding
to draw the pattern for it.

Fig. 45 is the shape of a piece of moulding drawn on the lines A K′, K′ K, and A L (Fig. 45a) its length. Obviously the pattern for the piece of moulding will be a rectangle marked with certain lines, upon which, with the given shape as a guide, the rectangle has to be bent and formed, so as, at its end, to present that shape.

Draw (Fig. 46) any line X X, and from any point A in it draw A Y at right angles to X X. From A along X X set off A L equal to A L (Fig. 45a), and from L draw L Y′ parallel to A Y. On A Y set off successively distances A B, B C, C D, D E, E F, F G, G H, H I, and I K, equal respectively to A B, B C, C D, D E, E F, F G, G H (F and G are any points in E H, the curved portion of the shape), H I, and I K of Fig. 45. From the points B, C, D, E, F, G, H, I and K draw lines

FIG. 45.

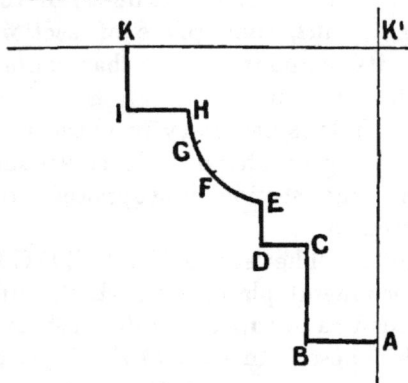

FIG. 45A.

FIG. 46

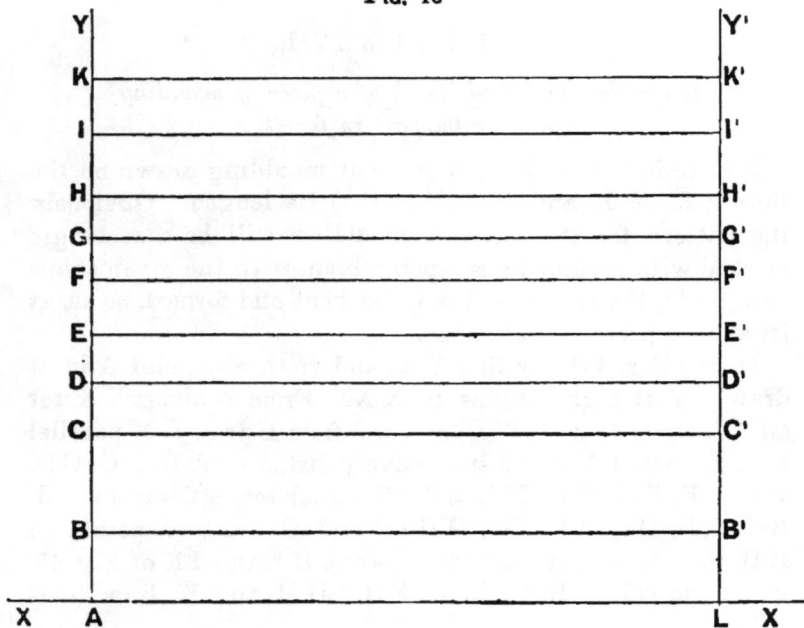

parallel to A L and terminating in L Y′ in the points B′, C′, D′, E′, F′, G′, H′, I′ and K′ respectively; A K K′ L will be the rectangle pattern required, and the several lines parallel to A L will be the lines by which the pattern is to be bent up and formed to the given shape.

In setting off on A Y the distances E F, F G, and G H, the workman must bear in mind that the distances he sets off are chords and not arcs, and that it may be necessary to make allowance for this.

PROBLEM XXVIII.

The shape of a moulding being given, to draw the shape at any angle of section of the moulding, and the pattern for a piece of the moulding cut at that angle. Also, conversely, the section of a moulding at any angle being given to draw the shape of the moulding.

The cases of this problem that usually occur in practice are where the plane of section contains the line of span of the moulding, but not the line of its depth, or contains the line of its depth, but not that of its span, and one particular example of where that plane contains neither line of depth nor line of span.

CASE I.—Where the plane of section contains the line of $\frac{\text{span}}{\text{depth}}$ of the moulding, but not the line of its $\frac{\text{depth}}{\text{span}}$.

(101A.) Here are seemingly two cases, they are however one and the same, as will be seen from this. Let $h\, g'\, f'\, e'\, d'\, c'\, b'\, a'$ (Fig. 47), drawn on the lines $a\, h$, $a\, \text{N}$, be the given shape of the moulding, one extremity of it falling on the line $a\, h$ at h, and the other at a' on $a\, \text{N}$. If of the lines $a\, h$, $a\, a'$, the line $a\, h$ represents in plan the surface to which the moulding applies (§ 98A), then $a\, h$ is the depth of the moulding, and $a\, a'$ its span. If $a\, a'$ represents that plan line, then $a\, a'$ is the depth of the moulding, and $a\, h$ is its span. Thus the lines $a\, h$, $a\, a'$, will be respectively depth and span, or span

x

and depth, according to which of them represents the surface
for the moulding, and the plane of section will be described

FIG. 47.

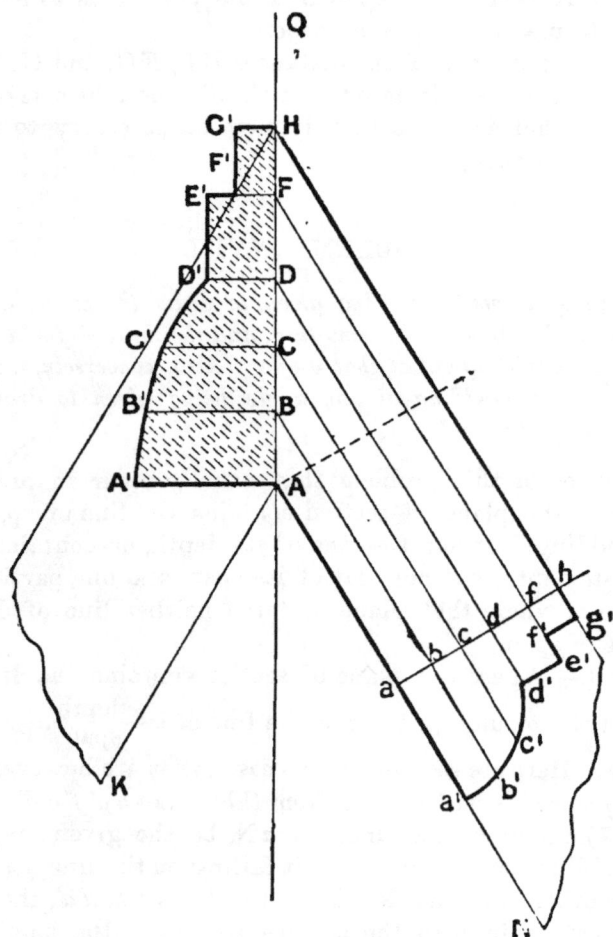

correspondingly. The working drawing for a section at an
angle is that which shows the angle.

(102A.) Producing the lines $a'a$ and $g'h$ indefinitely, lines
which are perpendicular to ah by construction, and by the

given nature of the moulding $a h$ H A will represent in plan, a piece of the moulding $h g' f' e' d' c' b' a'$. A plane of section of the moulding at any angle N A Q is represented by the line A Q cutting the indefinite lines a A and h H in the points A and H. This plane is perpendicular to the plane of the plan, but does not contain both its lines of span and depth. The plan line of a plane passing through the point A, and containing the lines both of span and depth, would be the line A Y.

(103A.) To get the shape of the moulding on the plane of A Q. Produce the line $e' f'$ (also perpendicular to $a h$ by the given nature of the moulding), cutting $a h$ and A Q in f and F respectively, and from the point d' draw a line parallel to the line just drawn, cutting $a h$ and A Q in d and D respectively. In $d' a'$, the curved part of the moulding, take any points e', b', and from these points draw lines c' C, and b' B, parallel to a' A, cutting $a h$ in c and b, and A Q in C and B, respectively. From A, B, C, D, F, and H, draw lines perpendicular to A H, and make A A', B B', C C', D D', F E', F F', and H G' equal respectively to $a a'$, $b b'$, $c c'$, $d d'$, $f e'$, $f f'$, and $h g'$. Join G' F' and E' D' by straight lines, and D' C' B' A' by an unbroken curved line; the line H G' F' E' D' C' B' A' will be shape required of the moulding on the plane of section A Q. This shape is often useful as a template.

To draw the pattern for the shape. Draw (Fig. 48) any line $a h$, and from a point a in it set off distances $a b$, $b c$, $c d$, $d e$, $e f$, $f g$ and $g h$, respectively equal to the distances $a' b'$, $b' c'$, $c' d'$, $d' e'$, $e' f'$, $f' g'$, and $g' h$ (Fig. 47). Through a draw a line perpendicular to $a h$; make a A equal to a A (Fig. 47), and from A draw a line parallel to $a h$ terminating at a point Y in an indefinite line drawn from h perpendicular to $a h$. Through b, c, d, e, f, and g, draw indefinite lines b B, c C, d D, e E, f F, and g G perpendicular to $a h$ and cutting the line A Y, then the rectangle a A Y h, with its guide lines between a A and h Y parallel to those lines, is the pattern of the length of moulding (see last Problem) a A Y h (Fig. 47). On b B, c C, and d D set off distances b B, c C, and d D equal

x 2

respectively to b B, c C, and d D (Fig. 47); on e E and f F distances e E and f F each equal to f F (Fig. 47) (the points e' and f' coinciding with each other in plan of the moulding), and on g G, and h H distances g G and h H, each equal to h G (Fig. 47) (the points g' and h' coinciding with each other in plan of the moulding). Through the points A, B, C, D, draw an unbroken curved line, and join D E, E F, F G, and G H by straight lines. Then A B C D E F G H will be the pattern required.

FIG. 48.

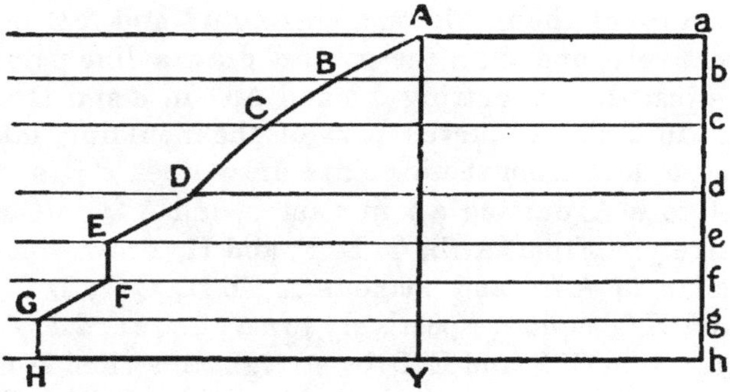

(104A.) The pattern for the piece of moulding cut at the angle N A Q (Fig. 47) is also found; it is a A H h of (Fig. 48); which pattern, if bent up on the lines d D, e E, f F, and g G, and rounded up between the lines a A and d D, will form the moulding required.

(105A.) In respect of the curve A B C D the caution given in the last problem should be borne in mind. In choosing division points on curves of mouldings, from which, by means of lines to find points for any required section, the workman must be guided by the length and style of a curve. Lines from angular points of a shape, such as h, g', f', e', and d', must always be drawn, as these not only give the angular points in the required section, but are lines on which the pattern of a piece of moulding must be angled' up to form the moulding.

The pattern piece of a moulding may be made of any desired length. Length is obtained by simply producing the lines A a, B b, C c, D d, E e, F f, G g, and H h. If length is not required, but the mitre-line end only of the pattern, this could be found by taking measurements from the line A Y (Fig. 47) instead of $a h$; the pattern thus obtained would be A B C D E F G H Y (Fig. 48).

(106A.) What has been done in this first Case should be noted by the student. The angular points of the moulding, and other chosen points are projected on to one of the lines, $a h$ (Fig. 47) on which the shape is drawn, and thence on to the line in which the moulding is cut. The plane of the cut, whatever its angle, being perpendicular to the surface to which the moulding is applied, span lengths are not affected, but remain unaltered. Depth lengths, however, such as $g' f'$, $e' d'$ (Fig. 47), and the perpendicular length from d' to the line $a a'$ are altered, and the true lengths that these become in the shape of the moulding in the plane of the cut appears in the line H A as H F, F D, and D A respectively.

CONVERSE PROBLEM.—*The section of a moulding at any angle being given, to draw the shape of the moulding.*

Let A D' E' F' G' H (Fig. 47) be section of a moulding N A H h cut at the angle N A Q. Divide A'D' into any number of equal parts, here three, in the points B' and C', and from the points A', B', C', D', and E' draw lines perpendicular to A Q, cutting that line in the points A, B, C, D, and F. The point F' is, by the nature of the section, a point in E' F; and the line G'H is already perpendicular to A Q. From any point a in N A draw $a h$ perpendicular to N A, cutting H h in h, and from B, C, D, and F draw lines parallel to N A, cutting $a h$ in b, c, d and f. From a on A N set off $a a'$ equal A A'; from b, c and d set off $b b'$, $c c'$, and $d d'$ respectively equal to B B', C C, and D D'; from f set off $f f'$ and $f e'$ respectively equal to F F' and F E'; and from h set off $h g$ equal to H G'. Join a' to d', through b' and c', by an unbroken curved line, and join $d' e'$, and $f' g'$. Then $a' b' c' d'$ $e' f' g' h$ will be the shape required.

CASE II.—Where the plane of section contains neither line of depth nor line of span of a moulding.

As stated above, only one particular example is to be dealt with; it is useless to here treat the case generally, seeing that other examples very seldom indeed occur in practice. The example being special, it is worked as a whole in Problem XXXII. succeeding.

PROBLEM XXIX.

The shape of a moulding being given, to draw the pattern for joining two pieces of it at any angle.

Let *h g' e' d' a'* (Fig. 47), drawn on the lines *a h, a a'*, be the given shape of the moulding, and K A N the angle at which the two pieces of it are to meet. Bisect the angle K A N (Problem VIII., p. 10), then A Q is the direction of the mitre line, and the two pieces are represented by K A H and *a* A H *h*. As there is the same section or cut on each piece, the problem now becomes that of drawing the pattern for the piece of moulding, say *a* A H *h*, cut at the angle A H *h*. This is done by means of the problem preceding.

(107A.) A method used in practice for finding the line A Q, that is for bisecting the angle K A N, is to set off from A equal distances along A K and A N, to join the points thus found by a straight line, and then by means of a square applied to this line to draw a line perpendicular to (square with it) and passing through the point A. This is a perfectly correct procedure.

PROBLEM XXX.

To draw the pattern for an aquarium (or other) base formed of a moulding of one and the same design.

Let A' B' C' D' (Fig. 50) be the plan of the inside edge of the aquarium (or other) base, and E F (Fig. 49), the shape of the moulding, drawn on the lines G F and G E; G E representing the surface on which the aquarium stands. Bisect (see Problem VIII., p. 10) each of the angles of the plan (working lines of bisection not shown) by the lines A′ A, B′ B, C′ C, D′ D. Through either of the angular points, here D′, draw a line perpendicular to A′ D′; and on it set off D f' equal to G E (Fig. 49). Through f' draw a line parallel to A′ D′, and intersecting A′ A and D′ D in A and D respectively. Through D and A draw lines parallel to D′ C′,

and intersecting C′ C and B′ B in C and B, and join C B. Then A B C D will be the plan of the outer edge of the aquarium stand or other base, and A′ A, B′ B, C′ C and D′ D the plans of the mitre joints (§ 101A).

FIG. 49.

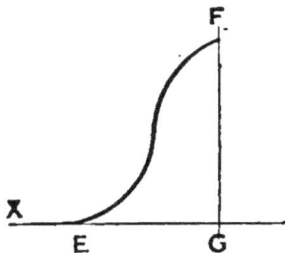

If the plans of the outside edge of the aquarium (or other) base be given or first drawn, as A B C D, the plan of the inside edge will be thus found. Bisect, first of all, the angles at A, B, C, and D. Next draw a line (not shown in the Fig.) perpendicular to one of the edges, say A D, on its inner side, from any point in it (the line f'' D′ is such a line, if f' be regarded as not a particular point in A D but as any point in it); on this line set off a distance equal to E G (Fig. 49), and through the extreme point of this draw a line parallel to A D, intersecting the bisecting lines of the angles at D and A in the points D′ and A′. Through D′ and A′ draw lines parallel to D C, intersecting the bisecting lines of the angles C and B in the points C′ and B′ and join C′ B′. Then A′ B′ C′ D′ is the plan of the inner edge of the stand.

The problem is now essentially completed, and is simply that of drawing the pattern for the E F (Fig. 49) moulding on either mitre plane, which is done by Problem XXIX. It will be a help to the student, however, to work the problem in detail right through. We already have the line D′ f' if the inner edge was first drawn; if the outer edge was first drawn, let fall now from D′ a line perpendicular to A D, meeting this line in f'; then D′f', being parallel to the line, not shown, drawn perpendicular to A D on which the distance E G was set off, will be itself equal to E G. Now on D′f', D′ A′, draw $f'd''$ the required shape of the moulding, disposed towards D′f' as to the surface on which the moulding stands;

FIG. 50.

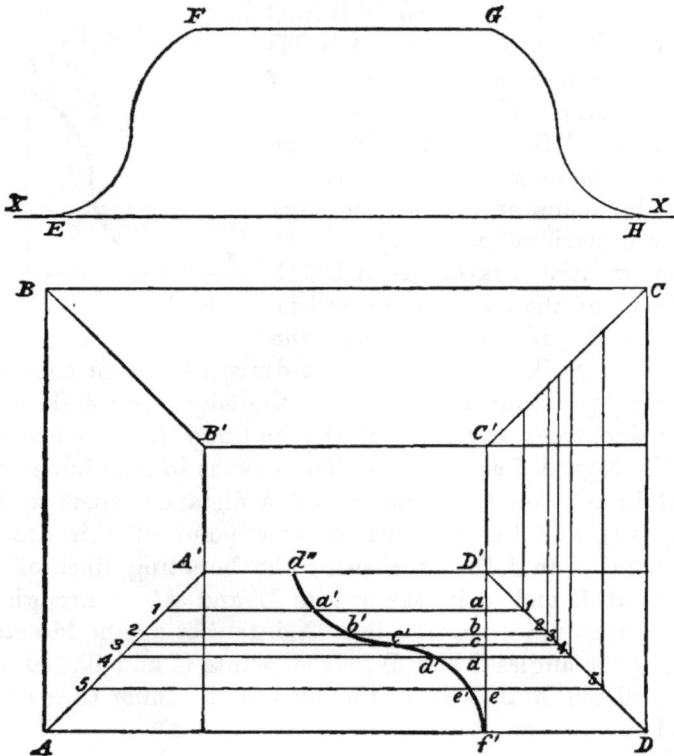

and divide it into any number of parts, equal or unequal (see § 105A, p. 308). The division is here into six equal parts, in the points a', b', c', d', and e', through which points draw lines 1 to 1, 2 to 2, 3 to 3, 4 to 4, and 5 to 5 parallel to A D, and terminating in A′ A and D′ D, and from points 1, 2, 3, 4, and 5, in the line D' D draw lines parallel to D C and terminating in C' C.

To draw the pattern for the A' D' A D portion of the base. Draw (Fig. 51) any line K L; and from any point D' in it

FIG. 51.

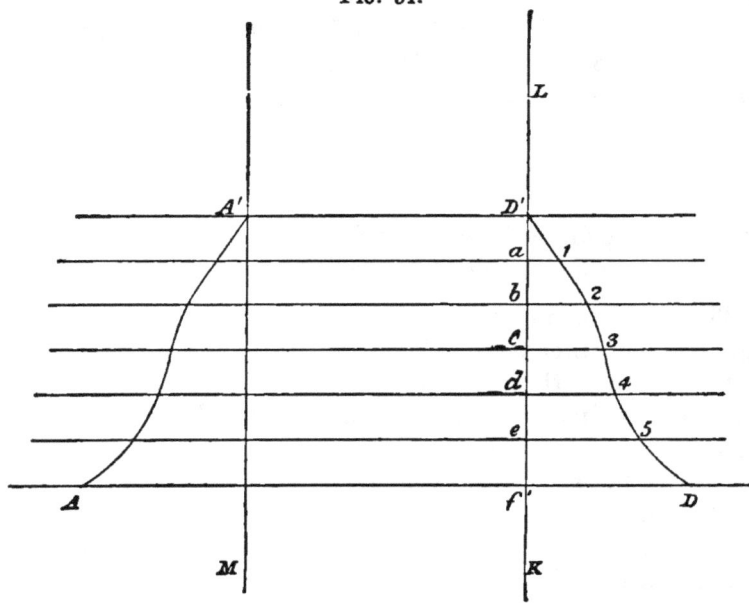

set off D' a, a b, b c, c d, d e, and e f equal respectively to the distances d" a', a' b', b' c', c' d', d' e', and e' f round the curve d" c' f (Fig. 50); and through the points D', a, b, c, d, e, and f draw lines perpendicular to K L. Make a 1 equal to a 1 (Fig. 50), and b 2, c 3, d 4, e 5, and f D respectively equal to b 2, c 3, d 4, e 5, and f D (Fig. 50); and through the points D', 1, 2, 3, 4, 5, and D draw an unbroken curved line. Now from D' set off D' A' equal to D' A' (Fig. 50), and through A' draw A' M parallel to K L. From the points in A' M, where the lines a, b, c, d, e, and f cut A' M, set off distances to the left of A' M corresponding to the distances a 1, b 2, c 3, d 4, e 5, and f D to the right of the line K L, and through the points there found draw an unbroken curved line.   Then A' D' A D will be the pattern required.

To draw the pattern for the D' C' D C (Fig. 50) portion of the base.   It will be at once seen from the plan that this differs from the A' D' A D portion, only in that the distance D' C' is less than the distance A' D'; and thus that, if the lines A' M and L K (Fig. 51) are brought closer together, so that A' D' is equal to D' C' (Fig. 50), the pattern so obtained will be the pattern required.

PROBLEM XXXI.

To draw the pattern for an aquarium (or other) base formed of mouldings not of one and the same design.

The problem now before us is that of fitting to a given aquarium or other body, which has to stand in a given position, a position, that is to say, of which the dimensions are fixed, a base made up of mouldings.   In the last problem there are no restrictions as to the dimensions of the base.

(108A.) Let A' B' C' D' (Fig. 52) be the plan of the aquarium or other body to which a moulding base has to be fitted, and A B C D the plan of the position the body has to stand in. Join A' A, B' B, C' C, and D' D; these lines last drawn will be the plans of the mitre joints (§ 101A).   Looking at the plan, it is at once seen that the pieces of moulding on the B' A' and C' D' sides of the body, though equal to each other, are not the same dimensions as the pieces, also equal to each other, on the B' C', and A' D' sides.   An imperative condition of two pieces of moulding properly joining is that the sections of the pieces shall be the same on the mitre-plane (§ 97A). That is to say, that here, for the pieces A' D' D A, C' D' D C to properly join notwithstanding the difference of their dimensions, their sections on the mitre-plane of D' D must be the same.   When two pieces to be joined are of one and the same design, this condition of same section on the mitre-plane is fulfilled when the mitre line bisects in plan the angle that the pieces of moulding, as joined, make with one another (§ 97A). That condition has now to be fulfilled when, the two pieces being unequal in dimensions, the mitre does *not* in plan bisect the angle made with each other by the pieces; the line D' D, for example, does not bisect the angle A' D' C'.

FIG. 52.

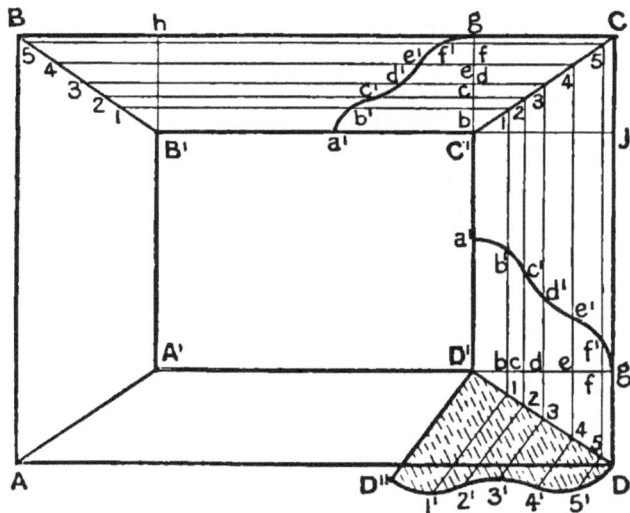

The shape for one of the mouldings may be given or chosen; let it be, say, that for the moulding piece D' C' C D, or the piece B' A' A B, which is equal to it; and let that shape be *g e' c' a'*, drawn on the lines D' *g* (a continuation of the line A' D'), and D' *a'* (part of D' C'). Regarding D' *g* as the line of span of the moulding, D' *a'* is the line of its depth, that is, of the distance the moulding reaches up on the side of the aquarium or other body. The depth of all four pieces of moulding round the body is the same; it is not affected by the want of correspondence between the plan A' B' C' D' and the plan A B C D. From C' on C' B' set off this depth, C' *a'*, equal to D' *a'*, and draw C' *g* perpendicular to C' B' and meeting the plan line B C of the outer edge of the piece of

moulding B' C' C B in a point g. The line C' g then is the
span of this moulding, and we shall now be able to draw on
the lines C' g, C' a', the shape for it.

Divide g e' c' a', the chosen shape for the moulding D' C' C D,
into any number of parts, here six, equal or unequal, in the
points b', c', d', e', f'. Through these points draw lines parallel
to C D, terminating in the mitre plan-lines C' C and D' D,
and cutting D' g in b, c, d, e, and f; the drawn lines being
the lines 1 to 1, 2 to 2, 3 to 3, 4 to 4, and 5 to 5.   From the
points 1, 2, 3, 4, and 5 on the line C' C draw lines parallel to
B C, terminating in B' B, and cutting C' g in b, c, d, e, and f.
From these last points b, c, d, e, and f, set off distances b b',
c c', d d', e e', and f f', respectively equal to the distances of
same letters from the line D' g, and through the points a' (on
C' B'), b', c', d', e', f', and g on C B draw an unbroken curved
line; this line will be the shape for the moulding B' C' C B.

FIG. 53.

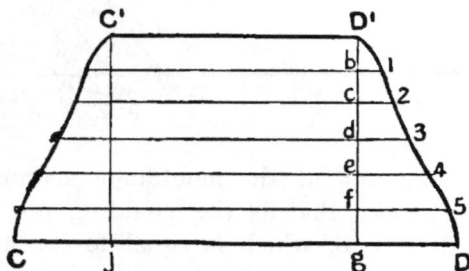

FIG. 53.

Having the shapes of each of the mouldings, to draw the
patterns for the pieces.   First that for the D' C' C D pieces.
Draw (Fig. 53) any line D' g, and from the point D' set off
distances D' b, b c, c d, d e, e f, and f g, equal respectively to
the distances a' b', b' c', c' d', d' e', e' f', and f' g of the shape
drawn on the lines D' g, D' a (Fig. 52), and through the points
D', b, c, d, e, f, and g draw lines perpendicular to D' g.   Make
g D equal to g D (Fig. 52), and make f 5, e 4, d 3, c 2, b 1 respectively equal to f 5, e 4, d 3, c 2, and b 1 from the line D' g
(Fig. 52); and from the point D' through the points 1, 2, 3,
4, 5 to D draw an unbroken curved line.   Now make g j

equal to $g\,j$ (Fig. 52), and through j draw $j\,C'$ parallel to $g\,D'$. Set off $j\,C'$ equal to $g\,D'$, and then obtain points for and draw the curve $C'\,C$ exactly as with the curve $D'\,3\,D$, the two curves being alike. The pattern required will be $C'\,D'\,D\,C$.

Fig. 54.

To draw the pattern for the $B'\,C'\,C\,B$ (Fig. 52) piece of moulding. Draw (Fig. 54) any line $C'\,g$, and from the point C' set off distances $C'\,b$, $b\,c$, $c\,d$, $d\,e$, $e\,f$, and $f\,g$, equal respectively to the distances $a'\,b'$, $b'\,c'$, $c'\,d'$, $d'\,e'$, $e'\,f'$, and $f'\,g$ of the shape drawn on the lines $C'\,g$, $C'\,a'$ (Fig. 52), and from the points C', b, c, d, e, f, and g draw lines perpendicular to $C'\,g$. Make $g\,C$ equal to $g\,C$ (Fig. 52), and make $f\,5$, $e\,4$, $d\,3$, $c\,2$, $b\,1$ respectively equal to $f\,5$, $e\,4$, $d\,3$, $c\,2$, and $b\,1$ from the line $C'\,g$ (Fig. 52); and from the point C' through the points 1, 2, 3, 4, 5 to C draw an unbroken curved line. Now make $g\,h$ equal to $g\,h$ (Fig. 52), and through h draw $h\,B'$ parallel to $C'\,g$. Set off $h\,B$ equal to $g\,C$, and then obtain points for and draw the curve $B'\,B$ exactly as with the curve $C'\,3\,C$; or copy the curve $B'\,B$ from the curve $C'\,3\,C$, the two curves being alike. The pattern $B'\,C'\,C\,B$ will be the pattern required.

The section or shape of either piece of moulding at the mitre plane is obtained as in Problem XXVIII., by drawing (Fig. 52) through the points D', and 1, 2, 3, 4, 5 on the line $D'\,D$, lines perpendicular to that line, and making $D'\,D''$, $11'$, $22'$, $33'$, $44'$, and $55'$, equal respectively to $D'\,a'$, $b\,b'$, $c\,c'$, $d\,d'$, $e\,e'$, and $f\,f'$, and drawing from the point D'' through the extremities of these lines $11'$, $22'$, $33'$, $44'$, and $55'$ to D an unbroken curved line. This section, however, is not needed for the problem.

PROBLEM XXXII.

*To draw the pattern for an inclined (raking) moulding, to mitre
or join with a given horizontal moulding.*

(109A.) This is the case referred to in Problem XXVIII.,
of section of a moulding by a mitre plane which contains
neither the line of depth of the moulding, nor the line of
its span.　It is easy to work because the mitre plane, so far
as the given horizontal moulding is concerned, contains both
of those lines.　The case is not one, the student must notice,
of joining two pieces of moulding *of the same shape* at any
angle, one of the pieces being inclined, but of so fitting to
one shape of moulding a moulding *of another shape* that the
plane of their junction shall contain the line of intersection
of the two surfaces to which the mouldings are to apply.
As a typical instance of the problem, suppose the mouldings
to be gutters, fitting say on a dormer window, or on some
part of a roof where one gutter is inclined and the other
horizontal.　The walls or surfaces (see Fig. 55) to which
the mouldings fit are in this instance assumed to be perpendicular to each other, they may however be at any
angle ; this does not affect the work of the problem.　The
plane of junction of the mouldings, however, *must contain the
line of junction of the surfaces of the walls.*

Let f', e', d', c', b', a (Fig. 55A) drawn on the lines $A'f'$, $A'a$
be the given shape of the horizontal moulding, the lines
$A'a$ being part elevation line of the surface to which the
horizontal moulding applies ; it may in fact be regarded as
representing the line of junction of the two flat faces or
surfaces of the walls to which the mouldings are to fit.　The
shape given is not chosen as a good one for a gutter only,
but rather as a useful shape to illustrate principles.　Hence
also the term mouldings is employed instead of gutters, all
gutters being mouldings (§ 100A, p. 303).　Perpendicular to
$A'a$ draw any line X X, produce $A'a$ indefinitely, as a N A,
and from any point A in the produced line draw A K at an

angle in plan equal to that at which the two pieces of moulding are to meet, now a square or right angle as above stated. Bisect the angle K A N by a line A Q, through f' draw f' M F perpendicular to X X, and intersecting A Q in F, and let M be the point in f' F in which a line from any point N in A' A and perpendicular to it cuts f' F, and

Fig. 55.

through F draw a line parallel to A K. Next choose division points in the curve $d'a$ as most convenient (§ 105A), the points here taken being c' and b'; from the points b', c', and d' draw lines, each perpendicular to X X, to meet the line A Q, the lines namely b' B, c' C, d' D, and through B, C, and D draw lines parallel to K A. Then N M F A will be the plan of the horizontal moulding, and F K A that of the inclined moulding, and A F the plan of the mitre-joint (§ 101A).

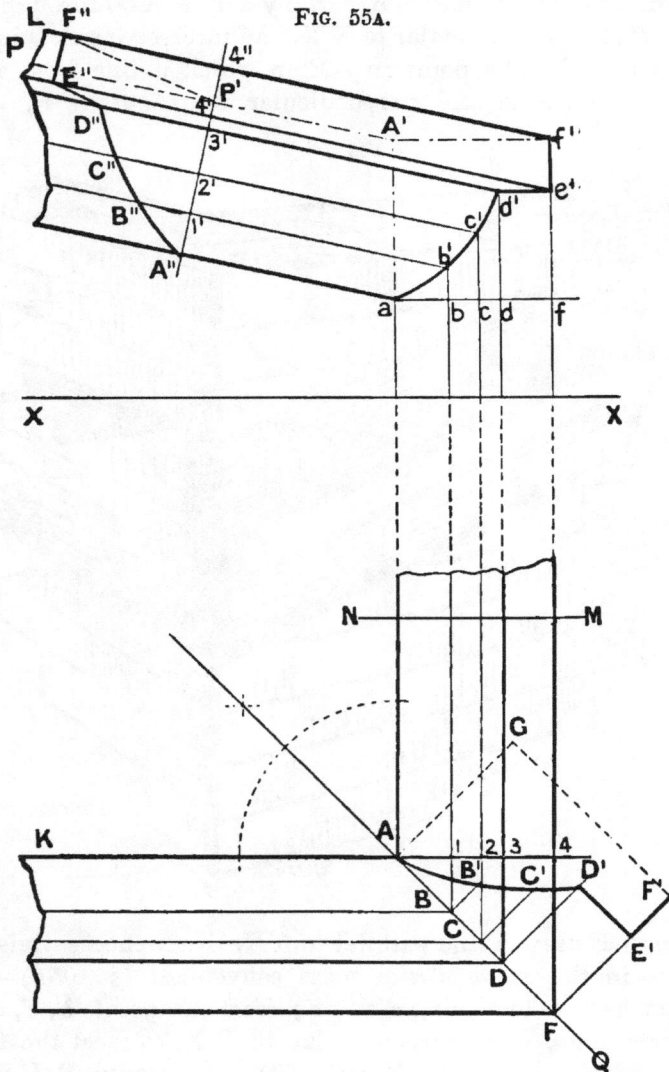

FIG. 55A.

To draw the elevation of the inclined or raking moulding. From A′ draw A′ P at an angle to X X equal to the inclination of the moulding, and from the points f', e', and d' (§ 105A), and from c', b', and a, draw lines parallel to A'P, the f L a (L being a point in the line drawn from f') is the elevation required.

To draw the shape of the raking moulding. From any point A″ in the line from a parallel to A'P draw a line perpendicular to aA″, meeting fL in 4″ and cutting the lines drawn parallel to A'P from e', d', c', and b' in the points respectively 4′, 3′, 2′, and 1′, and continue the line K A to cut the lines b'B, c'C, d'D, and f'F in the points respectively 1, 2, 3, and 4. As the inclination of the moulding does not affect span dimensions, set off 1′ B″ equal to 1 B, 2′ C″ equal to 2 C, 3′ D″ equal to 3 D, and 4′ E″, 4″ F‴ each equal to 4 F, the points f and e' coinciding with each other, in the point F, in plan. Joining the points F‴, E″, D″, C″, B″, A″ as shown, gives the shape for the inclined moulding, the shape that will mitre accurately in mitre plane A Q with the horizontal moulding.

To draw the pattern for the horizontal moulding N M F A. This is done by Problem XXVIII.; in detail it is as follows :— Draw (Fig. 56) any line Af, and from A set off distances A b' b' c',c' d', d' e', and e' f' respectively equal to a b', b' c', c' d' d' e', and e' f (Fig. 55A), and through the points A, b', c', d', e', and f', draw lines perpendicular to Af. Make f F, and e' E each equal to 4 F (Fig. 55A), and d' D, c' C, b' B, equal respectively to 3 D, 2 C, and 1 B (Fig. 55A). From A through the points B, C, and D draw an unbroken curved line, and join D E, E F. Then A B C D E F is the pattern required. The line N M corresponding to the line N M from any point N in A A' (Fig. 55A) is drawn to indicate that the moulding may be of any length.

To draw the pattern for the inclined moulding F K A. Draw (Fig. 57) any line A″ 4,″ and from A″ set off A″ 1′, 1′ 2′, 2′ 3′, 3′ 4′, 4′ 4″ equal respectively to A″B″, B″C″, C″D″, D″E″, E″F‴ (Fig. 55A), and through the points A″, 1′, 2′, 3′, 4′, and 4″, draw lines perpendicular to A″ 4″. Make A'a equal to A″ a (Fig. 55A), and 1′b', 2′ c', 3′ d', 4′ e', and 4″f'

Y

respectively equal to $1'b'$, $2'c'$, $3'd'$, $4'e'$, and $4''f'$ (Fig. 55a). Through the a, b', c', d', draw an unbroken curved line, and join $d'c'$, $e'f'$; then $ab'c'd'e'f'$ is the pattern required.

In the instance where the mouldings form a gutter a straight piece equal in depth to a A' (Fig. 55a) is necessary, with the horizontal moulding, to form a b·ck to the gutter to fit against the wall (see Fig. 55). To the raking gutter

Fig. 56.

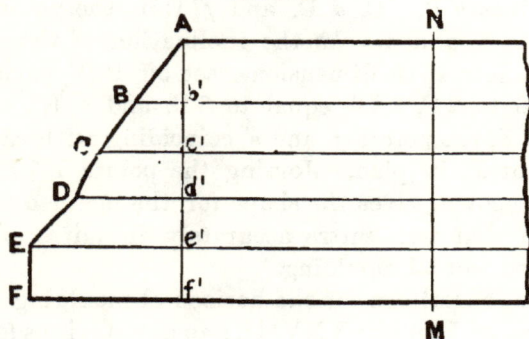

Fig. 57

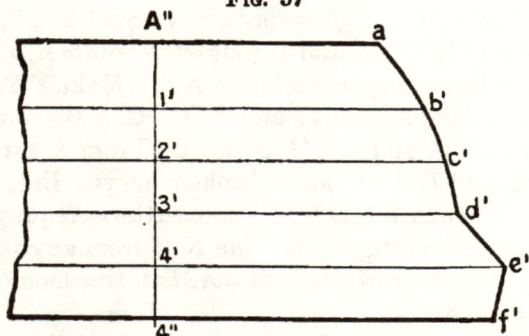

also a back piece is required (see Fig. 55). The "wall-line," against which the top of the gutter would have to fit, is A'P. The straight piece for the back would therefore be equal in depth to P'A" (Fig. 55a), P' being the point of intersection of A'P with A"4''. The highest point at the back of the gutter is P', the highest point in front is F''; the line F‴ P′ is drawn to make this clear; and the workman in drawing the pattern must be careful not to use the line A′ P except as it has been used in the working of the problem.

Although not needed, it will be useful to show how to obtain the mitre section of the mouldings at A F (Fig. 55A). Draw through point a a line a f perpendicular to A′ a, cutting b′ B, c′ D, d′ D, and e′ F in the points b, c, d, and f. Through A, B, C, D, and F draw lines perpendicular to A F, and set off distances F F″, F F′, D D′, C C′, B B′, equal respectively to f f′, f e′, d d′, c c′, and b b′. Join F′ E′, E′ D′, and from D′ through C′ and B′ to A draw an unbroken curve; the section sought is F′ E′ D′ C′ B′ A′. The lines A G and G F′ represent the lines of depth and span of the horizontal moulding on the plane of junction of the walls or surfaces to which the gutter fixes.

The three problems that immediately follow are examples of penetrations, and in each problem there is intersection of a cone. One example of penetrating bodies has already been dealt with in Problem IV., p. 251, and patterns for cones cut by a flat plane are treated of in Problem XV. (p. 276), XVII. (p. 279), and XIX. (p. 282). The problems referred to should be again worked by the student before proceeding with Problem XXXIII.; especially should he take the course here advised before he attempts Problem XXXV.

PROBLEM XXXIII.

To draw the necessary patterns where a conical body (part-cone) penetrates a flat surface, but is not square with it.

Examples in point would be where a pipe attaches to a flat surface of a boiler; fixes vertically on the slant of a roof; or, to take a domestic article, forms a spout to a toilet can.

Let D″ G″ G′ D′, E F G D, (Fig. 58) be respectively elevation and plan of a body, the side of which, G″ G′ in elevation,

Fig. 58.

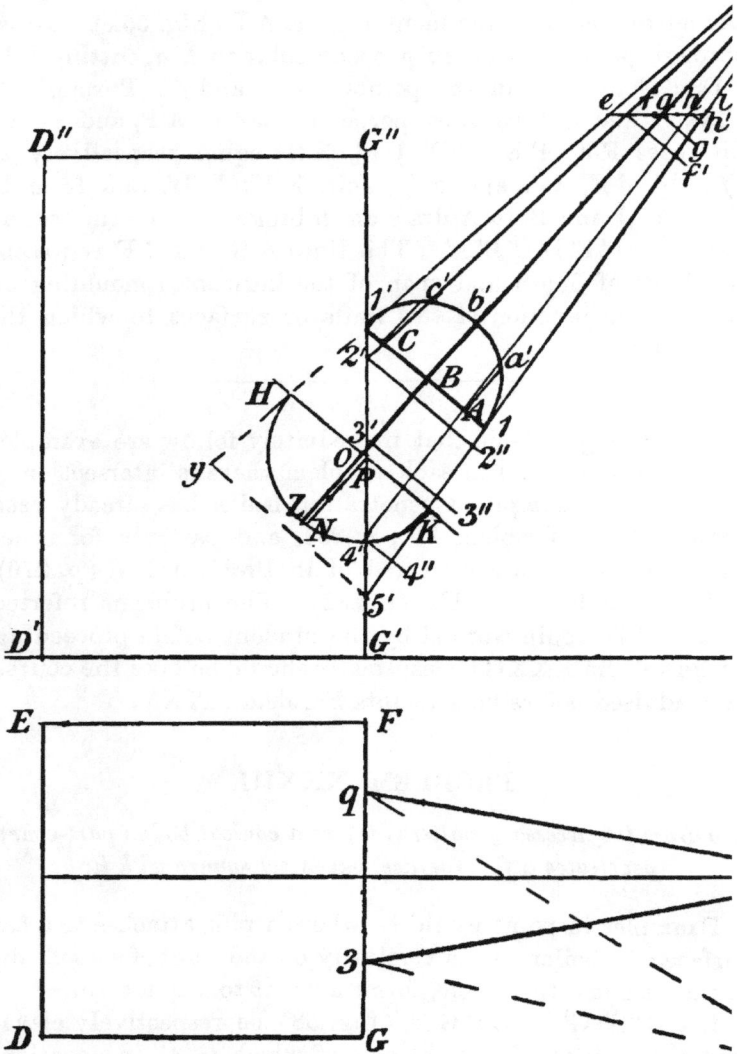

F G in plan, is flat, and let 1′ *e i* 5′ be in elevation the part-cone penetrating the flat side. The plane of the free end *e i* of the part-cone is here taken to be horizontal, the flat side G″ G′ being vertical. The apex of the cone is *v*′. In the plan the free end of the part-cone is omitted, but the cone, *q v* 3, is shown. As will be seen from the plan and elevation, the cone is out of square *in one direction only* with the flat side of the body. The case in which the cone is out of square *both ways* with the flat surface, that is to say, where it would show in elevation as in the Fig., but in plan as *q* V 3, is not taken, because it comparatively seldom occurs in practice.

The necessary patterns in the present problem are, first, a pattern for the part-cone 1′ *e i* 5′, and secondly a pattern of the curve of penetration 1′ 5′ in elevation, *q* 3 in plan. The following further construction is required.

From *v*′ along *v*′ 5′ set off *v*′ 1 equal to *v* 1′, and join 1′ 1; the line 1′ 1 will represent in edge elevation a circular section of the part-cone. On 1′ 1 describe a semicircle 1′ *b*′ 1, which divide into any convenient number of equal parts (here, to avoid the confusion arising of many lines, four), in the points *c*′, *b*′, *a*′. Through *c*′, *b*′, and *a*′ draw lines perpendicular to 1′ 1, intersecting it in C, B, and A respectively. Join *v*′ C, and produce it to meet G″ G′ in point 2′; similarly join *v*′ B, *v*′ A and produce them to meet G″ G′ in points 3′ and 4′ respectively. Also through 2′, 3′, 4′, *f*, *g* and *h* draw lines parallel to 1′ 1, intersecting *v*′ 5′ in 2″, 3″, 4″, *f*′, *g*′ and *h*′ respectively.

To draw the pattern for the part-cone.

With any point *v*′ (Fig. 59) as centre and radius equal to *v*′ 1′ (Fig. 58) describe an arc; at about the centre of it take a point 1, and from *v*′ through point 1 draw an indefinite line. (It is here assumed that the seam of the part-cone will correspond with the line 1′ *e*, Fig. 58.) Now set off right and left of *v*′ 1 distances 1 *a*′, *a*′ *b*′, *b*′ *c*′, *c*′ 1′ equal respectively to the curve distances of same lettering in Fig. 58. Through each of the points *a*′, *b*′, *c*′, and 1′, draw from *v*′, indefinite lines. Make *v*′ 5′ equal to *v*′ 5′ (Fig. 58), and right and left of *v*′ 5′ set off *v*′ 4″, *v*′ 3″, *v*′ 2″ equal respectively to the distances *v*′ 4″,

v' 3″, v' 2″ (Fig. 58). Also make v' i equal to v' i (Fig. 58), and right and left of v' 5′ set off distances v' h', v' g', v' f', v' e equal

Fig. 59.

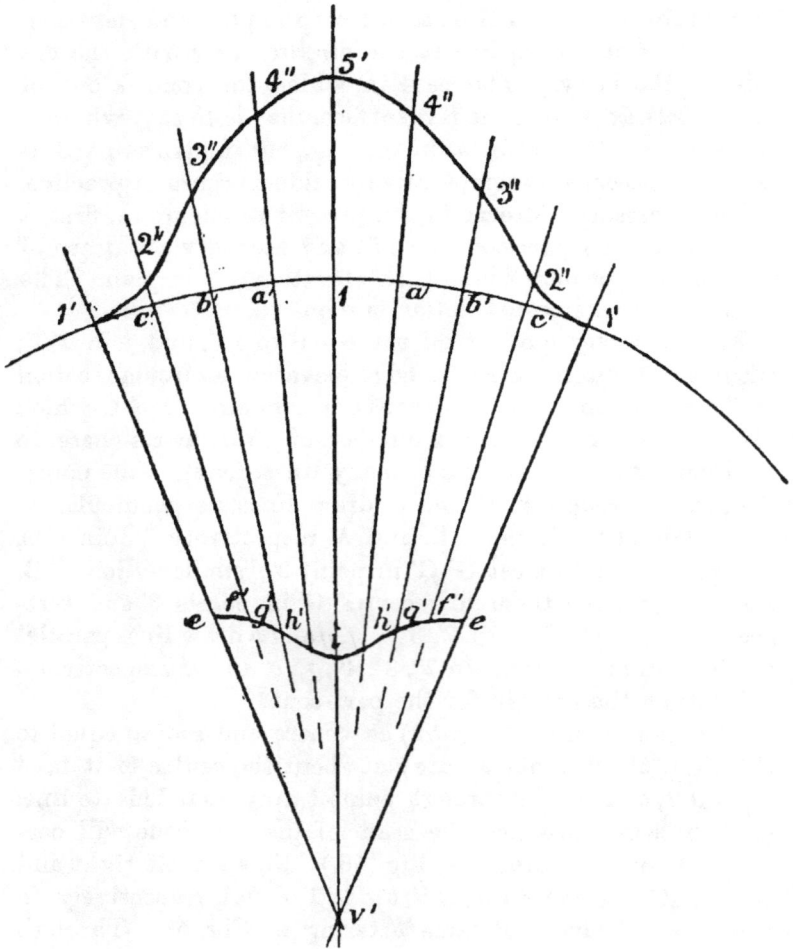

respectively to v' h', v' g', v' f', and v' e (Fig. 58). Through the points 1′, 2″, 3″, 4″, 5′, 4″, 3″, 2″, 1′ draw an unbroken curved line. Also through the points $e, f', g', h', i, h', g', f', e$ draw an unbroken curved line. The figure $e\,i\,e\,1'\,5'\,1'$ will be the pattern required.

To find the shape of the hole of penetration.

The case being that of a cone $y\,v'\,5'$ cut by a plane $G''\,G'$ which cuts all the generating lines of the cone but is not parallel to its base, the curve of penetration is an ellipse (see § 93A, p. 271–3). The line $1'\,5'$ is the length or major axis of the ellipse. Methods of describing an ellipse, length and width being known, have been given at pp. 15–18, and in Problem XXXVII., p. 343. The width of the ellipse is therefore what now has to be found. As the minor axis (width) of an ellipse crosses the major axis at its central point, bisect $1'\,5'$ in P; the point P is then the minor axis in elevation; and we can readily find the length of the line of which it is elevation by drawing the half-base of the cone in which the point lies. Produce $v'\,3'$, and through the point P draw a line parallel to $3''\,3'$, intersecting $v'\,y$ and $v'\,5'$ in H and K respectively and the produced $v'\,3'$ in O. With O as centre and O H as radius describe a semicircle intersecting $v'\,3'$ produced in Z; then H Z K will be a half base of the cone, turned round, on H K as axis, into the plane of the paper. The line Z O will be a semi-diameter of this base, and the line P N which passes through the point P parallel to Z O and meeting the semicircle in N, will be the half width of the ellipse where P is elevation of the width; will be, that is, half the minor axis required.

If the student will now refer to Problem XIV., p. 274, he will find that this Problem is exactly what has now been explained to him, the semicircle H Z K of Fig. 58 corresponding to the semicircle on the line A G of Fig. 22, p. 275, and the line P N to the line E F of that Fig. If the division of the semicircle $1'\,1$ has not been into an even number of parts, the procedure to find P N would be precisely that of finding E F in the Problem referred to.

PROBLEM XXXIV.

*To draw the patterns where a conical body (part-cone) penetrates
a cylindrical body, the axes of cone and cylinder being in
one plane but not at right angles to each other.*

Let D′ E′ D E, D F E (Fig. 60) be respectively elevation and
part plan of a cylinder, and let 1′ v′ 5′ be part-elevation, and
E v 3 part-plan of the penetrating cone.

The plane of the free end e i of the part-cone is here again
horizontal, the cylinder being vertical; but may be at any
angle, this affecting only the e i end of the part-cone pattern.

As in the last problem there is needed a pattern for the
penetrating part-cone, and the shape for the hole, 1′ 3′ 5′ in
elevation, E 3 in half-plan, in the penetrated cylinder. The
following further construction is required.

Set off from v′ a distance v′ 1 equal to v′ 1′ and join 1′ 1; the
line 1′ 1 will represent in edge elevation a circular section of
the cone. On 1′ 1 describe a semicircle 1′ b′ 1, which divide
into any convenient number of equal parts (here, to avoid the
confusion arising of too many lines, four), in the points c′, b′, a′
Through c′, b′, and a′ draw lines perpendicular to 1′ 1, intersecting it in C, B, and A respectively. Join v′ C and produce
it; similarly join v′ B, v′ A and produce them. From the
points C, B, and A draw lines perpendicular to v D (plan) and
meeting it in the points r, s, t respectively, and produce each
of these lines. From r along C r produced set off r c equal
to C c′, and join v c and produce it to intersect E F in point 2;
then v 2 is the plan of the line v′ 2′. From s along B s produced set off s b equal to B b′, and join v b and produce it to
intersect E F in point 3; then v 3 is the plan of the line v′ 3′.
From t along A t produced set off t a equal to A a′, and join
v a and produce it to intersect E F in point 4; then v 4 is the
plan of the line v′ 4′; it falls, the student should notice, inside
the line v 3. The line v E of the given half-plan of the cone
is the plan of the line v′ 1′ and also of v′ 5′. The points E,

2, 4, 3, are plans of points in the curve of penetration. Through point 3 draw a line perpendicular to D E to inter-

FIG. 60.

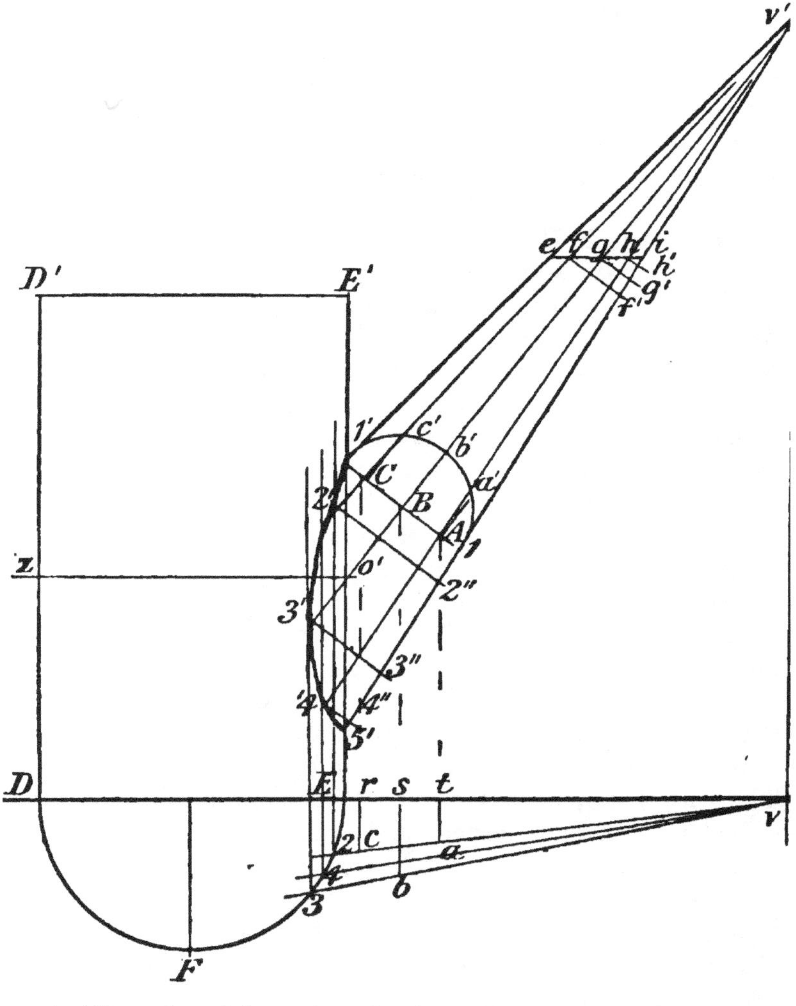

sect v' B produced in point 3'; through point 4 draw a line perpendicular to D E to intersect v' A produced in point 4';

and through point 2 draw a perpendicular to D E to intersect
v' C produced in 2'.   The points 1', 2', 3', 4', and 5' are the elevations of points in the curve of penetration, and joining
these points by an unbroken line will give the curve in
elevation. Through the points 2', 3', and 4' draw lines
parallel to 1' 1 and intersecting v' 5' in the points 2", 3", 4"
respectively, and through the points f, g, and h draw lines
ff', gg', hh' parallel to 1' 1.   Also, parallel to D E draw a line
Z O' to measure distances from.   The line is here drawn
through the point of intersection of v' B produced and E' E,
but this is not of necessity ; it might be drawn through any
other point in E' E.

In the previous problem neither plan points were found nor
plan lines drawn corresponding to the points and lines just
described, for the reason that such points and lines afford no
aid at all in the drawing of the shape of the hole of penetration.

To draw the pattern for the part-cone.   With any point v'
(Fig. 61) as centre and radius equal to v' 1' (Fig. 60) describe
an arc ; at about the centre of it take a point 1, and from v'
through 1 draw an indefinite line.   (As in the previous
problem it is assumed that the seam of the part-cone will
correspond with the line 1' e, Fig. 60.)   Now, proceeding
point for point exactly as in the previous problem, obtain the
figure $e i e$ 1' 5' 1', which will be the pattern required.

To draw the shape of the hole of penetration.   Essentially
the method of doing this has been already given in connection with Problem III., p. 246, the problem of the ' slanting T.'
The case there, is that of penetration of a cylinder by a
cylinder.   Whether the penetration, however, is by a cylinder
or by a cone makes no difference as to finding the shape
required.   For the convenience of the student the working is
again here given.   Draw (Fig. 62) two indefinite lines Z O,
1' 5' intersecting at right angles in O' ; and from O' on Z O
set off right and left a distance O' 2 equal to the distance
E 2 (Fig. 60) ; to the actual curve distance E 2 on the
cylinder, that is.   Similarly, set off right and left of O' the

distances 2 4, 4 3 equal to the distances 2 4, 4 3 respectively
(Fig. 60); and through each of the points 2, 4 and 3 draw

FIG. 61.

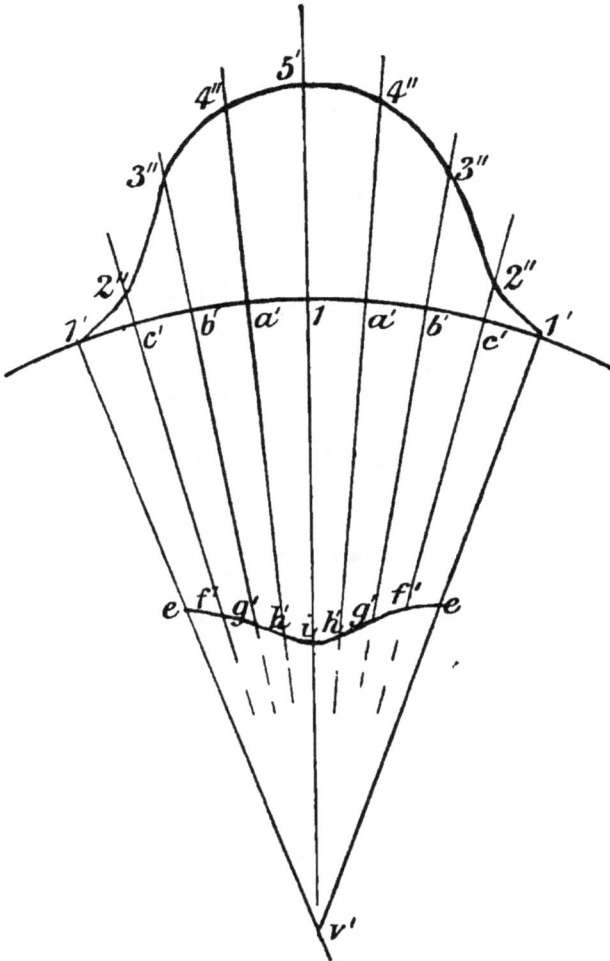

lines perpendicular to Z O. From O' on 1' 5' set off O' 1' and
O' 5' equal to O' 1', O' 5' (Fig. 60) respectively; also, above

Z O, from each point 2 set off a distance 22′ equal to the perpendicular from 2′ to Z O′ (Fig. 60) and, below Z O, from each
point 3 set off a distance 33′ equal to the perpendicular

FIG. 62.

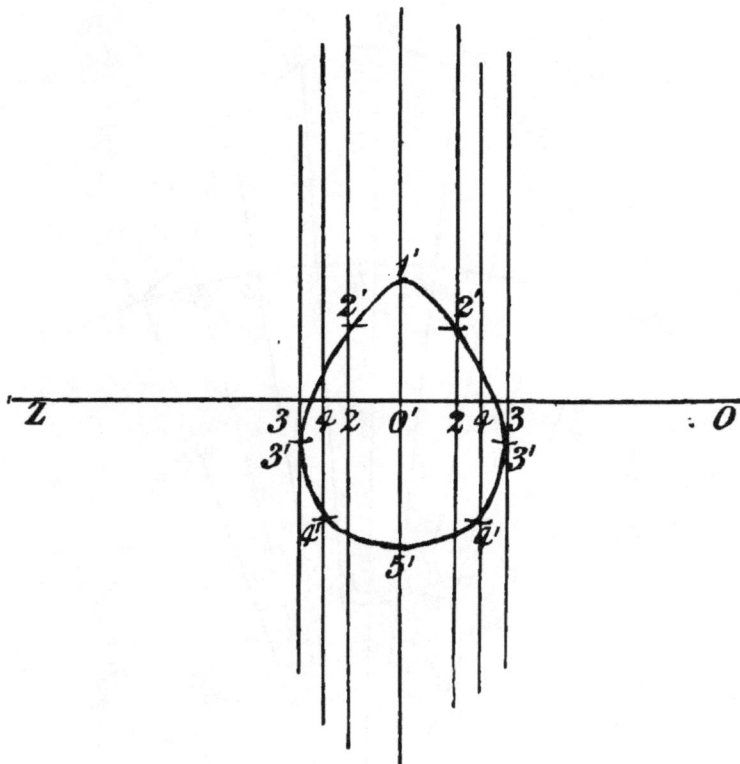

distance from 3′ to Z O′ (Fig. 60), and from each point 4 a
distance 44′ equal to the perpendicular distance from 4′ to
Z O′ (Fig. 60). Through the points thus obtained draw an
unbroken line; this will be the shape required.

PROBLEM XXXV.

To draw the patterns where a conical body (part-cone) penetrates a conical body (part-cone), the axes of the two cones being in one plane, but not at right angles to each other.

Let O' K' V K (Fig. 63) represent a cone frustum, V' being the apex of the complete cone, and V K Q part-plan ; and let 1' *e i* 5' be an elevation of the part-cone penetrating O' K' V K, with *v'* for its apex. The Fig. shows the cones in position ; the line V K produced is the plan line of the axis of the penetrating cone, the point *v* being the plan of *v'* its apex. The plane of the *e i* end of the penetrating cone is horizontal as before, but may be at any angle, this affecting only the *e i* end of the part-cone pattern.

The necessary patterns are a pattern for the penetrating part-cone, and a pattern for the penetrated cone with the shape of the hole of penetration in position upon it. The following further construction, very closely following that of the previous problem, is required.

Set off from *v'* a distance *v'* 1 equal to *v'* 1' and join 1' 1 ; the line 1' 1 will represent in edge elevation a circular section of the cone. On 1' 1 describe a semicircle 1' *b'* 1, which divide into any convenient number of equal parts (here four), in the points *c'*, *b'*, *a.'* Through *c'*, *b'* and *a'*, draw lines perpendicular to 1' 1 intersecting it in C, B, and A respectively. Join *v'* C and produce it ; similarly join *v'* B, *v'* A and produce them. From the points C, B, and A let fall lines perpendicular to V *v* and intersecting it, and produce those lines.

From the point where the perpendicular from A intersects V *v* set off on the perpendicular a distance equal to A *a'*, and join *v* with the point *a* thus found and produce it. Working in like fashion find points *b* and *c* and draw lines *v b*, *v c* and produce them. Now divide the quadrant of the part-plan of the penetrated cone into any convenient number of equal

parts (here four), in the points L, M, and N; join L V, M V,
N V; and from the points L, M, and N erect perpendiculars,
intersecting V v in the points L', M', and N' respectively.
Join L' V', M' V', N' V', intersecting v' C produced in the
points l' (2' in Fig.), m' and n' respectively. (The point
2' is a point to be presently obtained. It lies, however, so
close to l', that l' must be represented by it.) From n', m',
and 2' draw lines perpendicular to V v and produce them
to intersect V N, V M, and V L in the points n, m and l respectively. Also from the point where v' C produced intersects V' K let fall a perpendicular intersecting V v (the
perpendicular is not shown); and through the points n, m,
l, and the intersection point just found draw an unbroken
curved line. This curve is the plan of the intersection
of the part-cone O' K' K V with the plane of v' n', and the
point 2 where the curve intersects v c produced is the plan
of a point in the curve of penetration of the cones. From
point 2 erect a perpendicular, meeting v' n' in the point 2';
this point will be the elevation of point 2 in plan, and we
now have two points in the elevation of the curve of penetration, namely the points 1' and 2'. The finding of a
third point is similar in the working to the finding of point
2, and is as follows. From the points x' and y' where the
line V' B produced intersects V' L' and V' M' respectively,
let fall perpendiculars to V v and produce them to intersect
V L and V M respectively in the points x and y. From the
point where v' B produced intersects V' K let fall a perpendicular intersecting V v (the perpendicular is not shown);
and through the points y, x, and the intersection point just
found, draw an unbroken curved line. This curve is the
plan of the intersection of the part-cone O' K' K V, with the
plane of v' y', and the point 3 where this curve intersects v b
produced is the plan of the third point in the curve of penetration, the elevation of the point being the point 3' where a
perpendicular from point 3 meets v' y'. For a fourth elevation
point the working is again similar. From the points (not
lettered) where the line v' A produced meets the lines V' L,

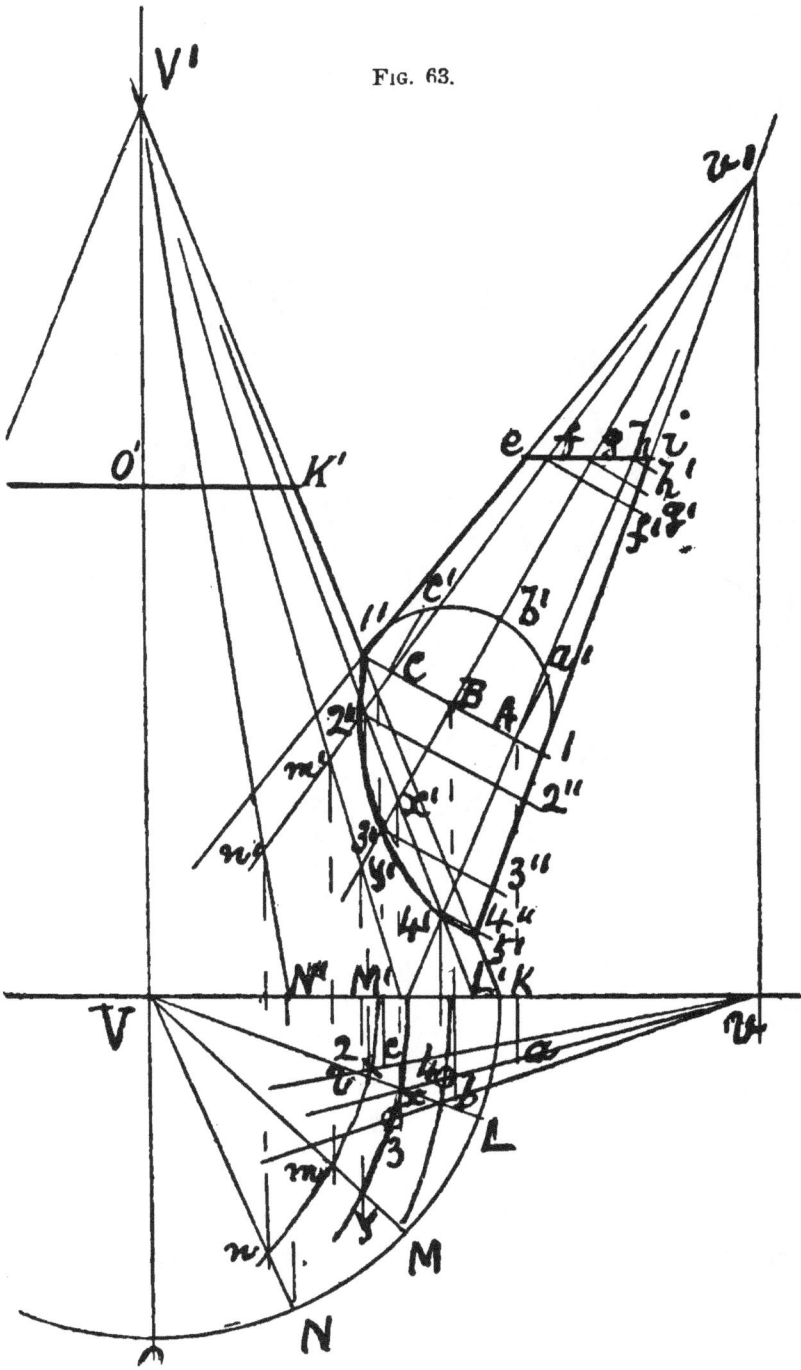

FIG. 63.

V′ M′ respectively, let fall perpendiculars to V v and, producing them, obtain points of intersection with V L and

FIG. 64.

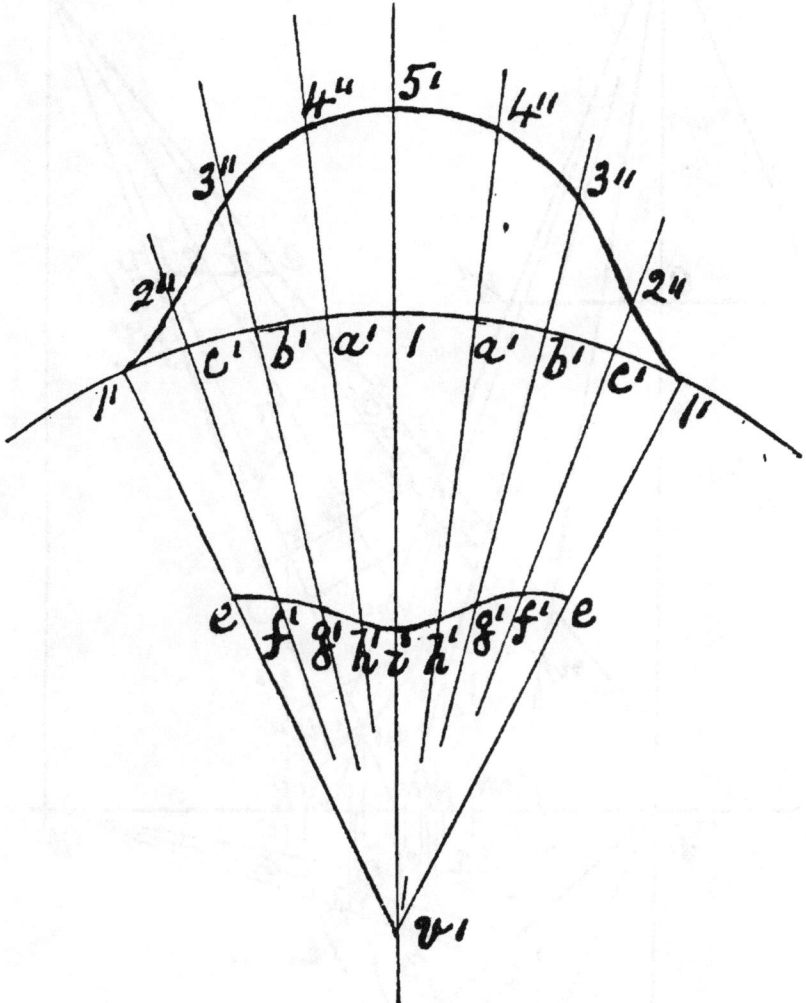

V M respectively. From the point where v' A produced intersects V′ K, let fall a perpendicular (the perpendicular is not shown), and obtain a point of intersection in V v. Through the intersection points in V L, VM, and V v thus found, draw an unbroken curved line. This curve will be the plan of the intersection of the part-cone O' K' K V with the plane of v' A produced, and the point 4 where this curve intersects $v\,a$ produced is the plan of the fourth point in the curve of penetration, the elevation of the point being the point 4' where a perpendicular from point 4 meets v' A produced. A fifth point in the elevation is the point 5'. From the point 1', through the points 2', 3', and 4', to 5' draw an unbroken curved line; this curve is the elevation of the curve of penetration. Now through the the points 2', 3', and 4' draw lines parallel to 1' 1, meeting the line v' 5' in the points 2'', 3'', and 4'' respectively; also similar lines through the points f, g, and h meeting v' 5' in f', g', and h' respectively.

To draw the pattern for the part-cone.

With any point v' (Fig. 64) as centre and radius equal to v' 1' (Fig. 63) describe an arc; at about the centre of it take point 1, and from v' through 1 draw an indefinite line. (As in the previous problem it is assumed that the seam of the part-cone will correspond with the line 1' e, Fig. 63.) Now, proceeding point for point exactly as in the previous problem, obtain the figure $e\,i\,e$ 1' 5' 1', which will be the pattern required.

To draw the shape of the hole of penetration.

In order to make the working clear, and avoid complication of lines, part of Fig. 63 is reproduced in Fig. 65, with which we now have to deal. Join V' with each of the points 2', 3', and 4' and produce V' 2', V' 3', V' 4' to intersect V K in H' F', and G', respectively. Through F', G', and H' draw F' F, G' G and H' H perpendicular to V K and intersecting the quadrant K F Q in F, G, and H respectively. Through the points 2' 3', and 4' draw lines 2' r, 3' s, and 4' t parallel to V K and intersecting V' K in the points r, s and t respectively.

Next, with V' (Fig. 66) as centre and radius V' K (Fig 65) describe an arc Q Q, and in it take any convenient point K, and right and left of K set off distances K H, H G, G F equal

z

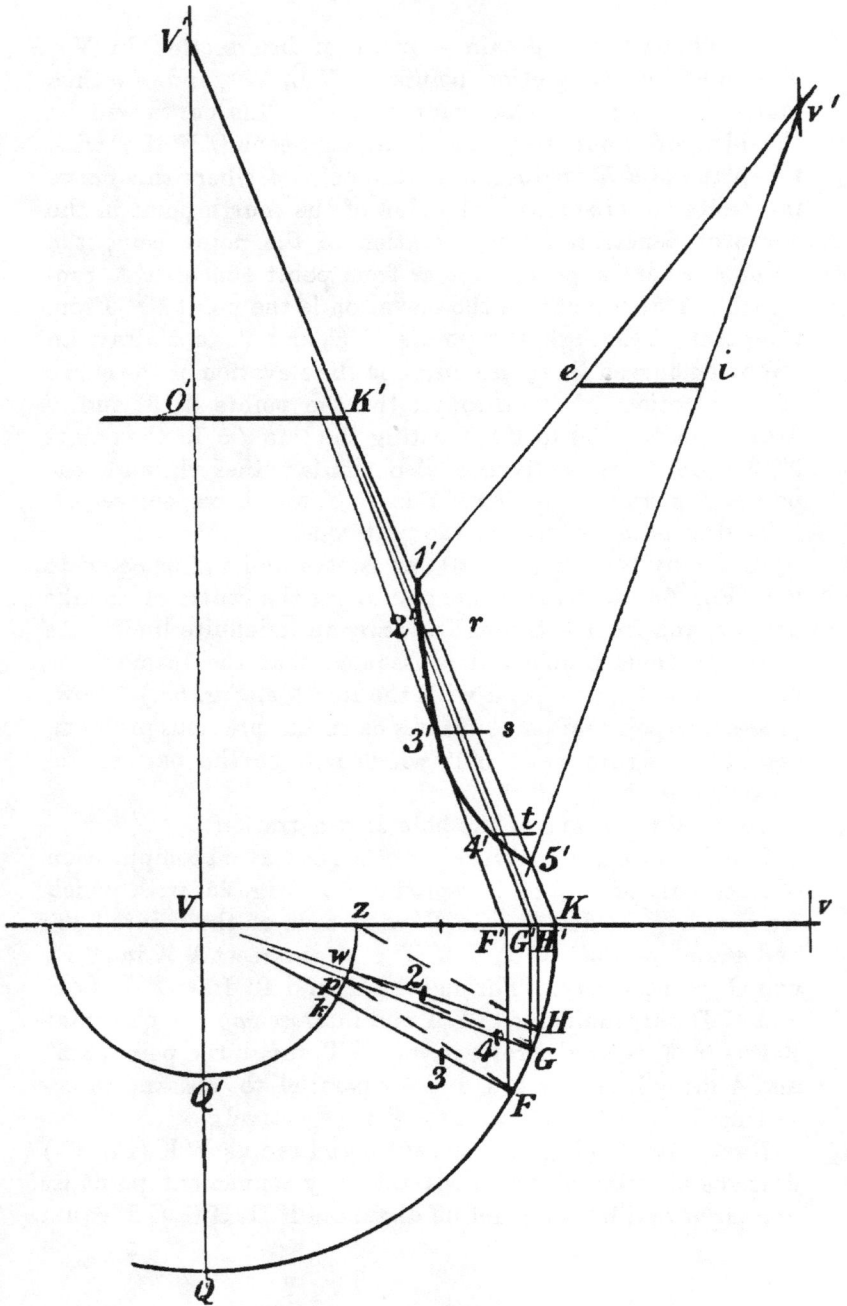

Fig. 65.

respectively to the curve distances K H, H G, G F (Fig. 65).
Join V′ K and right and left of V′ K join the points H, G, and

Fig. 66.

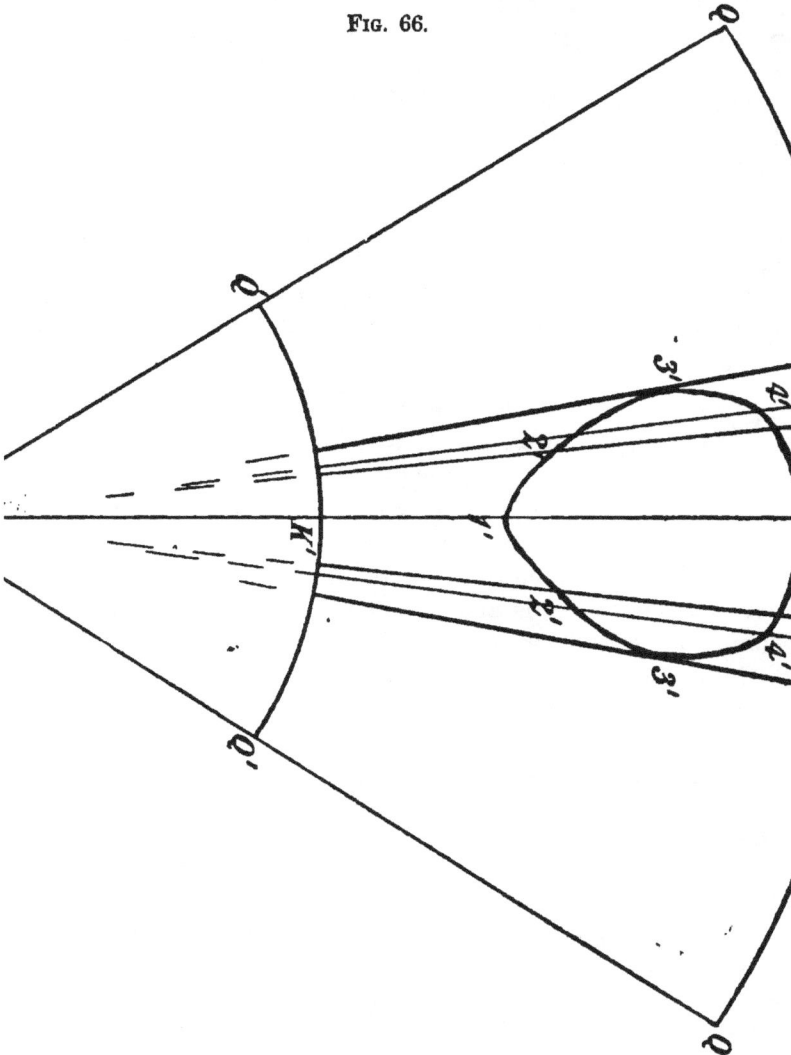

F to V′. On V′ K from K set off distances K 1′, K 5′ equal
respectively to K 1′, K 5′ (Fig. 65); and make the distances

G 4′, H 2′, F 3′ right and left each of them of V′ K, equal
respectively to the distances K *t*, K *r*, K *s*. Join the points
1′, 2′, 3′, 4′, 5′, 4′, 3′, 2′, 1′ by an unbroken curved line; this
will give on a portion of the pattern for the penetrated cone
the required shape of the hole of penetration.

If on Q Q from the point F right and left of V′ K a distance
F Q be set off equal to the curve distance F Q (Fig. 65) and
each of the points Q be joined to V′, and if further with V′ as
centre and radius V′ K′ (Fig. 65) an arc Q′ Q′ be described
intersecting the lines V′ Q in the points Q′, Q′, then the figure
Q′ Q Q Q′ with. the figure 3′ 5′ 3′ 1′ within it will be the
pattern of half the penetrated cone-frustum with, in proper
position with respect to the pattern, the shape of the hole
of penetration.

PROBLEM XXXVI.

To draw, **without long radii**, *the shape of the hole of
penetration.*

Large work is now supposed, and consequently that the
apex of the cone-frustum is inaccessible. Having, however,
the necessary dimensions as to frustum and penetrating
cone (either the ends and height of the frustum, or ends and
slant), we can draw (see Problems VI. and VII., pp. 37 and
39) the part elevation O′ K′ K V (Fig. 63) of the frustum
and its part plan V K Q, and also the elevation 1′ *e i* 5′ of the
penetrating cone in position. The procedure to draw the
hole of penetration by short radius starts with the plan
points of the curve of penetration. The method of finding
these (points 2, 3 and 4 of last problem), differs from that
there described only in the way of obtaining the lines from
L, M and N intersecting O′ K′. The lines in Fig. 63 were
drawn to the cone apex : with our frustum, now, we have no
apex. Draw, therefore, with V as centre, and O′ K′ as radius
a quadrant intersecting the lines V L, V M and V N (none
of the construction here described is shown; the student
will, however, have no difficulty in mentally following on
Fig. 63 what has to be done); and, from the points of inter-

section, erect perpendiculars to intersect O' K'. Joining these points of intersection in O' K' with L', M' and N' will give the lines required.

The plan points having been found, draw (Fig. 65), (the part reproduction of Fig. 63) from V through the plan points lines intersecting the quadrant K Q in H, F and G, V H being through point 2, V F through 3, and V G through 4. The quadrant z Q' in Fig. 65 is the quadrant the student was just above instructed to mentally draw in Fig. 63, and this quadrant intersects the lines now drawn in w, p and k respectively. Join H z, G w, F p.

Fɪɢ. 67.

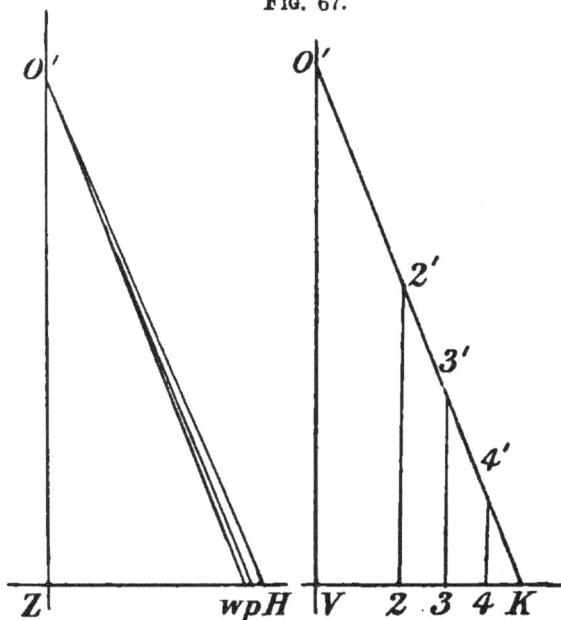

Next (Fig. 67) draw two lines Z O', Z H, perpendicular to each other and intersecting in Z; make Z O' equal to V O' (Fig. 65), Z H equal to z H (Fig. 65), Z w equal to G w (Fig. 65), and Z p equal to F p (Fig. 65); and join O' H, O' p, O' w; then O' H, O' w and O' p are the true lengths of the diagonals represented in plan by z H, G w and F p respectively. Now (Fig. 67) draw two lines V O', V K perpen-

Fig. 68.

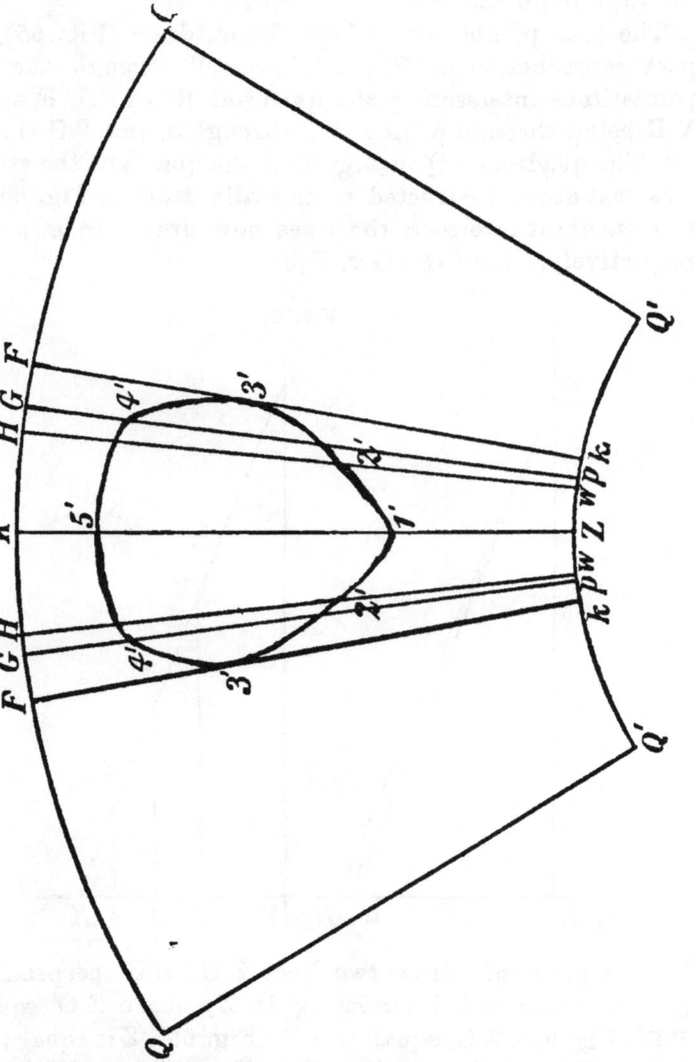

dicular to each other and intersecting in V, make **V O'** equal to V O' (Fig. 65), make **V K** equal to z K (Fig. 65), and join

O' K. Then O' K is the true length of z K, w H, p G, k F, and Q' Q (Fig. 65). Make V 2, V 3, and V 4 equal respectively to w 2, k 3, and p 4 (Fig. 65), and through the points 2, 3, and 4 draw lines perpendicular to V K and intersecting O' K in the points 2', 3', and 4' respectively. This gives the points 2', 3', and 4' in their true positions on their generating lines, positions corresponding respectively with the positions of the points r, s, and t on V' K (Fig. 65) as obtained in long radius method, last Problem.

To set out the hole of penetration.

Draw (Fig. 68) a line K Z equal to O' K (Fig. 65), and with Z and K as centres, and radii respectively equal to O' H (Fig. 67) and K H (Fig. 65), describe right and left of K arcs intersecting in H, and with H and Z as centres and radii O' K (Fig. 67), and $z w$ (Fig. 65), describe right and left of Z arcs intersecting in w. Also with w and H as centres and radii O' w (Fig. 67), and H G (Fig. 65), describe right and left of K arcs intersecting in G ; and with G and w as centres and radii O' K (Fig. 67) and $w p$ (Fig. 65), describe right and left of Z arcs intersecting in p. With p and G as centres and radii O'p (Fig. 67) and G F (Fig. 65), describe right and left of K arcs intersecting in F ; and with with F and p as centres and radii K O' (Fig. 67) and $p k$ (Fig. 65), describe arcs intersecting in k. Join right and left of K Z the points H w, G p, F k and Q Q. On K Z set off K 5' and K 1' equal respectively to K 5' and K 1' (Fig. 65) ; on each line H w set off H 2' equal to K 2' (Fig. 67), on each line G p set off G 4' equal to K 4' (Fig. 67), and on each line F k set off F 3' equal to K 3' (Fig. 67). Now through the points 1', 2', 3', 4', 5', 4', 3', 2' to 1' draw an unbroken line, and this will give the shape of the hole required. Also through F, G, H, K, H, G, F draw an unbroken curved line, and through k, p, w, Z, w, p, k. The arcs F Q and the arcs k Q' are obtained as shown in Problem VIII., p. 41. Join the extremes of these arcs by the lines Q Q' ; we then have complete the pattern of the half-frustum O' K' K V with the hole of penetration in position upon it.

PROBLEM XXXVII.

To draw an oval, its length and width being given.

For long narrow ovals this method is better than that of
Problem XII. p. 13.

Fig. 69.

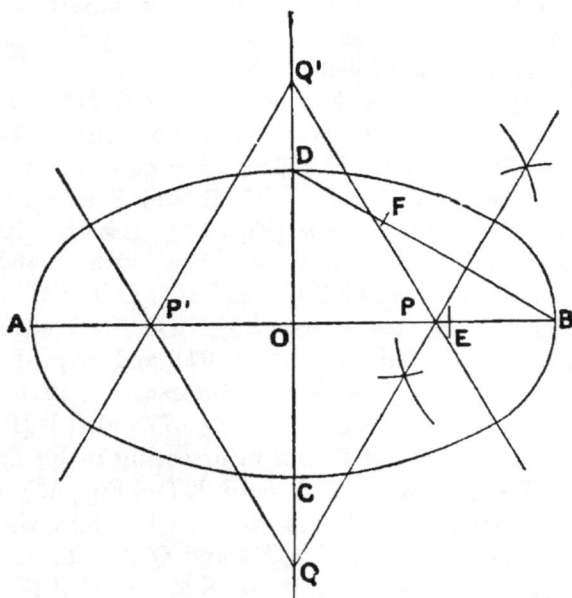

Draw (Fig. 69) two lines A B, C D, perpendicular to one
another, and intersecting in O.   Make O A and O B each
equal to half the length, and O C and O D each equal to half
the width, of the oval.   From O mark off O E equal to O D.
Join D B; make D F equal to E B, and bisect B F by a line
perpendicular to it, and cutting the axes A B, and C D produced, in points P and Q respectively; then P will be the
centre for an end curve, and Q the centre for a side curve.
Make A P' equal to E B, and O Q', on C D produced, equal
to O Q; join Q' P', Q' P, and Q P'.   With Q as centre and
radius Q D describe an arc, meeting the produced lines Q F'
and Q P; with Q' as centre and same radius describe an arc

meeting the produced lines Q' P' and Q' P; and with P B as radius and P and P' successively as centres, describe arcs to meet those just drawn; this will complete the oval.

PROBLEM XXXVIII.

To draw a parabola (§ 94A, p. 273), *the axis and width (see Problem XVII., p. 279) being given.*

Draw (Fig. 70) two lines T S and A E, at right angles and meeting in S. Make S T equal to the given axis, S E equal to

FIG. 70.

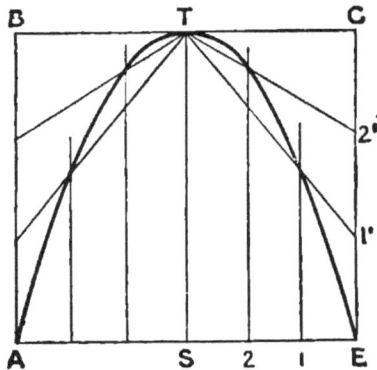

half the given width (double ordinate, see Fig. 25, p. 279), and S A equal to S E. Through E and A draw E C and A B perpendicular to A E, and through T draw B C parallel to A E and cutting A B and E C in points B and C respectively. Divide S E into any number of equal parts (here three) in the points 1 and 2, and divide E C into the same number of equal parts that S E is divided into, in the points 1' and 2'. Through points 1 and 2 draw lines parallel to T S, and join 1' T and 2' T; then the intersection of T 2' with the line drawn through point 2 will be a point on the parabola, as also will be the intersection of 1' T' with the line through point 1. Through T and the points just obtained, draw to E an unbroken curved line; repeat the construction on the other side of T S, and the required parabola is completed.

PROBLEM XXXIX.

To draw an hyperbola (§ 95A, p. 273), *the length, major axis and width* (*double ordinate*) *being given.*

Fig. 71.

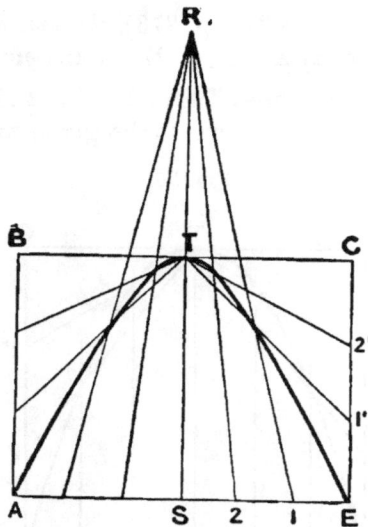

Draw (Fig. 71) two lines R S and A E at right angles and meeting in S.   Make R T equal to the given major axis, S T equal to the given length of the hyperbola (see Fig. 27, p. 282).   Make S E equal to half the given width (double ordinate), and S A equal to S E.   Through E and A draw E C and A B perpendicular to A E, and through T draw B C parallel to A E and cutting A B and E C in points B and C respectively.   Divide S E into any number of equal parts (here three) in the points 1 and 2, and divide E C into the same number of equal parts that S E is divided into, in the points 1' and 2'.   Join 1 R, 2 R, 1' T and 2' T; then the intersection of 1 R and 1' T will be a point in the hyperbola, as will also be the intersection of 2 R and 2' T.   Through T and the points just obtained draw to E an unbroken curved line; repeat the construction on the other side of S R, and the required hyperbola is completed.

Book IV.

CHAPTER I.

Pipe Bending.

Before taking up problems on the construction of bends from flat sheet-metal, it will not be outside the scope of this volume to devote some little space to a consideration of the methods to employ, and tools to use, in bending ordinary metal tubing, as this is a knowledge oftentimes wanted in the sheet-metal workshop, and is by no means common knowledge there. The workman who is unacquainted with these methods and tools finds great difficulty in giving to a straight piece of tubing a desired curvature; in, that is, so shaping it, that the completed bend shall be free from ridges, dents, or 'kinks,' and not flattened in the throat; the perfection of a bend, of course, being when the internal diameter or 'bore' of the tube is the same throughout.

What has to be done in the way of tube bending in the ordinary workshop must be done with the appliances that are usually found there, or can easily be made there. The tools necessary for the production of bends, specially smooth and circular in section, such as those of wind musical instruments, cannot be made in the ordinary sheet-metal workshop; it will be well, however, to also give a brief description of the working of these bends.

The method of bending a pipe or tube varies with the metal of it, its diameter, and the curvature required. Ordinary gas barrel, of the smaller sizes, may readily be bent without any special preparation to curves, even of small radius. A piece of iron, preferably round iron, being gripped in a vice, horizontally or vertically as most convenient, and used as a fulcrum or bending post, the barrel may be shaped round it to the curve required. A slight curvature should first be given, then, changing the place of contact of barrel and fulcrum, a further curvature; and so on. A piece of wood, if of sufficiently large diameter to bear the strain it will be put to, may be similarly used. If the barrel is of so large diameter that its bending against a piece of iron in the manner described is beyond the workman's strength, it should be made red-hot where the bend is to come. Thus softened, the workman will find the bending comparatively easy. In bending iron barrel in this way there is always a slight flattening in the throat of the bend. In such work, however, as iron barrel is generally used for, this flattening is not of consequence; the thickness of the barrel and the toughness of the metal prevent any great amount of flattening.

The procedure necessary to bend tubes made of metals softer than iron—of copper, brass, zinc, tin, lead, or 'compo,' for instance—is less simple, and to prevent buckling, puckering or flattening, in the throat of a bend, such tubing must be 'loaded.' The substances used for loading are various, and the choice between them depends upon the metal of which the tube to be bent is made, and greatly upon the finish and symmetry required in the particular work to be done. Lead, resin, pitch, and resin and pitch in equal parts, are the substances most in favour amongst workmen. When 'set' after being melted and run into tubing, they are found to bend without breaking as the bending of a tube progresses, and to offer the needful resistance to change of section of it. A spiral spring not closely coiled, or a piece of cane or solid rubber, may often be advantageously used to 'load' soft metal pipe with. Either of these loadings can be pulled through the made bend, and will serve again and again as loading. A tightly-rolled piece of paper will often serve, even for brass tubing; if used for brass tubing, it can be burnt out if need be. On an emergency, and if the ends of the piece of tube to be bent are tightly corked or otherwise sealed up, and the look of the finished work is not of particular importance, a tube may be loaded with sand, or even with water. The reason why sand and water, as loading, are suitable only for an emergency, is that the plugging at one of the tube ends often gives way in the course of the bending.

The melting-point of lead is 323° C., and, as a loading substance for brazed brass tube, lead has this disadvantage—that when melting it out of a bent tube, there is danger lest any weak spot in the brazed seam should crack or open up, and special care needs to be taken to warm up the tube slowly and equally in melting out the lead because of this. When the lead has been run out of a bent tube, little particles of lead often remain in the tube adhering to the surface. To dislodge these, the tube should be again warmed up, to a temperature a little higher than that of the melting point of lead, and the open end struck smartly on the bench, or with a piece of wood, the tube being held with a pair of pliers, or otherwise as may be convenient. Resin or pitch, as loading substances, leave behind a thin adherent film after being melted out of a tube. This must not be forgotten when choice of a loading substance has to be made. If it is imperative that the inner surface of a bent tube should be clean, then neither of these substances can be used with a tube of soft metal, as the film has to be *burnt* off, which would mean spoiling the tube. They may be used with copper or brass tubing, when, for the reason that the throat of the bend need not be perfectly circular in section, it is desired that the loading substance shall not offer any great resistance to bending.

Brass or copper drawn tubing should be annealed before being bent. In loading a piece of pipe with either lead, pitch, or resin, two or three layers of brown paper should be wrapped round one end of it and securely tied. If lead is the loading material, the tube should be rigidly fixed vertically, with its closed end embedded in sand, so that molten lead may not run out to do mischief. The lead is poured from an ordinary plumbers' ladle. And in loading with pitch and resin the tube should at least rest and be secured with its closed end on some solid substance, to prevent leaking out of the hot pitch or resin.

The appliances commonly used in bending light tubes are shown in Figs. 1 and 3. In Fig. 1, B C is a channel-piece of stout metal, say brass, of semicircular section. It may be made by cutting a tube into halves longitudinally, and giving to one of the pieces the desired B C curve, by stretching with blows of a hammer each of the longitudinal edges of the piece. To this channel-piece, at one end of it, a semicircular strap D is soldered or riveted, or both, so as to form at this end of the channel a complete ring-piece. A B C is a stout piece of metal, shaped, one edge of it, to the

Fig. 1.

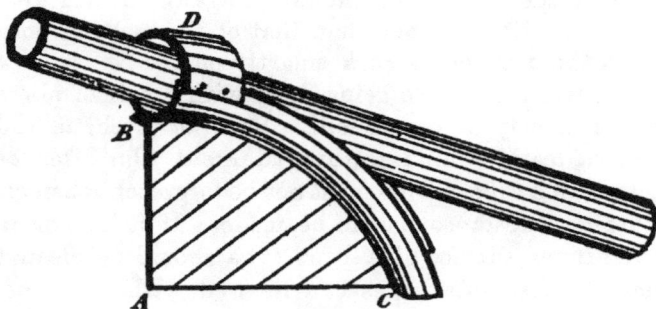

B C curve, and firmly soldered to the throat of the channel piece. The tool in use is gripped in a vice by its A B C piece, one end of the pipe to be bent is placed in the ring-piece D, as shown in the Fig., and the bending is effected in the forcing of the pipe by hand pressure into the B C channel.

Should there be a dent in the straight piece of tube to which curvature has to be given, it may be removed as shown in Fig. 2. A brass or steel ball being forced into the tube, so as to rest on the edge formed by the dent, may be made to pass right through by striking it smartly and repeatedly with a second tube of slightly smaller diameter than the dented tube. Slight dents in a lead-loaded tube or bend may be worked up by means of light blows of a round-faced hammer on either side of the dent; this treatment will force the dent outwards.

FIG. 2.

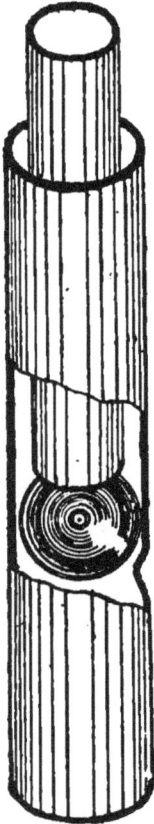

The appliance shown in Fig. 3 is in all respects the same as that represented in Fig. 1, except that the channel piece B C is of V section instead of semicircular section. Zinc tubes used for 'bead' for baths and other articles are bent without being loaded; and in bending these unloaded tubes a V-shaped

FIG. 3.

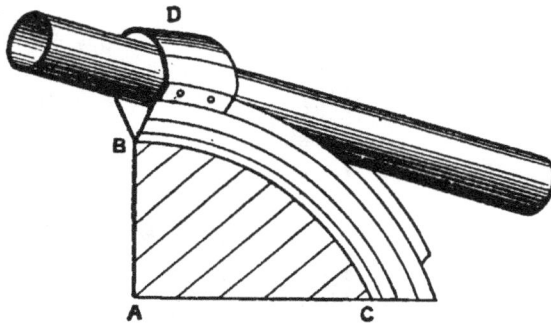

channel-piece answers best, for the reason that the flat sides of it give support to the tube, and help to keep its circular section and to prevent flattening in the throat.

In Figs. 4 to 9, appliances for bending stout tubes are represented. Fig. 4 shows a piece of hard wood with a hole through it, in which is a piece of pipe ready for bending. The hole is just large enough to admit the tube, and the edges of the hole on either side of the wood are rounded, so that in bending the tube there are no sharp edges in the wood to cause dents in the tube. In using the tool, the wood block is fixed in a vice.

Fig. 5 represents an appliance for bending pipes into coils.

A circular block of wood is cut away, part of it, so as to form a shouldered shank, as seen in the Fig., the shank being squared to fit a hole in a bench, or other convenient place. An iron staple is driven into the block, or an iron plate to which a staple has been fixed is screwed on to it. The illustration shows how to start a bend. In making coils, the diameter of the circular portion of the block is that of the hollow of the coil.

FIG. 4.

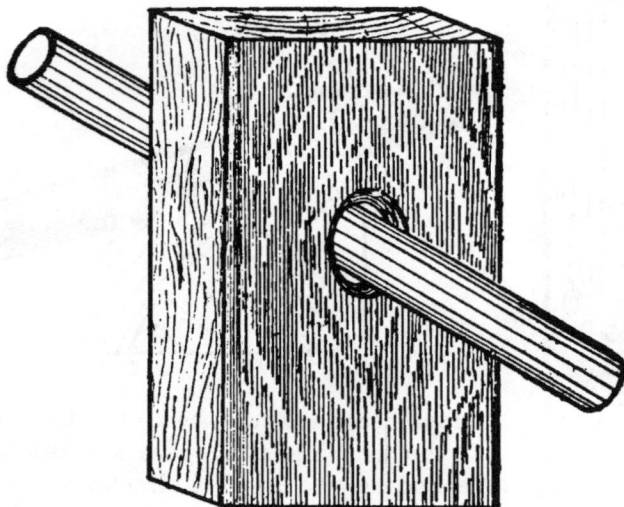

For making sharp bends in small tubes, the arrangement represented in Fig. 6 is employed. Two iron staples are driven into a stout wooden block, a small movable steel mandrel A A is passed through them, and fixed by wedges B B. The loaded tube is put under the mandrel and bent against it to the curve required.

When a number of pieces of tube of equal diameter, which have been bent to one curve, are required to be of true circular section, and to have their outside surfaces smooth and free from marks, moulds of cast brass are made use of, in which the tubes can be placed for the necessary manipulation. The moulds are in two pieces and alike in general

FIG. 5.

FIG. 6.

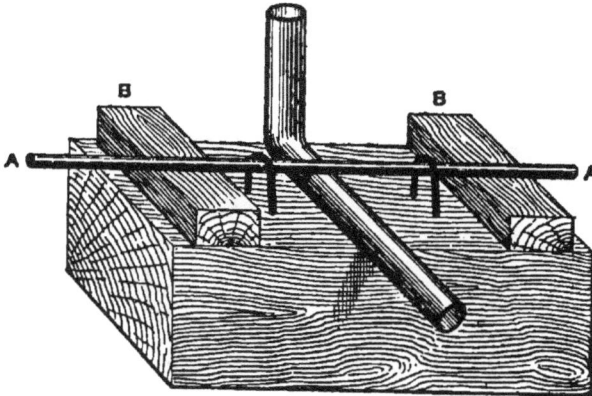

shape, and each piece is channelled; the channelling being such, that when the two pieces of a mould are put together

2 A

the channels form a hollow chamber which will hold one of
the pieces of bent tube. The grooves are made quite smooth,
and each groove of a mould is of the same section throughout,

FIGS. 7 and 8.

Sectional Elevation. Section on line A.B

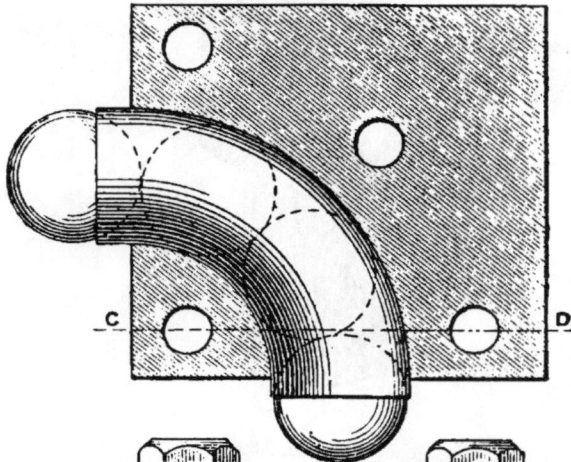

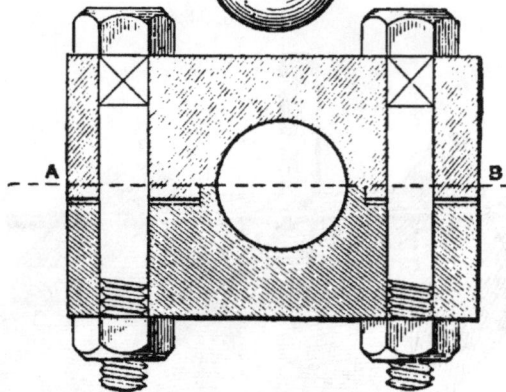

Cross Section on line C. D.

that section being a semi-circle. In the section (Fig. 8) of
two half-moulds bolted together, it will be seen that the
hollow formed by the two grooves is circular; the circle is

of the same diameter as the outside of the tube that is being worked. Fig. 7 is a section of the upper of the two half-moulds shown in Fig. 8, a section on the line A B; the mould is for a quadrant bend. The through-holes, shown in Fig. 7, are for the bolts by which the mould-halves are fixed together. In the Fig., a piece of tube is represented lying in position in the half-mould. The section, Fig. 8, is in the plane of the line C D of Fig. 7, which plane passes longitudinally through the middle of the bolts that go in the holes on the line C D.

The moulds are thus used. A piece of tube as rough-bent is placed in the groove of a half-mould, the other half-mould is then placed in position upon it, and the two pieces are firmly bolted together. Brass or steel balls (steel by prefer-ence) turned quite smooth and true, and of diameter equal to the inside of the tube being worked, are then forced through the tube. One ball having been placed in the tube, it is driven well in by light blows with the mallet. A second ball is placed in the tube on the first, and the mallet again used. In malleting the second ball into the tube, the first ball is driven still further in. A third ball is now applied and dealt with as the first two were, and so on to whatever number of balls are necessary to go right through the tube, the number of balls being determined by the length of the quadrant. A glance at Fig. 7, in which the balls are shown, will make the working quite clear. As the balls fit very closely, all the irregularities that the tube presents are forced outwards against the smooth sides of the mould, and the pressure of the substance of the tube by the balls against the mould groove being continuous, a perfectly smooth exterior surface is given to the tube as well as a true circular section. Should there have been any convexities outside the tube, these will have been already forced down when the tube was tightened up in the mould. The balls should be lightly dusted with black lead before using.

A semicircular bend is worked in like fashion. Fig. 10 is a

section of two half-moulds for such bend bolted up.   Fig. 9 is
a section of the upper of the two half-moulds shown in Fig. 10,
a section on the line A B.   The through-holes of Fig. 9 are

FIGS. 9 and 10.

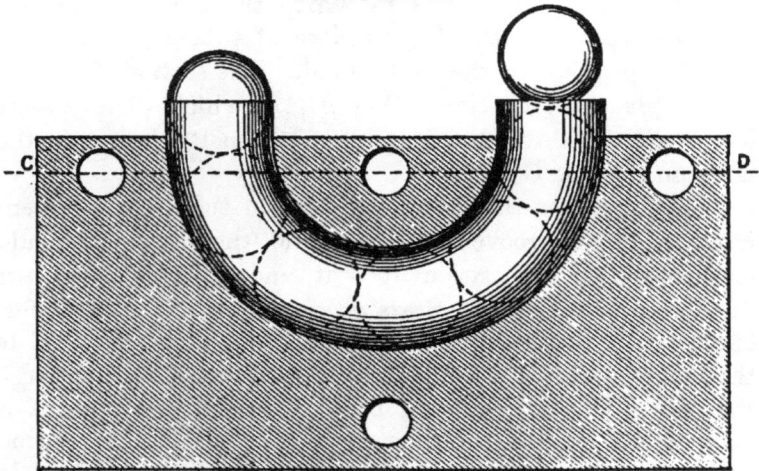

Sectional Elevation. Section on line A.B.

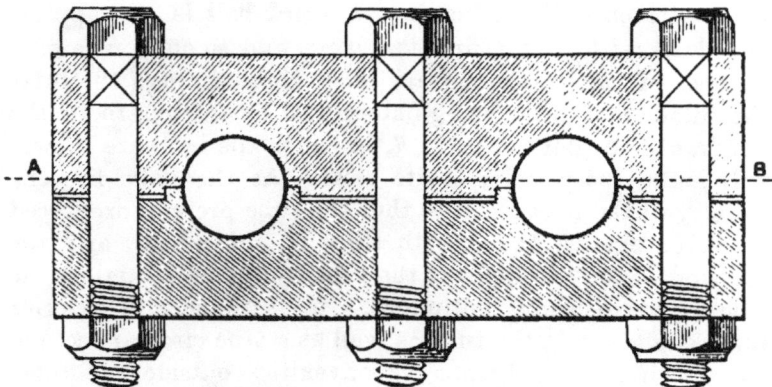

Cross Section on line C.D.

the bolt holes.   A semicircular piece of tube lying in the
groove of the half-mould is seen in the Fig.   The section,
Fig. 10, is in the plane of the line C D of Fig. 9, which plane

passes longitudinally through the middle of the bolts that go in the holes on the line C D. The manipulation of a piece of tube, rough-bent into a semicircle and placed in a mould such as represented, is exactly the same as has been explained for a quadrant tube.

Manifestly, a piece of the same tube can be bent into an arc greater than a semicircle with the semicircular mould. The rough bend, as required, being first made, and a semicircular piece of it having been smoothed, the bolts would be loosened and the tube shifted so as to bring a further portion of the rough bend into the hollow of the mould. This being then again bolted up, a driving through of the steel balls, like to that which has been described, will smooth out the further arc. Similarly, an arc greater than a quadrant, or a semicircular arc, or an arc greater than a semicircle, can be worked in the quadrant tool. An arc greater than a semicircle can also be manipulated in the semicircular tool; and with one or other of the tools, or with both. S curves, compounded or two quadrants of two semicircles, or of semicircle and quadrant, can be readily fashioned. And the grooves of the moulds made use of need not be of one radius, though they must always be of one section. An S, for example, could be made up of a large semicircle and a small one.

It may be said in respect of all the pipe bending that has been so far considered, that the result depends more on the tool than on the workman, or, at any rate, as much on the tool as on the workman. Bending *lead* pipe, however, is quite another matter, and has not been specially referred to. Here the result depends far more on the workman than on the tool, and the bending of lead pipe is an important part of the sheet-lead worker's (plumber's) training, and calls for its own tools. The general particulars of pipe bending that have been given will enable a workman in sheet-metal, other than lead, to bend a piece of lead pipe on occasion when a job he is engaged upon may require it, which is all that can fittingly be taught in a book such as this, which is for the most part geometrical.

Bends which are to be of such ' bore,' metal, and thickness, that they cannot, by any of the methods that have been described, be conveniently formed from pipes that correspond in diameter, metal, and thickness, have to be made up out of flat sheet-metal, in pieces, two or more.   Various problems in such bends now follow on.

CHAPTER II.

PATTERNS FOR PIPE-BENDS.

A SPECIAL prefatory word is necessary to these problems. Bodies of compound curvature, such as those now under consideration, will not develop into the flat; patterns in the flat for these, therefore, can only be approximate, and the sheet-metal after being cut to a pattern, must be worked into the desired shape by extension and alteration of surface, where required, with the hammer. It may be said of the patterns for such bodies as we are now considering, that they are geometrical compromises, and that the fewer the pattern-pieces for any one such body the greater the compromise. The direction of compromise in each particular instance is decided by experience. As a general rule, a pattern should be started if possible from a central line, and the hammer work left for the surfaces outside of this line. Especially should this course be followed if there is in the body for which pattern has to be made a central line which lies in a flat plane—in a plane, that is, like the surface of the metal plate on which the pattern is to be laid out; because then that line will not be disturbed (will be a neutral line) in the subsequent manipulation. The full meaning of these remarks will be more apparent as we proceed.

PROBLEM I.

To draw the pattern for a bend of equal circular section.

Six cases will be treated of—five in this problem, and one in the problem following.

And, the 'quadrant' bend being a very common one, and an extreme instance, it will be taken as typical. For any other length of arc, however, the working will be exactly similar.

CASE I.—The bend to be made up of two like pieces, each of them semicircular in section. That is, the seams of the bend are to follow the curve of it, one in the middle of the throat, the other in the middle of the back. Manifestly but one pattern-piece is required.

First draw a side elevation and a semicircular section of the half-bend for which pattern is required, as follows. Draw (Fig. 12) two lines Q 1″, Q 7″ at the same angle to each other as that which the given arc for the bend subtends: here an angle of 90°, that being the angle subtended by a 'quadrant.' On Q 1″ set off Q 1 equal to the given radius for the inner arc of the bend, and with Q as centre and Q 1 as radius describe the quadrant 1 7; this quadrant will be, in elevation, the inner curve or throat line of the half-bend. Now set off along Q 1″ from point 1 a distance 1 1″ equal to the diameter of the given bore of the bend, and with Q as centre and Q 1″ as radius describe the quadrant 1″ 7″; this quadrant will be, in elevation, the outer curve or back line of the half-bend; and the whole figure 1″ 7″ 7 1 will be a side elevation of it. Bisect 1″ 1 in A, and through A at right angles to Q 1″ draw a line A O. From point A on the under side of Q 1″ set off on the line just drawn a distance A O equal to A 1, through O draw a line d D parallel to Q 1″, and with O as centre and O A as radius describe the semicircle d A D; this will be a semicircular section of the half-bend. Now divide each of the quadrants A d, A D into any, the same, number of equal parts (here three) in the points b, c and B, C respectively, and with Q as centre and radius Q A describe the quadrant A P.

With what line should the laying-out of the pattern be started? Draw (Fig. 11) a front elevation A P 7″ 7 1 of the half bend, and continue the line A P of it indefinitely both

ways to X and X. The plane of the paper, X X X X (Fig. 12),
being supposed to be the flat surface of the sheet metal on

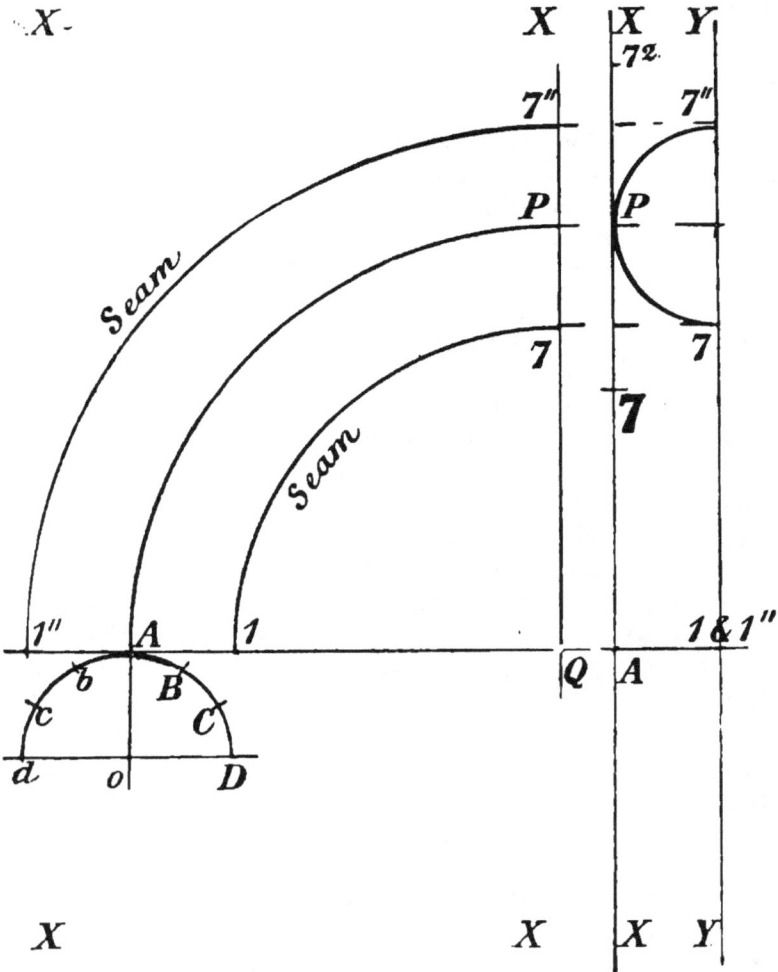

Fig. 12.   Fig. 11.

which the pattern is to be drawn, the line X X in Fig. 11
shows this surface edgeways, the line A P, part of the line

X X (that is, lying wholly in the plane X X), being the
quadrant line A P (Fig. 12). Now the line A P is the line
formed by the contact points of *all* the semicircular sections
of the half-bend ; and if the pattern for the half-bend is fully
drawn on the plane X X X X, Fig. 12 (the figure 1″ 7″ 7 1
in Fig. 12 is not pattern, but elevation, it must be remembered), this line A P of it remains on that plane right
through from the drawing of the pattern for the half-bend
to the completion of it by shaping up—is, in fact, a ' neutral '
line. The line being thus a constant both as to form and
position, and also being central, presents itself as the line
with which the drawing of the pattern should be started.
The other quadrant lines of the side elevation (the lines
7″ 1″, 7 1) are still quadrant lines in the finished half-bend,
but the lines for them that were drawn on the plane X X X X,
the flat surface of the sheet metal, *have not remained on that
plane*; they have in the shaping up been transferred to the
plane of division of the bend, to the plane that, appearing
edgeways in Fig. 11, shows there as the line Y Y. To make
this quite clear. The extreme points 7², **7** (Fig. 11) of the
line 7² P **7**, which is the *pattern* line for the 7″ P 7 end
(Fig. 12) of the half-bend, are also the extreme points 7″, 7
of, respectively, the quadrants 7″ 1″, 7 1 ; and as this line
becomes, after the shaping up of the pattern, the curve
7″ P 7 (Fig. 11), it is evident that the points 7², **7** on the
X X plane have been transferred to the plane Y Y. And
so with all the rest of each of these two quadrants. The
quadrant of which P is an extremity has alone remained
in the X X plane.

To draw the pattern for the half-bend.

Draw (Fig. 13) at right angles to each other two lines
Q 1″, Q 7″. From Q set off Q A equal to Q A (Fig. 12), and
with Q as centre and radius Q A describe the quadrant A P.
This line, as we just above found, is the neutral line with
which to start the pattern. Bisect A P in point *a* ; join Q *a*
and produce it ; and from *a* towards Q set off one after another
the distances *a* B, B C, C D equal respectively to the distances

A B, B C, C D (Fig. 12), and, away from Q, also successively, the distances *a b, b c, c d* equal respectively to the distances A *b, b c, c d* (Fig. 12). Through point *d* with radius Q *d* describe the arc 1″ 7″ terminating in the points 1″ and 7″ in Q 1″ and Q 7″ respectively, and through point D with Q D as

Fig. 13.

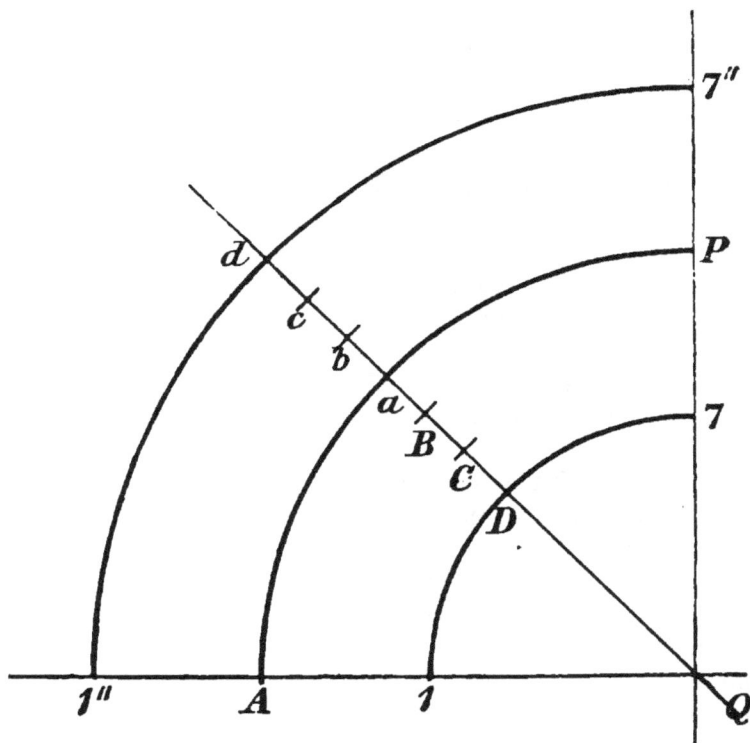

radius describe the arc 1 7 terminating in the points 1 and 7 in Q 1″ and Q 7 respectively; then 1″ *d* 7″ 7 D 1 will be the pattern required. Bent up, stretched and hollowed judiciously with the hammer, and trimmed, it will shape into the half-bend of the elevation and section.

Bends thus made cannot be recommended where any con-

siderable internal pressure is to be resisted ; as, on account
of the great amount of stretching required at the throat of
the bend, the metal becomes thin and weak there.   In a few
small sizes, half-bends stamped up are now commercial
articles, two stampings making up the complete bend.

CASE II.—The bend is to be made up of two pieces, with the
    seams following the curve of the bend, centrally on its
    opposite sides.   Two pattern-pieces will be required.

Draw (Fig. 14) two lines Q 1″, Q 7″ at right angles to
each other (the bend being quadrant) and intersecting in Q.
On Q 1″ set off Q 1 equal to the given radius of the bend, and
with Q as centre and Q 1 as radius describe the quadrant 17 ;
this quadrant will be in elevation the throat of the bend.
New set off along Q 1″ from point 1 a distance 11″ equal to
the diameter of the required bore of the bend, and with Q as
centre and Q 1″ as radius, describe the quadrant 1″ 7″; this
quadrant will be in elevation the outer curve of the bend ;
the figure 1″ 7″ 7 1 will be a side elevation of the bend.
Next draw through the points 1″, 1 and perpendicular to
Q 1″, two lines 1″d, 1 D; through any point d in 1″d draw
d D parallel to Q 1″ ; with O the mid-point of d D as centre,
and O d as radius describe the circle d a D g; and through
O draw a line g A perpendicular to Q 1″, cutting the circle
just described at g and a and meeting Q 1″ at A.   The circle
a d g D will be a circular section, anywhere, of the bend ; the
semicircle a D g will be a section anywhere of its throat half,
and the semicircle a d g of its back or ‘ saddle’ half.   Now with
Q as centre and Q A as radius describe the quadrant A P ;
this quadrant will be the seam of the bend, in elevation ;
divide the quadrant 1 7 into any number of equal parts (here
six) in the points 2, 3, 4, 5, 6 ; and from Q draw lines through
these points (the lines are only partly drawn in the Fig.),
meeting the quadrant A P in the points 2′, 3′, 4′, 5′, 6′, and
the quadrant 1″ 7″ in the points 2″, 3″, 4″, 5″, 6″ ; which will
be a division of the quadrants A P and 1″ 7″, each of them,
into also six equal parts.   Further, divide the arcs a d, d g,

g D, D *a*, each of which is one quarter of the circumference of the circular section of the bend, into, each of them, any (the same) number of equal parts (here 3) in the points *b* and *c* for the arc *a d*, and the points *e* and *f*, F and E, C and B for the arcs *d g*, *g* D, and D *a* respectively.

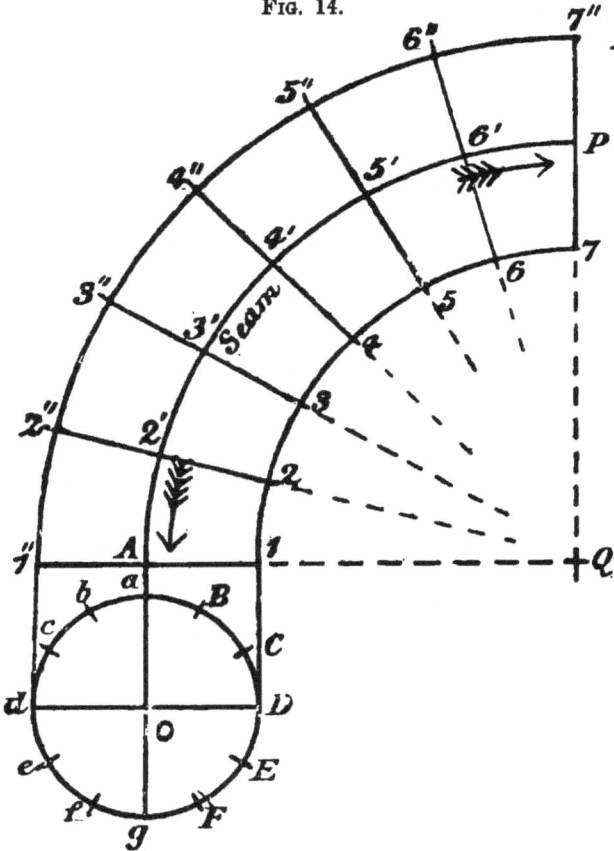

FIG. 14.

One pattern-piece will be required for the throat portion A 4′ P 7 4 1 (*a* D *g* in section) of the bend, and another for the saddle portion A 4′ P 7″ 4″ 1″ (*a d g* in section) of it.

To draw the pattern for the A 4′ P 7 4 1 portion.

The line to commence it upon is easily decided. If the finished throat portion of the bend were laid on a flat surface, it would rest with its A 4′ P edge (or opposite corresponding edge) on that surface. But either of these edge lines is an extremity line, and quite unsuited, therefore, to start a pattern with. And there is no line whatever that can be drawn in the flat for this throat half pattern that would lie against a surface in the shaped up throat piece; no line that would be 'neutral,' that is. This being the case, manifestly the next best thing to do is to start the pattern with a line of single curvature only, that is a line which, in making up from the flat to the half-round, bends wholly in one plane. The only line which answers to this condition is the central line 1 4 7; with this line then, developed of course into a straight line, we will begin the required pattern.

Draw (Fig. 15) an indefinite straight line 1 7; through any point 4 in it draw at right angles a line 4′ 4′. From point 4 to the left of the line 4′ 4′ set off successively the distances 4 3, 3 2, 2 1 equal respectively to the distances 4 3, 3 2, 2 1 (Fig. 14), and to the right of 4′ 4′ the distances 4 5, 5 6, 6 7 equal respectively to the distances 4 5, 5 6, 6 7 (Fig. 14); the straight line 1 7 (Fig. 15) will then be equal in length to the quadrant curve 1 7 of Fig. 14. Now from the point 4 above the line 1 7, set off one after another the distances 4 C, C B, B 4′ equal respectively to the distances D C, C B, B a (Fig. 14), thus making the line 4 C B 4′ equal in length to the arc D a (Fig. 14); and, below 17 the distances 4 E, E F, F 4′ equal respectively to the distances D E, E F, F g (Fig. 14), thus making the line 4 E F 4′ equal in length to the arc D g (Fig. 14). The whole line 4′ 4′ will then be equal in length to the semicircle a D g, which is the semicircular section of the throat half of the bend. Through each point 4′ draw indefinite lines parallel to 1 7; from the point 4′ on the B C side of 17 set off one after another distances 4′ 3′, 3′ 2′, 2′ A equal respectively to the distances 4′ 3′, 3 2′, 2′ A (Fig. 14), and, also successively, the distances 4′ 5′, 5′ 6′, 6′ P equal respectively to the distances 4′ 5′, 5′ 6′, 6′ P (Fig. 14); and from the point 4' on the E F side of 1 7, set off successively distances 4' 3', 3' 2', 2' A' equal respectively to the distances 4' 3, 3' 2', 2 A' (Fig. 14), and the distances, also one after another, 4' 5', 5' 6', 6' P' equal respectively to the distances 4' 5', 5' 6', 6' P (Fig. 14). The straight lines A P, A' P' of Fig. 15 will then be each equal in length to the curve line A P (Fig. 14), which is the line of the seam of the bend.

Fig. 15.

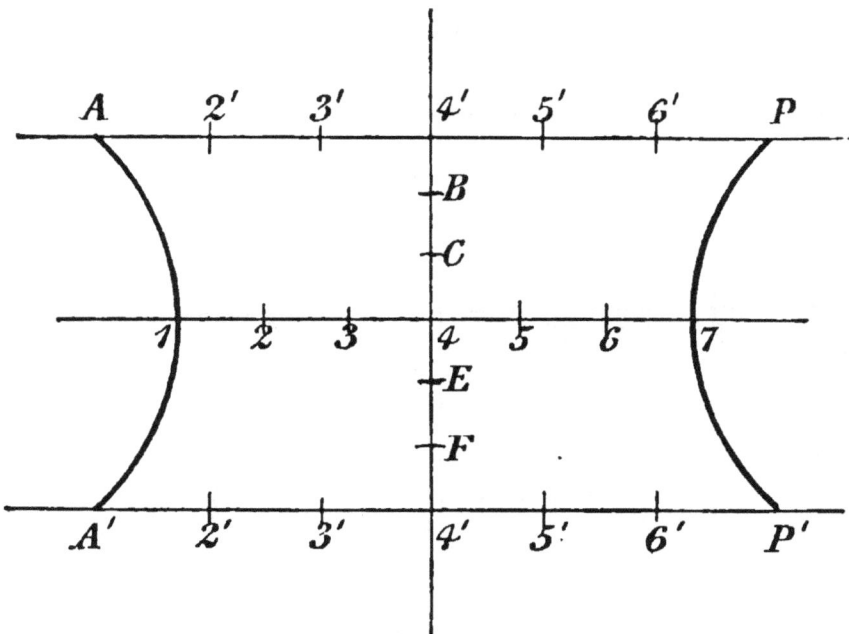

From A to A' draw an arc passing through the point 1, and from P to P' draw an arc passing through the point 7. The figure A 4' P 7 P' 4' A' 1 will be the pattern required. Bent up (the bending of the line 1 4 7 is of single curvature only), got into form by stretching where needed with the hammer, and trimmed, it will shape into the throat piece for the quadrant bend.

To draw the pattern for the A 4' P 7" 4" 1" (Fig. 14) portion of the bend. Just as for the throat-half we commenced the pattern with the central line 1 4 7 developed, we start the saddle-half pattern with the line 1" 4" 7" developed.

Draw (Fig. 16) an indefinite straight line 1" 7". On it, from any point 4", set off successively to the left of 4" the distances 4" 3", 3" 2", 2" 1" equal respectively to the distances 4" 3", 3" 2", 2" 1" (Fig. 14), and to the right of 4" successively the distances 4" 5", 5" 6", 6" 7" equal respec-

FIG. 16.

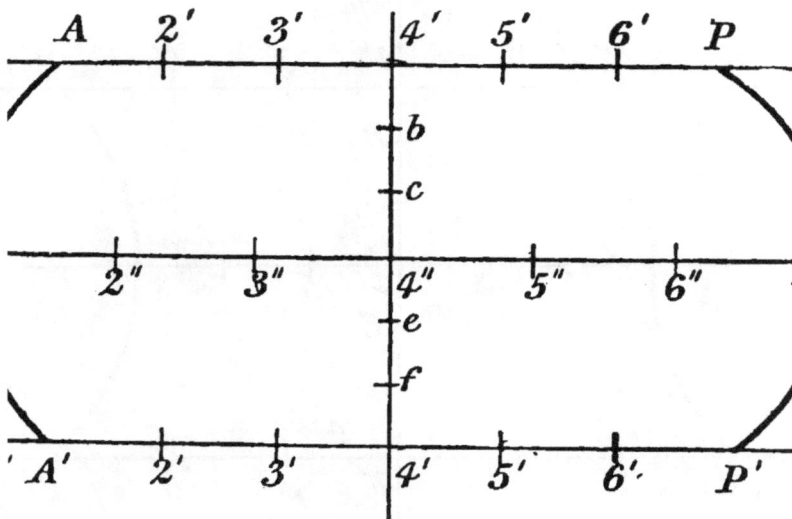

tively to the distances 4" 5", 5" 6", 6" 7" (Fig. 14); the whole line 1" 7" will then be equal to the quadrant curve 1" 7" (Fig. 14). Through point 4" draw a line 4' 4' perpendicular to 1" 7", and from the point 4" set off one after another the distances 4" c, c b, b 4' equal respectively to the distances d c, c b, b a (Fig. 14), thus making the line 4" c b 4' equal to the arc d a (Fig. 14); and, also successively, the distances 4" e, e f, f 4' equal respectively to the distances d e, e f, f g (Fig. 14), thus making the line 4" e f 4' equal to the arc d g

(Fig. 14). The whole line 4' 4' will then be equal in length to the semicircle $a\,d\,g$, which is the semicircular section of the back half of the bend. Through each point 4' draw indefinite lines parallel to 1" 7"; from the point 4' on the $b\,c$ side of 1" 7" set off one after another distances 4' 3', 3' 2', 2' A equal respectively to the distances 4' 3', 3' 2', 2' A (Fig. 14), and, also successively, the distances 4' 5', 5' 6', 6' P equal respectively to the distances 4' 5', 5' 6', 6' P (Fig. 14); and from the point 4' on the $e\,f$ side of 1" 7" set off successively distances 4' 3', 3' 2', 2' A' equal respectively to the distances 4' 3', 3' 2', 2' A (Fig. 14), and the distances, also one after another, 4' 5', 5' 6', 6' P' equal respectively to the distances 4' 5', 5' 6', 6' P (Fig. 14). The straight lines A P, A' P' of Fig. 16 will then be each equal in length to the curve line A P (Fig. 14), which is the line of the seam of the bend. From A to A' draw an arc passing through the point 1", and from P to P' draw an arc passing through the point 7". The figure A 4' P 7" P' 4' A' 1" will be the pattern required. Bent up (the bending of the line 1" 4" 7" is of single curvature only), got into form where needed with the hammer, and trimmed, it will shape into the saddle-piece for the quadrant bend.

Making thus, in two pieces, is of advantage when the seams have to be brazed, as the pieces when lying on a flat surface will butt up against each other.

CASE III.—The bend to be made up of four pieces; the seams, all of them, to follow the curve of the bend; one to be in the middle of the throat, one in the middle of the back, and the other two central on the opposite sides of the bend. Two pattern-pieces will be required, one for each half-throat of the bend and the other for each half-back of it.

This Case is a sort of combination of the two previous Cases, inasmuch as the seams come in throat, back *and* sides. Patterns for the pieces of which the bend is to be made up might be obtained either by dividing the half-bend

2 B

pattern of Case I. into two portions, or by halving each of
the patterns of Case II. Choice of which course to take
may at once be come to. In Case II. the patterns start on a

FIG. 17.

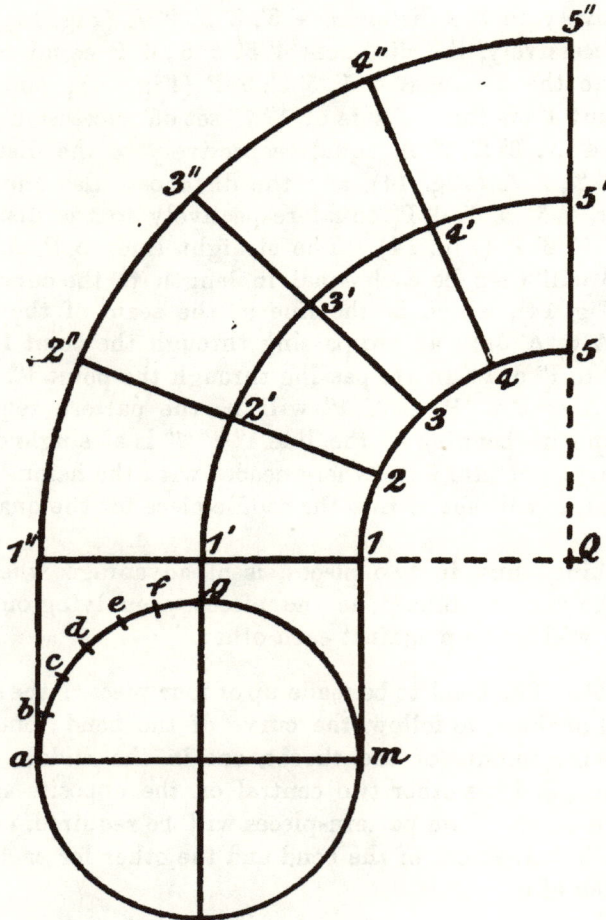

line which, though of only single curvature in the making
up of the bend, nevertheless *is* a line of curvature; in Case I.
they start upon a line which in the making up does not

undergo curvature at all; in Case II. there is no such line. From this it is evident that we had better work from the Case I. neutral line. Making-up bends in four pieces is necessary for large work principally, and it is desirable with such work to construct the half-throat pattern in the way that is shown just below. Let us now deal with the problem.

Draw (Fig. 17) a side elevation 1″ 5″ 5 1 of the bend and a circular section *a g m* of it, in the manner described in Case II. Also draw the central quadrant 1′ 5′; divide the arc 1 5

Fig. 18.

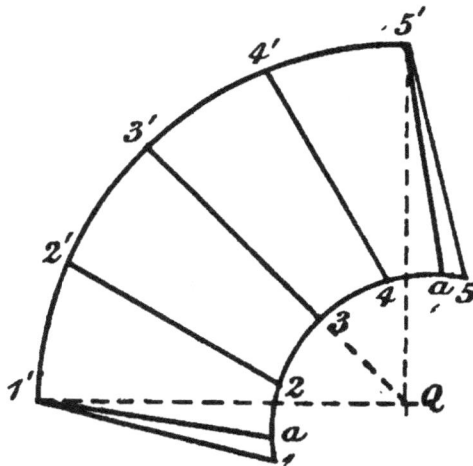

into any number of equal parts (here four) in the points 2, 3, 4; and draw lines from Q through these division points, cutting the central quadrant in the points 2′, 3′, 4′ respectively, and the outer curve of the elevation in the points 2″, 3″, 4″ respectively. Also divide the quadrant *a g* of the circular section into any number of equal parts (here six) in the points *b, c, d, e* and *f*.

To draw the pattern for the half-throat piece 1′ 3′ 5′ 5 3 1.

Draw (Fig. 18) two lines Q 1′, Q 5′ perpendicular to each other and intersecting in Q. With Q as centre and Q 1′

2 B 2

(Fig. 17) as radius, describe an arc 1' 3' 5' cutting Q 1', Q 5' in points 1' and 5' respectively. With this quadrant arc, equal to the quadrant 1' 3' 5' (Fig.17) we start the pattern. First divide the curve 1' 3' 5' into the same number of equal parts that the curve 1' 3' 5' (Fig. 17) is divided into, in the points 2', 3' and 4'. Now join the middle point of the quadrant, (in this

FIG. 19.

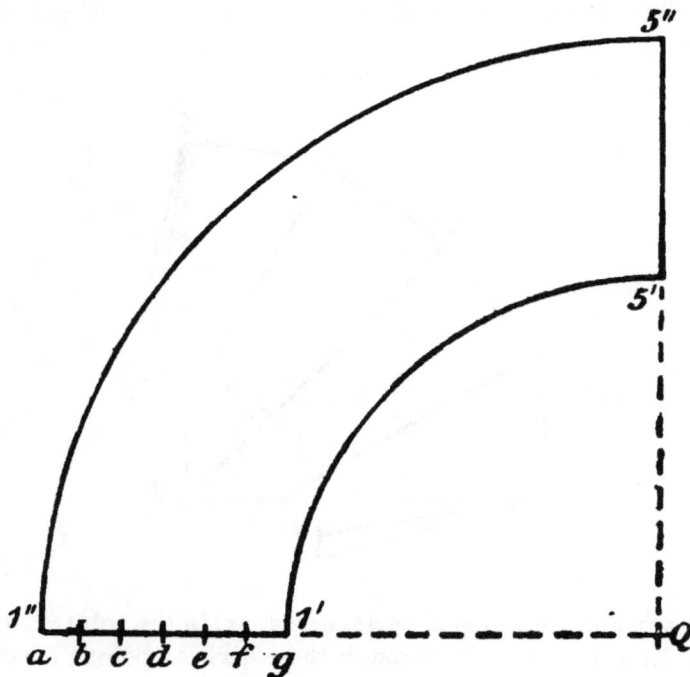

instance the mid-point coincides with the point 3', the arc having been divided into an even number of parts), with the centre Q, and from point 3' along 3' Q set off a distance 3' 3 equal to the length of the quadrant arc a g (Fig 17). Now with 4' and 3 as centres and radii respectively the quadrant-length a g and the distance 3 4 (Fig. 17), describe arcs inter-secting in point 4; and with 5' and 4 as centres and radii respectively the quadrant-length $a\,g$ and the distance 4 5 (Fig. 17) describe arcs intersecting in point 5. Working in exactly like manner mark points 2 and 1. Through the points 1, 2, 3, 4 and 5 draw an unbroken curved line, and join 4′ 4, 5′ 5, 2′ 2 and 1′ 1; we then have as pattern the figure 1′ 3′ 5′ 5 3 1. In practice, however, because of the stretching of the metal and consequent gain of surface coming of

FIG. 20.

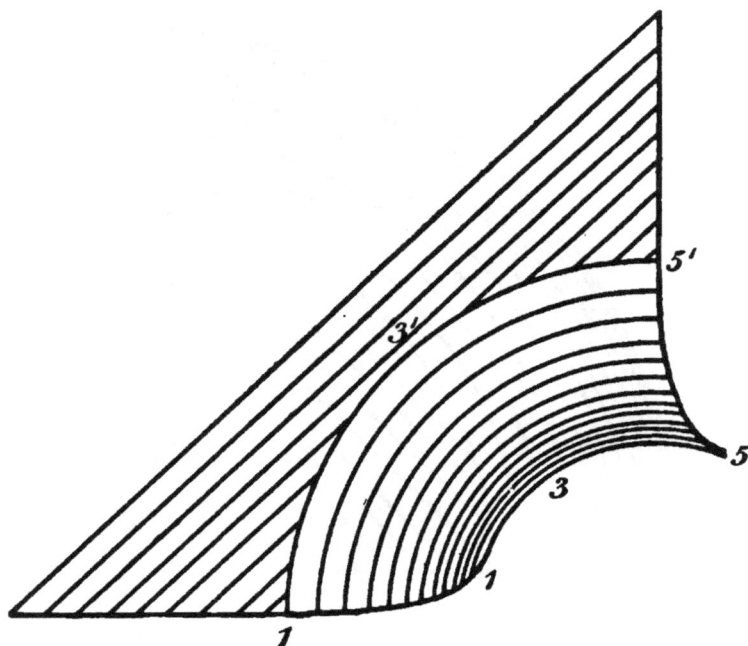

the working-in of the throat curve, it is necessary to cut off pieces 1 1′ a, 5 5′ a at the ends of the pattern, which thus becomes 1′ 3′ 5′ a 3 a. No rule can be laid down for marking these pieces; so much depends upon the individual workman, on experience, and on the metal that is being worked.

To draw the pattern for the 1″ 3″ 5″ 5′ 3′ 1′ portion of the bend.

Draw (Fig. 19) two indefinite straight lines Q 1″, Q 5″, and with Q as centre and radius Q 1′ (Fig. 17) describe the quadrant 5′ 1′ cutting Q 1″ and Q 5″ in the point 1′ and 5′ respectively.   From the point 1′ set off successively the distances 1′ f, f e, e d, d c, c b, b 1″ equal to the distances g f, f e, e d, d c, c b, b a (Fig. 17), and with Q as centre and radius Q 1″ describe the quadrant 5″ 1″; then the figure 1″ 5″ 5′ 1′ will be the pattern required.

FIG. 21.

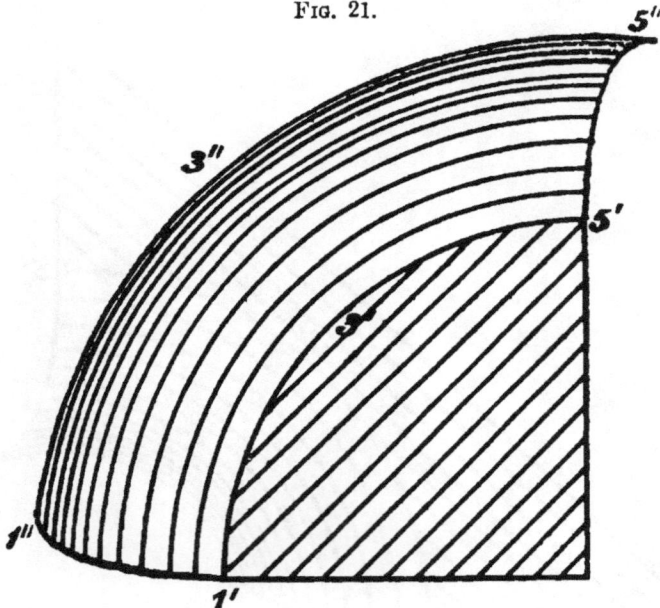

In Fig. 20 the half-throat pattern piece (Fig. 18) is represented as worked into shape.   A portion of the flat sheet on which the pattern is supposed drawn is shown as left attached, to illustrate that the 1′ 3′ 5′ curve of the pattern (Fig 18) is still intact when the shaping up by the hammer is completed.   In Fig. 21 the half-back pattern piece (Fig. 19) is represented with a portion of the flat metal sheet left attached, and similarly here the same curve, the 1′ 3′ 5′ curve, remains intact.

CASE IV.—The bend to be made in four pieces; two of these being part-sides and alike, the third being a piece for the throat, and the fourth a piece for the back; the seams coming two on each side of the bend.

Draw (Fig. 22) two lines Q 1″, Q 7″ at right angles to each other and intersecting in Q. With Q as centre and radius

FIG. 22.

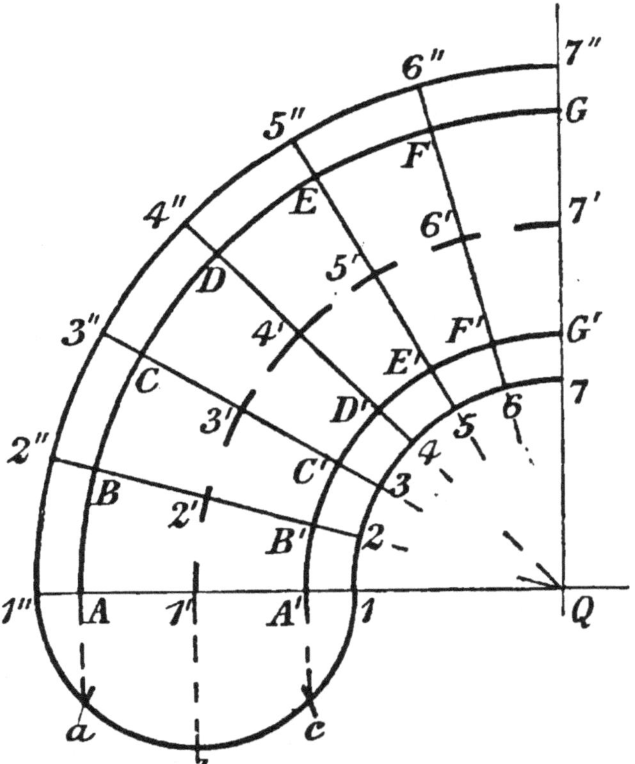

Q 1 equal to that of the given curve for the throat of the bend describe the quadrant 1 7. Along Q 1″ from point 1, set off a distance 1 1″ equal to the diameter of the circular section of the bend, and with Q as centre and Q 1″ as radius describe the quadrant 1″ 7″; then the figure 1″ 7″ 7 1 will be a side elevation of the bend. Bisect 1 1″ in 1′; from 1′ draw a line 1′ b perpendicular to Q 1″; and with 1′ as centre and radius 1⌐1 describe the semicircle 1″ b 1 cutting the line 1′ b in b. Now divide each of the quadrants 1″ b, 1 b into two equal parts in the points a and c respectively, and from a and

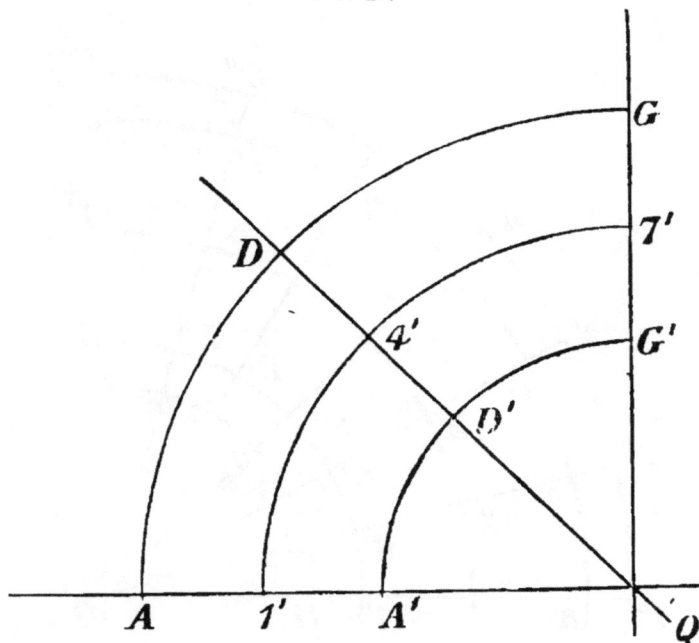

Fig. 23.

c draw a A, c A′ perpendicular to Q 1″ and cutting it in the points A and A′ respectively. With Q as centre and radius successively Q A, Q 1′, Q A′ describe the quadrants A G, 1′ 7′, A′ G′; the quadrants A G, A′ G′ will represent the seams of the bend. Divide the quadrant 1″ 7″ into any convenient number of equal parts (here six) in the points 2″, 3″, 4″, 5″, 6″, and join each of these points with Q, the line 2″ Q intersecting the quadrants A G, 1′ 7′, A′ G′ in the points B, 2′, and B′

respectively, and the lines 3″ Q, 4″ Q, 5″ Q, and 6 Q, intersecting these same quadrants similarly and giving the points
C, 3′, C′; D, 4′, D′,¡E, 5′, E′; and F, 6′, F′. The four pieces
for the bend will be as follows:—two part-side pieces alike,
represented each, in elevation, by the figure A D G G′ D′ A′,
and in transverse section by the line *a b c*; a throat piece, in
side elevation A′ D′ G′ 7 4 1, but of course, in actuality, double
that, extending, that is, from the seam A′ D′ G′ to the seam
on the opposite side of the bend, that corresponds with the
seam A′ D′ G′, (represented also in half-section by the line *c* 1);
and a back-piece, in half-section *a* 1″, and in side elevation
A D G 7″ 4″ 1″.

Fig. 24.

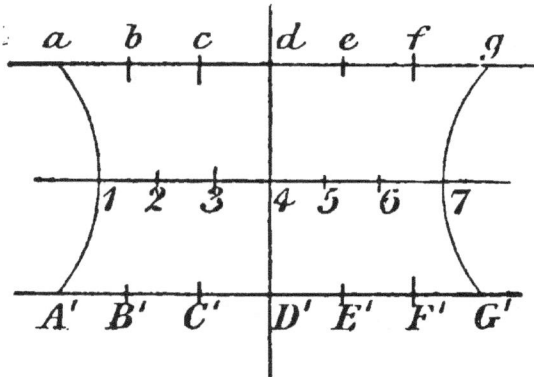

To draw the pattern for the side-piece.

Draw (Fig. 23) two lines Q G and Q A perpendicular to each
other and intersecting in Q. From A along Q A set off a
distance Q 1′ equal to Q 1′ (Fig. 22), and with Q 1′ as radius
describe the quadrant 1′ 4′ 7′. Also from 1′ set off distances
1′ A′, 1′ A equal respectively to the distances *b c*, *b a*,
(Fig. 22), with Q A′ as radius describe the quadrant A′ D′ G′,
and Q A as radius describe the quadrant A D G; the figure
A D G G′ D′ A′ will be the pattern required.

To draw the pattern for the throat piece.

Draw (Fig. 24) an indefinite straight line 1 7. From a point

4 about central in it set off to the left distances 4 3, 3 2, 2 1 equal
respectively to the distances 4 3, 3 2, 2 1 (Fig. 22), and, to the
right, distances 4 5, 5 6, 6 7 equal respectively to the distances
4 5, 5 6, 6 7 (Fig. 22).   Through point 4 draw an indefinite line
perpendicular to 1 7 ; above it set off 4 d and below it 4 D′
equal each to the curve distance 1 c (Fig. 22) ; and through d
and D′ draw indefinite straight lines parallel to 17.   Now to
the left of d D′ set off from d the distances d c, c b, b a, and from
D′ the distances D′ C′, C′ B′, B′ A′ equal, each set, to the distances D′ C′, C′ B′, B′ A′ (Fig. 22) ; and to the right of d D′ set

FIG. 25.

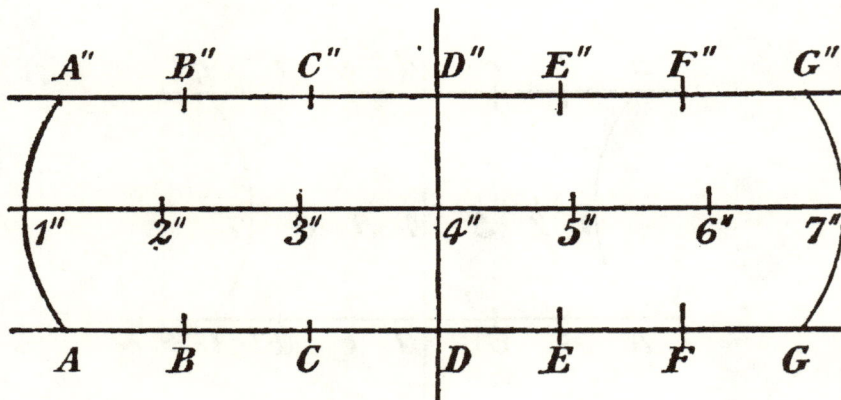

off from d distances d e, e f, f g and from D′ distances D′ E′,
E′ F′, F′ G′ equal respectively to the distances D′ E′, E′ F′, F′ G′
(Fig. 22).   Join the points a 1 A′ by an unbroken curved
line, also the points g 7 G′ ; the figure a g 7 G′ A′ 1 will be the
pattern required.

   To draw the pattern for the back-piece.

   Draw (Fig. 25) an indefinite straight line 1″ 7″.   From a
point 4″ about central in it set off to the left distances 4″ 3″,
3″ 2″, 2″ 1″ equal respectively to the distances 4″ 3″, 3″ 2″, 2″ 1″
(Fig. 22) and, to the right distances 4″ 5″, 5″ 6″, 6″ 7″ equal
respectively to the distances 4″ 5″, 5″ 6″, 6″ 7″ (Fig. 22).

Through point 4″ draw an indefinite line perpendicular to 1″ 7″; above it set off 4″ D″ and below it 4″ D equal each to the curve distance 1″ a (Fig. 22); and through D″ and D draw indefinite straight lines parallel to 1″ 7″. Now to the left of D″ D set off from D″ the distances D″ C″, C″ B″, B″ A″, and from D the distances D C, C B, B A equal, each set, to the distances D C, C B, B A (Fig, 22); and to the right of D″ D from D″ the distances D″ E″, E″ F″, F″ G″, and from D the distances D E, E F, F G, equal, each set, to the distances D E, E F, F G (Fig. 22). Join the points A″ 1″ A by an unbroken curved line; also the points G″ 7″ G; the figure A″ G″ 7″ G A 1″ will be the pattern required.

CASE V.—The bend to be made up of circular segments.

Draw (Fig. 26) two indefinite lines Q a′, Q 7′ at right angles to each other and intersecting in Q. With Q as centre and radius equal to that of the given quadrant for the throat of the bend describe the quadrant g′ 7 meeting the lines Q a′, and Q 7′ in the points g′ and 7 respectively. From g′ set off a distance g′ a′ equal to the given diamet r of the circular section of the bend, and with Q as centre and Q a′ as radius describe a quadrant a′ 7′ meeting Q a′ and Q 7′ in a′ and 7′ respectively; then g′ a′ 7′ 7 will be a side elevation of the bend. Now divide the quadrant a′ 7′ into as many equal parts as the bend is to have segments, here six; the figure g′ a′ a″ g″ shows one of the segments, the remaining segments are indicated by lines, not lettered. To find the pattern necessary for one of the circular segments is to find the pattern for all.

To find the segment pattern.

On the line g′ a′ describe the semicircle g′ d a′, which divide into any convenient number of equal parts (here six) in the points, f, e, d, c and b, and from each of these points draw lines perpendicular to g′ a′ and intersecting it in the points f′, e′, d′, c′ and b′. Bisect the arc a′ a″ and join the point of intersection with Q: also draw a′ a″ the chord of the arc a′ a″, and from the points b′, c′, d′, e′, f′ and

g' draw lines parallel to the line $a'\,a''$ meeting the line $a''\,g''$ in the points b'', c'', d'', e'', f'' and g'', and intersecting the line drawn from the bisection point of the arc $a'\,a''$ to Q, in the points A, B, C, D, E, F and G respectively.

FIG. 26.

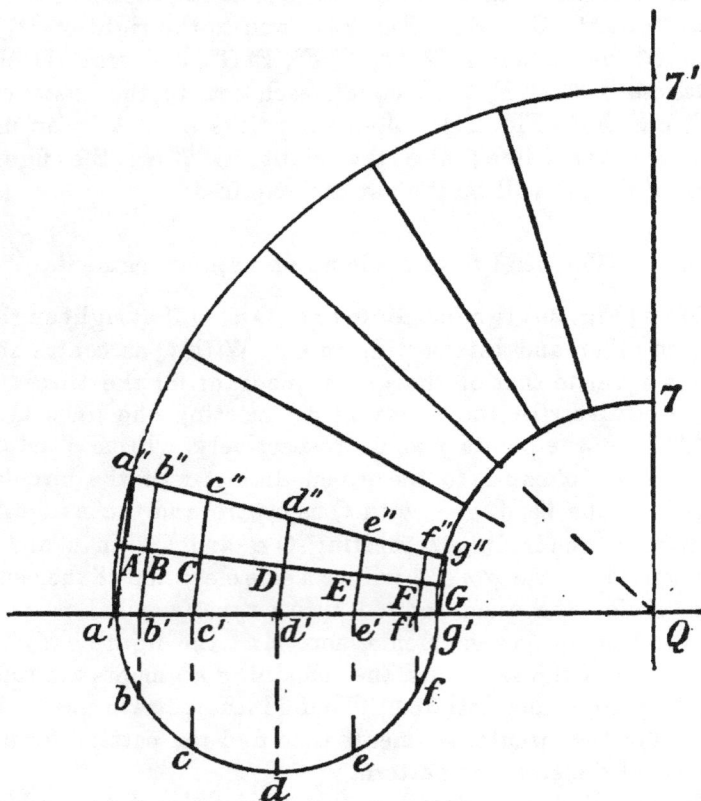

Next (Fig. 27) draw G A an indefinite straight line, and from any point G in it set off distances G F, F E, E D, D C, CB, B A equal to the distances $g'f$, fe, ed, dc, cb and ba' (Fig. 26) respectively. Through the point G draw a line $g'\,g''$ perpendicular to G A, and through the points F, E, D, C, B and A draw lines parallel to $g'\,g''$. Make G g', G g'' equal each to

G g' (or G g'') (Fig. 26), and similarly make F f' and F f'', E e' and E e'', D d' and D d'', C c' and C c'', B b' and B b'', A a' and A a'', equal respectively, pair for pair, to F f' (or F f''), E e', D d', C c', B b', and A a' (Fig. 26). From the point g' through f', e', d', c' and b' to a' draw an unbroken curved line; also from g''

FIG. 27.

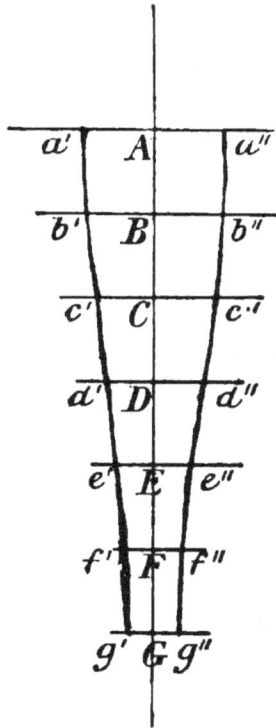

through f'', e'', d'', c'' and b'' to a''; the figure g' a' a'' g'' will be the pattern for the half of one of the circular segments of the bend. The like work repeated, but reversed in order, on the other side of the line a' a'', will give, in the entire figure so completed, the pattern for a whole segment of the bend.

PROBLEM II.

To draw the pattern for a lobster-back cowl.

Draw an indefinite line A O (Fig. 28), and from any point
O in it draw a perpendicular O T. With O as centre and
radius equal to the diameter of cowl required, describe an
arc A A′ T, cutting O T in T. Produce O T downwards, and
make O Q equal to the depth of the rim required; through
Q draw Q P, parallel to A O, and from A draw A P perpendicular to A O and cutting Q P in P; A O Q P is the elevation
of the rim. To draw the hood T S R O, draw T S perpendicular to O T, make T S and O R of the required dimensions
for the hood, and join S R. This completes the elevation
P A′ S R O Q of the cowl. Now divide the arc A T into the
same number of equal parts that the shell of the lobster-back
is to have segments. Here we take the number of segments
as three; A′ and G are the points of division of the arc.
Joining these points to O we have in G O T, A′ O G, and
A O A′ the elevations of the segments of the shell.

To draw the pattern for the segments.

On either of the lines A O, A′ O, G O, T O describe a semicircle (it is here described on A O), and divide it into any
convenient number of equal parts (here six) in the points
b, c, d, e and *f*. Through *b, c, d, e* and *f* draw lines perpendicular to A O, and cutting it in points B, C, D, E and F
respectively. With O as centre and radii O B, O C, O D,
O E and O F successively, describe arcs B B′ H, C C′ K, D D′ L,
E E′ M and F F′ N, cutting A′ O in the points B′, C′, D′, E′
and F′ respectively, and T O in the points H, K, L, M and
N respectively. We now find the pattern for the segment
A O A′; the pattern for that segment, the segments being all
equal, will be the pattern for all.

Bisect the arc A A′ in *k* and join *k* O. Join A A′, B B′, C C′,
D D′, E E′ and F F′ by straight lines, cutting *k* O in points
a′, b′, c′, d′, e′ and *f′* respectively. Next draw an indefinite
line *a′* O (Fig. 29), and on it set off *a′ b′, b′ c′, c′ d′, d′ e′, e′ f′*

and f' O, each equal to A b (Fig. 28), one of the equal parts
into which the semicircle A d O is divided, and through each
of the points a', b', c', d', e' and f' draw lines perpendicular to
a' O.   Make a' A', b' B', c' C', d' D', e' E' and f' F' respectively
equal to a' A', b' B', c' C', d' D', e' E' and f' F' (Fig. 28); also
make a' A, b' B, c' C, d' D, e' E and f'F respectively equal to

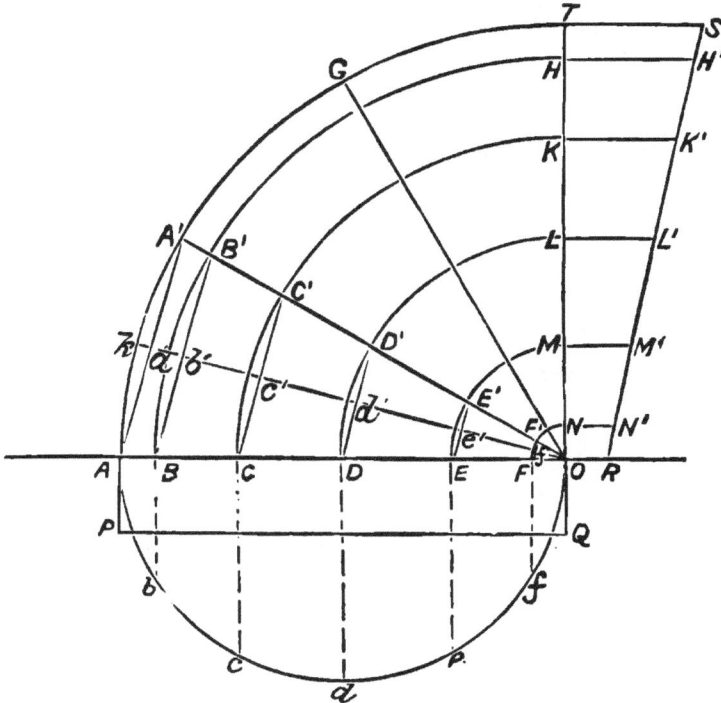

Fig. 28.

a' A, b' B, c' C, d' D, e' E, and f' F (Fig. 28).   Join A, B, C, D,
E, F and O, and also A', B', C', D', E', F' and O', by unbroken
curved lines, then A O A' will be half the pattern required,
which can be completed by drawing on A A' a figure exactly
the same as A O A'.

To draw the pattern for the hood T S R O.

Draw (Fig. 30) an indefinite line T O, and on it set off

T H, H K, K L, L M, M N and N O, each equal to A b (Fig. 28), one of the equal parts of semicircle A d O. Through each of the points T, H, K, L, M, N and O draw lines perpendicular to T O; and make T S, H H′, K K′, L L′, M M′, N N′ and O R respectively equal to T S, H H′, K K′, L L′,

FIG. 29.

FIG. 30.

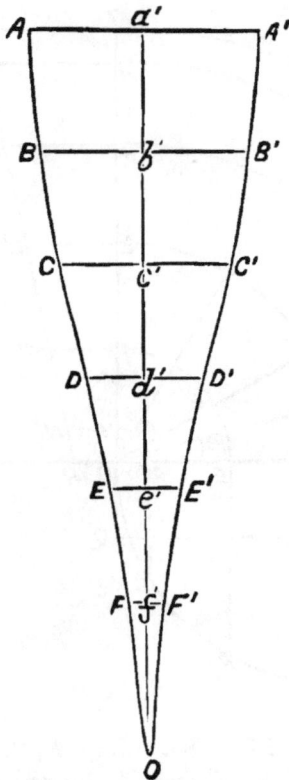

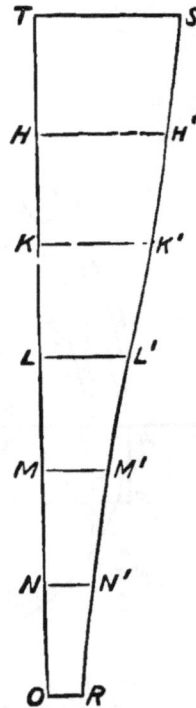

M M′, N N′ and O R (Fig. 28). Join S, H′, K′, L′, M′, N′ and R by an unbroken curved line; T S R O will be the pattern for one half of the hood; duplicating this on the line T S will complete the pattern.

If closer approximation is necessary the lines a' O (Fig. 29)

and T O (Fig. 30) should each of them be made equal to half
the circumfereuce of a circle having A O (Fig. 28) for its
diameter, and then divided into as many equal parts as semicircle A d O (Fig. 28) is divided into. In practice, for convenience in making up, it is necessary to cut off the point O
(Fig. 29) of the pattern. How much shall be cut off must be
left to the experience of the workman; it varies according to
the size of the "boss" used to cover the throat. Also, in
practice, to ensure neatness of fitting, the rim at O (Fig. 28)
is notched out a little. This, again, is a matter in which
experience must be the workman's guide.

The rim being cylindrical, its pattern needs no description
beyond saying that it is a rectangle, one side of which is
equal to the depth of the rim, and the other side equal to
the circumference of a circle having A O for its diameter.

Note.—In practice the usual sizes of rim A O Q P and hood
T S R O are as follows :—

Diameter of Cowl.	Depth of. Rim.	Size of Hood at Widest and Narrowest Parts.	
8 inches.	2 inches.	4 inches and 1 inch.	
9 ,,	2¼ ,,	4½ ,,	1⅛ ,,
10 ,,	2½ ,,	4¾ ,,	1¼ ,,
11 ,,	2¾ ,,	5 ,,	1½ ,,
12 ,,	3 ,,	5½ ,,	1¾ ,,

PROBLEM III.

*To draw the pattern for a round neck T-piece to connect
equal circular pipes.*

Draw (Fig. 31) two indefinite lines K g, 7′ a′ at right
angles to each other and intersecting in P, and set off from
P right and left along 7′ a′, and towards g on line K g distances
P 7′, P R and P 1′ equal each to the diameter of the pipe
with which the T is to connect. Through the point 1′

2 c

draw an indefinite line Q T' parallel to 7' *a'*, and through the points 7' and R draw lines H N, G S parallel to K *g* and intersecting Q T' in Q and Q' respectively. Bisect 7' Q in

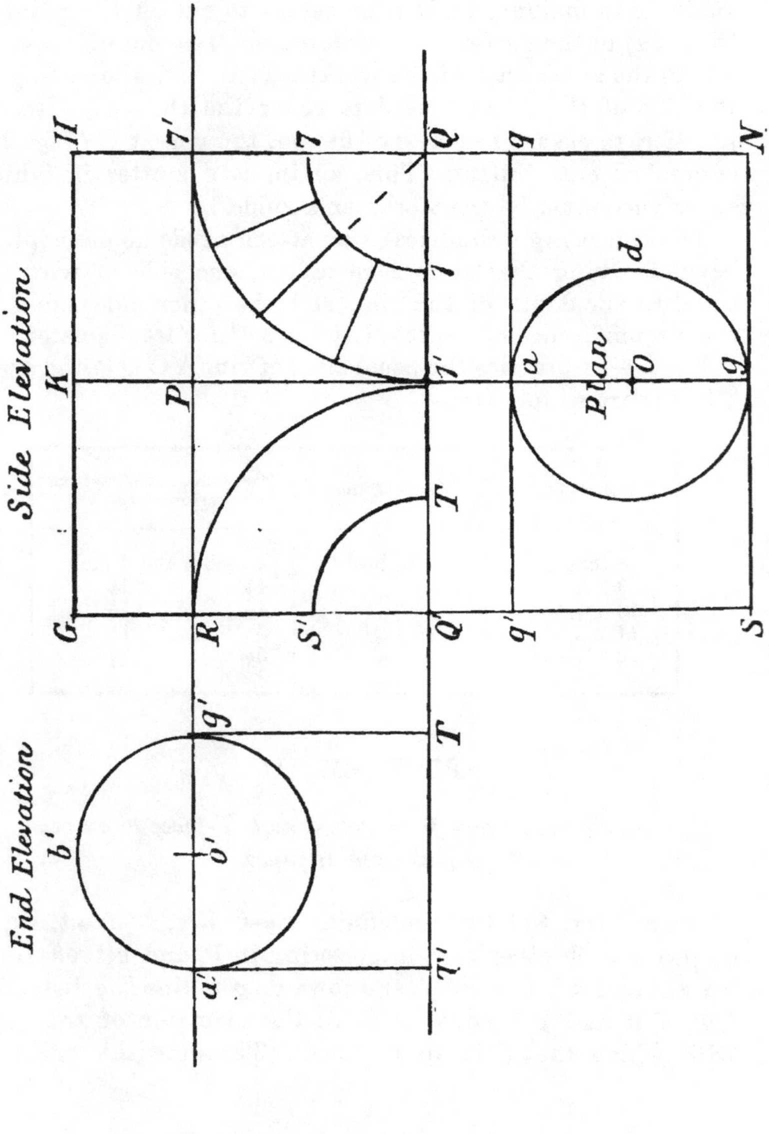

Fig. 31.

point 7, and as 7′ Q is equal to P 1′, each of the divisions 7′ 7, Q 7 will be equal to half the diameter of the pipe that the T connects with. From point 7′ along 7′ H set off 7′ H equal to 7′ 7, from point R on G S set off distances R S′, R G each equal to 7′ 7, and join G H. Now with Q and Q′ as centres and radius Q 7 describe the quadrants 7 1 and S′ T respectively, with Q as centre and radius Q 7′ describe the quadrant 7′ 1′, and with Q′ as centre and same radius describe the quadrant R 1′. Then T S′ G H 7 1 will be a side elevation of the round-neck T. On 7′ a′ take any convenient point O, and with O′ as centre and Q 7 as radius describe the circle a′ b′ g′ cutting 7′ a′ in the points a′ and g′. From points a′ and g′ draw lines a′ T″, g′ T perpendicular to 7′ a′ and meeting Q T′ in the points T′ and T respectively. Then T′ a′ b′ g′ T will be an end elevation of the T. On K g take any convenient point O, and with O as centre and Q 7 as radius, describe the circle a d g, cutting K g in the points a and g. Through point a draw a line q′ q, parallel to Q T′ and cutting G S and H N in points q′ and q respectively, and through g draw a line S N parallel to Q T′, and cutting G S in H N in S and N respectively. Then q′ a q N g S will be a plan of the T.

To draw the pattern for the R G H 7′ 1′ portion of the T.

Looking at the side and end elevations, it will be seen that the R G H 7′ portion of the T is a semicylinder of a′ b′ g′ section, and that the R P 7′ 1′ part of it lies wholly in the plane of a′ T″, and the portion on the opposite side of the T corresponding to R P 7′ 1′ wholly in the plane of g′ T. The pattern for the R G H 7′ 1′ piece of the T can consequently be obtained as follows.

Draw (Fig. 32) two lines R 7′, 1′ 1″ at right angles to each other and intersecting in P. From P along P 1″ set off P K equal to the developed length of the arc a d g (Fig. 31), that is equal to semicircumference of the circular pipe. Through K draw G H parallel to R 7′; and from P set off distances P 1′, P 7′, P R each equal to P 1′ (Fig. 31), and from K a distance K 1″ also equal to P 1′ (Fig. 31). Through the

2 c 2

points 1′, 1″ draw $q′q$ and S N parallel to R 7′; and through the points 7′ and R draw lines q N, $q′$ S parallel each to 1′ 1″, cutting $q′q$ in q and $q′$, and G H in G and H, and S N in S and N. With $q′$, q, N and S successively as centres

Fig. 32.

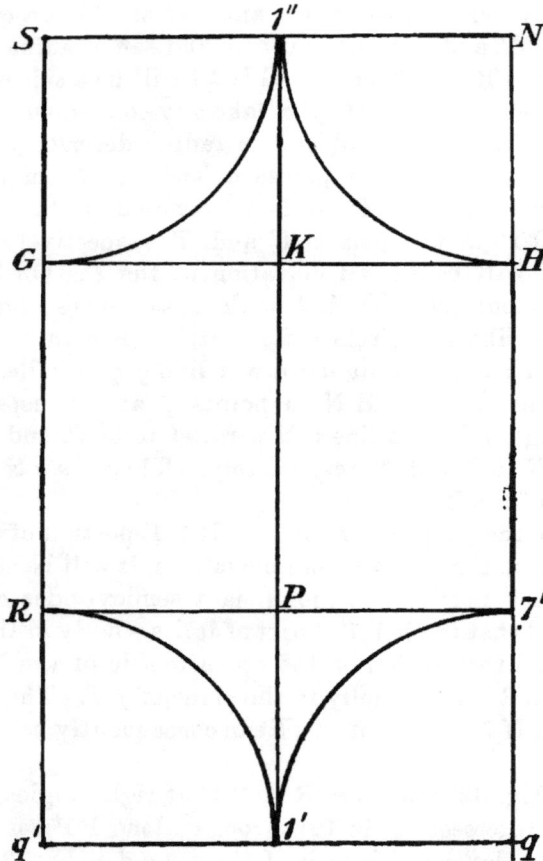

and radius $q′$ R describe the quadrants 1′ R, 1′ 7′, 1″ H and 1″ G. The figure 1′ R G 1″ H 7′ will be the pattern required.

Looking at the side elevation and plan in Fig. 31, it will

be seen that the 1' 7' 7 1, 1' R S' T, portions of the T are halves of bends of equal circular section (the 1' 7' 7 1 portion has drawn on it lines converging to Q in order to show the beginning of the construction of the pattern). In the case of bends, the quadrant curves 1' 7', 1' R, and corresponding quadrants on opposite side of the T would have been neutral lines (see p. 362). Here they are extreme curves of the throats of the T and of the portion of it for which we have just drawn the pattern. According to the dimensions of the pipe that is being dealt with, the throats may be made either in one or in two pieces. If in one piece the method of draw-ing the pattern is given in Case II. of Problem I. (p. 364); if in two pieces the method of procedure will be found in Case III., p. 369, of same Problem.

PROBLEM IV.

To draw the pattern for a curved elbow (bend) of square section, to join two unequal square pipes.

CASE I.—Where the throat of the bend is one of the angles of the square.

Let A D H E (Fig. 33) be a part side elevation of the bend, that is to say, its extreme lines, the A E end being the larger. The A E R T portion of the Fig. is unconnected with this Problem. Bisect A E in A', from A' draw A' a' perpendicular to A E and make A' a' equal to A' A, and join A a', a' E; then A a' E will be a half-section of the larger end of the bend. Similarly draw D d' H a half-section of the smaller end of it. To complete the side elevation, divide each of the curves A D, E H into any convenient number of equal parts (here three) in the points B, C, and F, G, respectively. Join B F and C G, and bisect these lines in B' and C' respectively. From point A' through points B' and C' draw to D' an unbroken curved line; the figure A D D' H E A' will be the complete side elevation of the bend. The bend will be made up of

four pieces; the seams will correspond with the lines of the
angles of the bend, that is, with A B C D, E F G H, A' B' C' D',

FIG. 33.

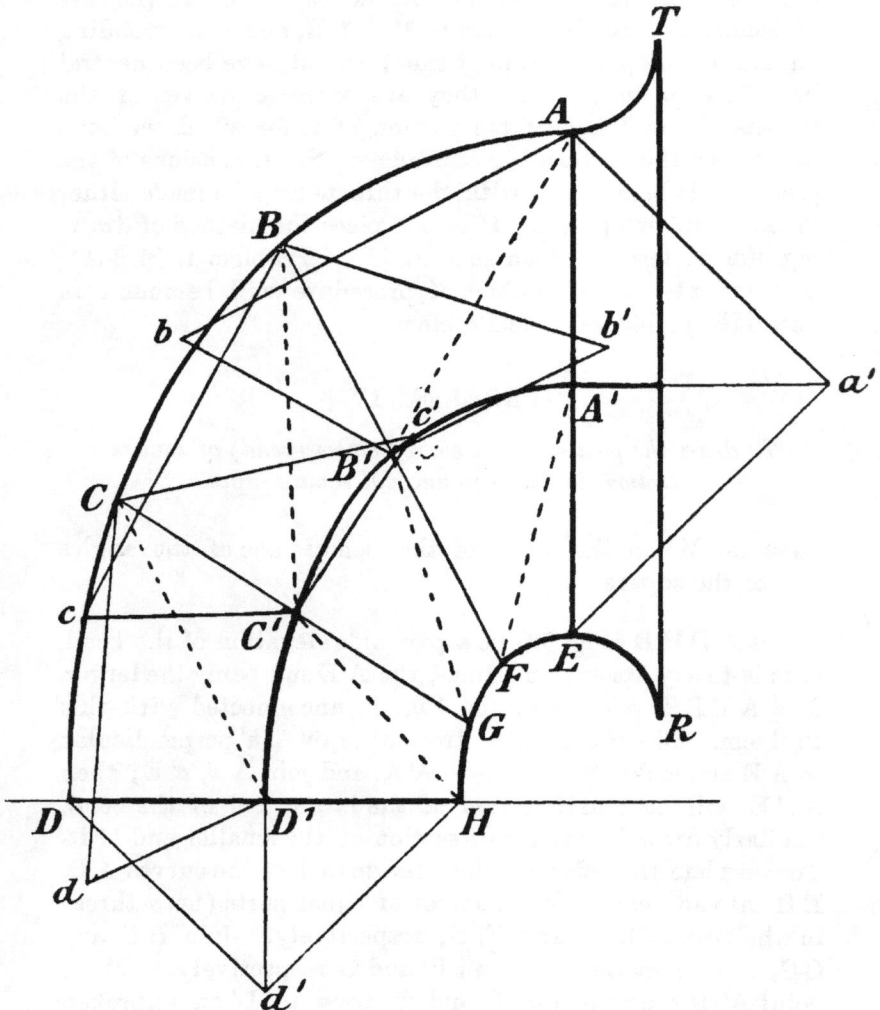

and the line not seen in the elevation that is opposite
to A' B' C' D'. Two patterns will be required, one for the

A B C D D′ C′ B′ A′ and corresponding piece, and one for the A′ B′ C′ D′ H G F E and corresponding piece.

To draw the pattern for the A B C D D′ C′ B′ A′ pieces.

We already have, in A a', and E a', the true lengths of A A′, and E A′ respectively; also, in D d', H d', the true lengths of D D′, and H D′, respectively. To find the true lengths of B B′, and of F B′, equal to it, through B′ draw B′ b' perpendicular to and equal to B B′, and join B b'; then B b' will be the true length required. To find the true lengths of C C′, and of G C′, equal to it, through C′ draw C′ c' perpendicular to and equal to C′ C, and join C c'; then C c' will be the true length required. Now join A B′, B C′, C D′ and find their true lengths. Through B′ draw B′ b perpendicular to A B′ and equal to B′ b', and join A b; through C′ draw C′ c perpendicular to B C′ and equal to C′ c', and join B c; and through D′ draw D′ d perpendicular to C D′ and equal to D′ d', and join C d; then A b, B c, and C d will be respectively the true lengths required.

Next draw (Fig. 34) a line A A′ equal to A a' (Fig. 33), and with A and A′ as centres and radii respectively equal to A b and A′ B′ (Fig. 33) describe arcs intersecting in B′; and with B′ and A as centres and radii respectively equal to B b' and A B (Fig. 33) describe arcs intersecting in B. With B and B′ as centres and radii respectively equal to B c and B′ C′ (Fig. 33) describe arcs intersecting in C′; and with C′ and B as centres and radii respectively equal to C c and B C (Fig. 33) describe arcs intersecting in C. Also with C and C′ as centres and radii respectively equal to C d and C′ D′ (Fig. 33) describe arcs intersecting in D′; and with D′ and C as centres and radii respectively equal to D d' and C D (Fig. 33) describe arcs intersecting in D. Join D D′; through D, C, B, A draw an unbroken curved line. Also draw an unbroken curved line through D′, C′, B′, A′. Then D C B A A′ B′ C′ D′ will be the pattern required.

The other lines drawn in Fig. 34 correspond to the lines of same lettering in Fig. 33; they are not a necessary part of the construction.

To draw the pattern for the A′ B′ C′ D′ H G F E pieces.

Join A′ F, B′ G, and C′ H ; and, to avoid confusion of lines, find their true lengths apart from the elevation. Draw (Fig. 35) from any point O the lines O A′ and O F perpendicular to each other, and equal respectively to A′ a′ and

FIG. 34.

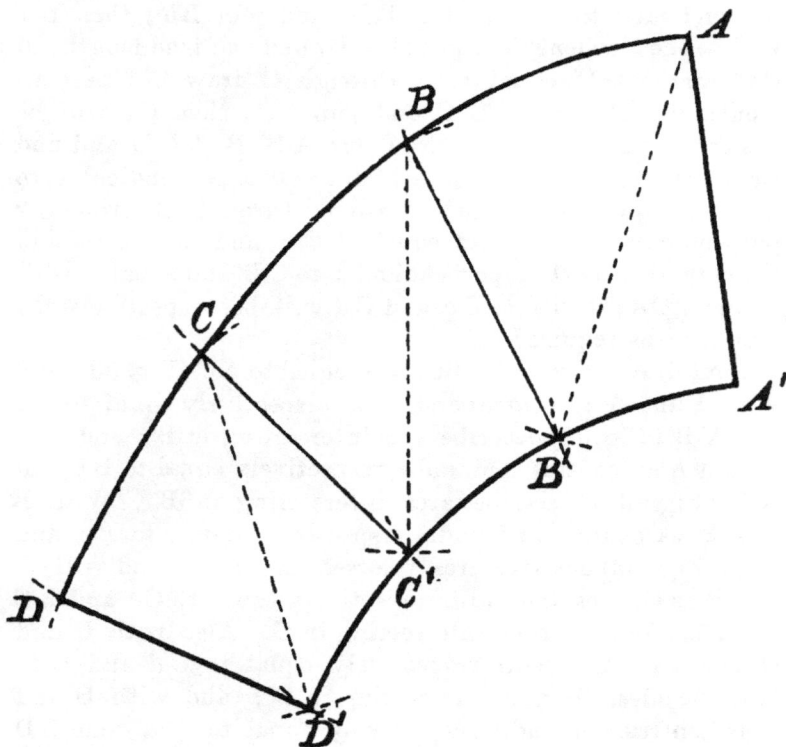

A′ F (Fig. 33), and join A′ F ; then A′ F will be the true length of A′ F (Fig. 33).   Now set off on ˙O A′ distances O B′, O C′, equal respectively to B′ b′, C′ c′ (Fig. 33), and on O F distances O G, O H, equal respectively to B′ G, C′ H (Fig. 33), and join B′ G, C H.   Then B′ G will be the true length of B′ G (Fig. 33), and C′ H the true length of C′ H

(Fig. 33). In E a', H d' (Fig. 33), we already have the true lengths of E A', H D' respectively, and in B b', C c', the true lengths of B' F and C' G.

Now draw E A' (Fig. 36) equal to E a' (Fig. 33), and with A' and E as centres and radii respectively equal to A' F (Fig. 35) and E F (Fig. 33) describe arcs intersecting in F; and with F and A' as centres and radii respectively equal to B b' and A' B' (Fig. 33) describe arcs intersecting in B'.

FIG. 35.   FIG. 36.

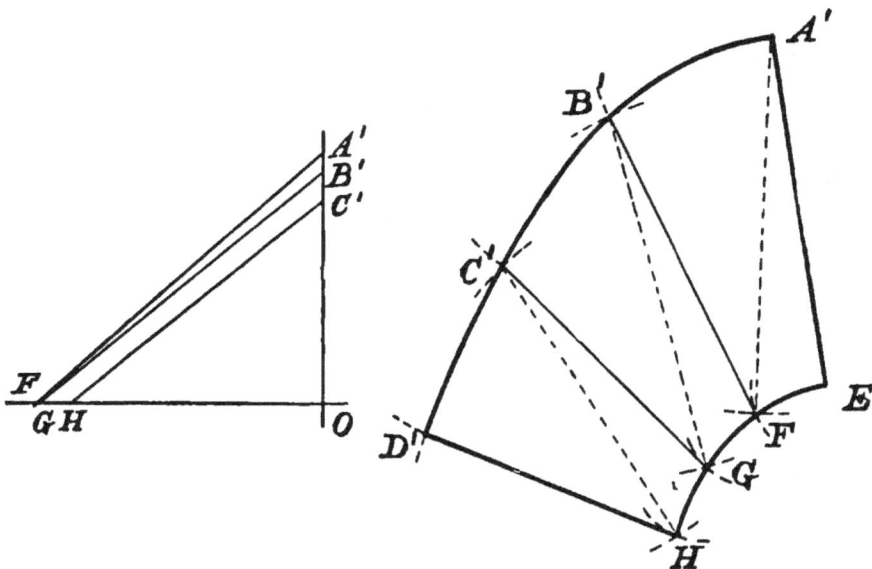

With B' and F as centres and radii respectively equal to B' G (Fig. 35) and F G (Fig. 33) describe arcs intersecting in G; and with G and B' as centres and radii respectively equal to C c' and B' C' (Fig. 33), describe arcs intersecting in C'. Also with C' and G as centres and radii respectively equal to C' H (Fig. 35) and G H (Fig. 33) describe arcs intersecting in H; and with H and C' as centres and radii respectively equal to H d' and C' D' (Fig. 33) describe arcs intersecting in D'. Join H D'; through D', C', B', A' draw an unbroken curved line. Also draw an unbroken curved line through H, G, F, E. Then D' C' B' A' E F' G H will be the pattern required.

As to the other lines drawn in Fig. 36, see the remarks made in connection with the preceding pattern. The patterns necessary for the formation of the bend A B C D D' H G F E A' are now completed.

CASE II.—Where the throat of the bend is one of the sides of the square.

Let A a e E, D d h H (Fig. 37) be the half sections of two square pipes to be joined, A a e E being half section of the larger, and D d h H half section of the smaller, and let A B C D H G F E be the side elevation of the bend to be constructed to join the two pipes. The bend will be made up in four pieces, and the seams will, as in Case I., correspond with the angles of the bend, that is, with A B C D, H G F E, and the lines not seen in the elevation, that are opposite to the mentioned lines. The back and throat of the bend being perpendicular to the plane of the paper, are represented respectively by the single curve lines. Three patterns will be required, one for the A D H E and opposite side, one for the back A B C D, and one for the throat E F G H.

To draw the pattern for the A D H E piece.

This pattern is in the main a repetition of the side of the bend as seen in the elevation, that is A D H E is the pattern required. A slight extension, however, should be made to the pattern beyond the lines A E and D H, seeing that whilst A D H E, the *elevation* of the bend, is in the plane of the paper, A D H E, the *side* of the bend, is not in that plane but at an angle to it.

To draw the pattern for the back A B C D of the bend.

Divide each of the curves A B C D, E F G H into any convenient number of equal parts (here three) in the points B, C, and F, G respectively, and join B F, C G. Now draw (Fig. 38) any line A D, and from any point A in it set off succes-

sively distances A B, B C, C D, equal respectively to the
distances A B, B C, C D round the curve A B C D (Fig. 37),

Fig. 37.

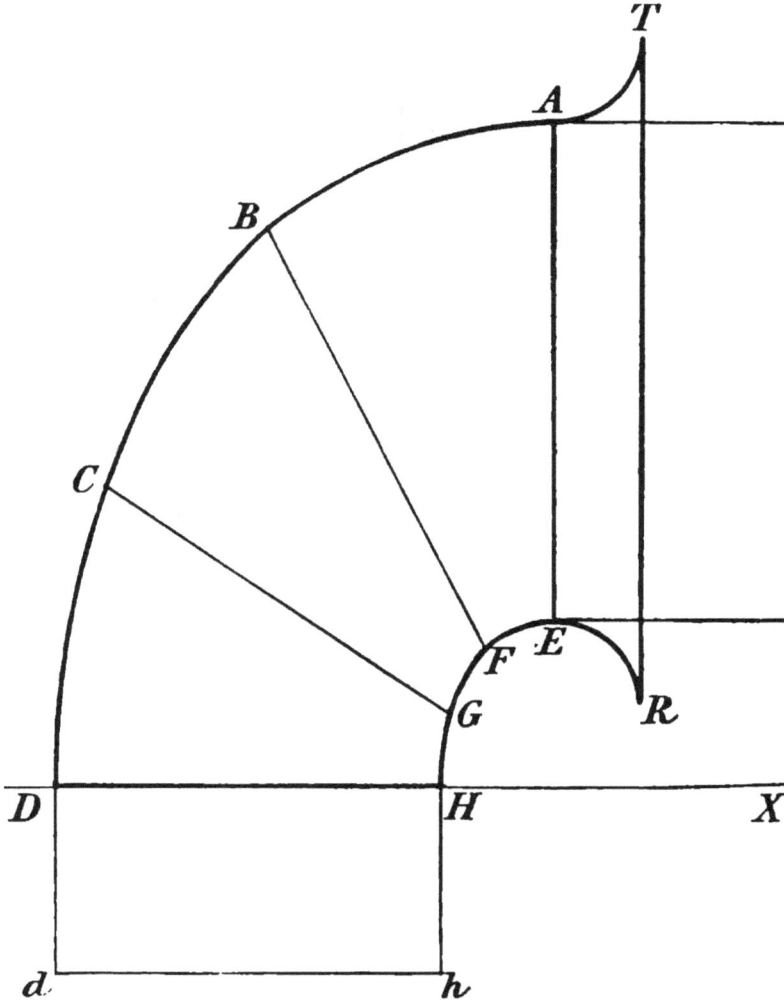

and through the points A, B, C and D draw lines perpendicular to A D. Set off right and left of A D, a distance A a

equal to A *a* (Fig. 37); B *b* equal to half B F (Fig. 37); C *c*
equal to half C G (Fig. 37); and D *d* equal to D *d* (half

FIG. 38.

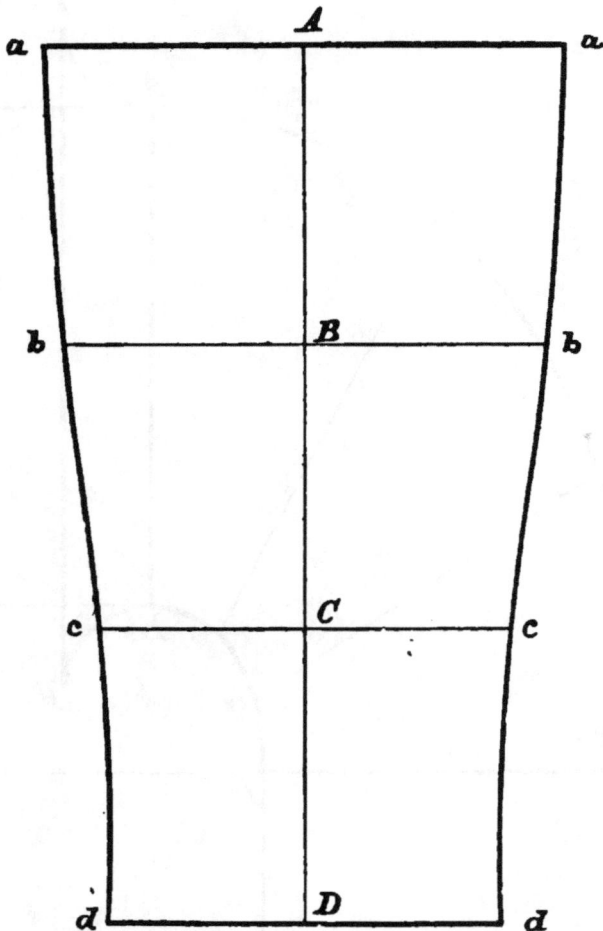

D H) (Fig. 37).   Through each set of points *a, b, c, d,* draw
an unbroken curved line.   Then the enclosed figure *a d d a* is
the pattern required.

To draw the pattern for the throat E F G H of the bend.

Draw (Fig 39) any line E H, and from any point E in it set off successively distances E F, F G, G H equal respectively to the distances E F, F G, G H round the curve E F G H (Fig. 37), and through the points E, F, G and H draw lines perpendicular to E H. Set off right and left of E H, a distance

FIG. 39.

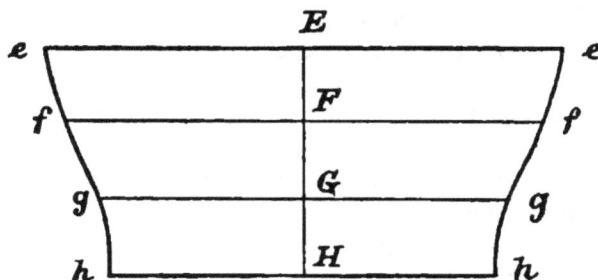

E e equal to E e (Fig. 37); F f equal to half B F (Fig. 37); G g equal to half C G (Fig. 37); H h equal to H h (half D H) (Fig. 37). Through each set of points e, f, g, h, draw an unbroken curved line. The enclosed figure e h h e is the pattern required. The patterns necessary for the formation of the bend A B C D H G F E are now complete.

PROBLEM V.

To draw the pattern for a ventilator of square section.

CASE I.—Where the throat of the ventilator is one of the angles of the square.

Let T A C D H F E R (Fig. 33) represent the ventilator in side elevation; the seams of it are to be at the angles. The portion A D H E is a curved tapering elbow or bend of square section. On D H draw D d' H, half section of the square at D H, and on A E draw A a' E half section of the square where the mouthpiece is attached. These half sections may be regarded as the half sections of two unequal square pipes, and the drawing of the pattern for the A D H E portion of the ventilator now becomes that of the setting out a pattern for a square tapering bend, the throat of which is one of the angles of the square, to join the two unequal square pipes whose half sections are D d' H and A a' E. Divide each of the curves A C D, H F E into any convenient number of equal parts (here three) in the points B, C and F, G respectively, and join B F and C G cutting the elevation curve A' D' in the points B' and C'; the lines B F and C G will be respectively bisected in the points B' and C'. The further procedure for the pattern will be exactly as shown in Case I. of preceding Problem.

To complete the ventilator. It finishes off with a moulding, of which the curve A T is a transverse section. The student will have no difficulty in dealing with this mouth of the ventilator; he will find every instruction in Problem XXX., p. 310.

PROBLEM VI.

To draw the pattern for a tapering circular bend.

The methods given are approximate, as indeed all ordinary practical methods must be for problems of this description.

CASE I.—The bend to be made up of circular segments.

Let K' K B' B H H' Q' Q' (Fig. 40) represent in elevation a tapering circular bend. (The B T R H portion of the Fig. is unconnected with this Problem.) Divide each of the curves B K', and H Q' into as many equal parts as it is determined there shall be segments, here say three parts, the points of division being B', K, and H', Q respectively. Joining B' to H', and K to Q, we have in B B' H' H, B' K Q H', and K K' Q' Q elevations of the three segments of which the ventilator is to be made up.

To draw the pattern for the segment B B' H' H.

Bisect the arcs B' B in a and H' H in h, and join $a\,h$; also join B' B, intersecting $a\,h$ in b. On $a\,h$ describe a semicircle $a\,e'\,h$ and divide it into any number of equal parts (here six) in the points c', d', e', f' and g'. From these points draw lines

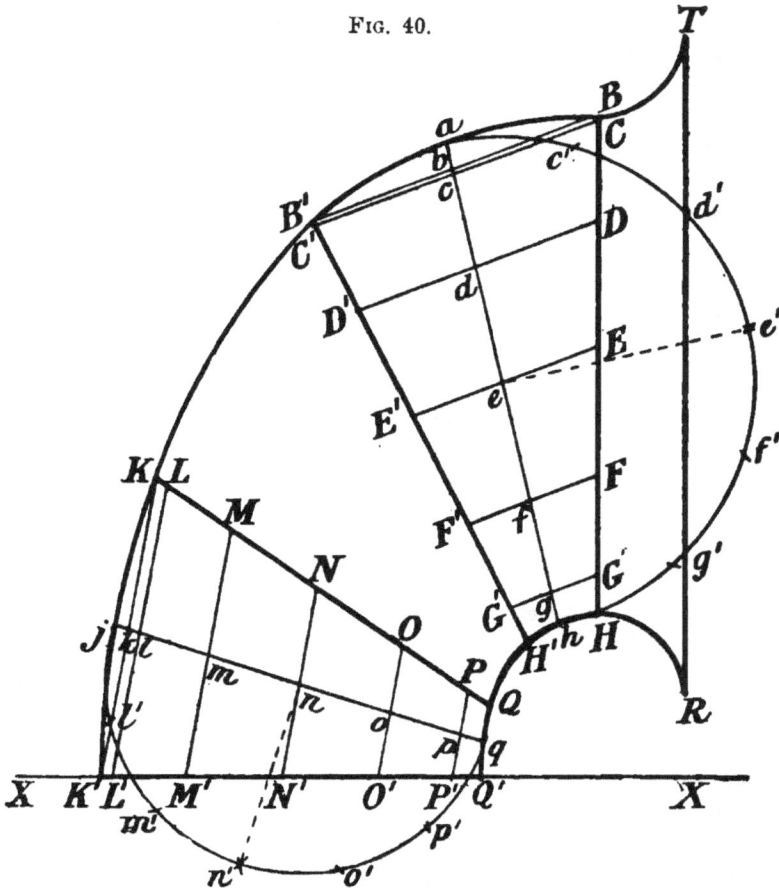

Fig. 40.

perpendicular to $a\,h$, cutting $a\,h$ in the points c, d, e, f and g respectively (to avoid confusion these lines are not shown in the figure), and through c, d, e, f and g draw C' C, D' D, E' E, F' F, and G' G parallel each to B' B.

Next draw (Fig. 41) an indefinite line $b\,h$, and set off on it
distances $b\,c$, $c\,d$, $d\,e$, $e\,f$, $f\,g$ and $g\,h$, each equal to $a\,c'$ (Fig. 40),
that is, to one of the divisions of the semicircle $a\,e'\,h$, and
through b draw B′ B at the same angle to $b\,h$ that B′ B
(Fig. 40) is to $b\,h$. Now through c, d, e, f, g and h (Fig. 41)
draw C′ C, D′ D, E′ E, F′ F, G′ G and H′ H parallel each to
B′ B. Make b B, c C, d D, e E, f F, g G and h H equal respectively to b B, c C, d D, e E, f F, g G and h H (Fig. 40).
Also make b B′, c C′, d D′, e E′, f F′, g G′ and h H′ equal respectively to b B′, c C′, d D′, e E′, f F′, g G′ and h H′ (Fig. 40).
Joining B, C, D, E, F, G and H, and B′, C′, D′, E′, F′, G′ and
H′ by unbroken curved lines gives in B H H′ B′ the pattern
required.

To draw the pattern for the segment K′ K Q Q′.

Bisect the arc K′ K in j and Q′ Q in q, and join $j\,q$; also
join K′ K, intersecting $j\,q$ in k. On $j\,q$ describe a semicircle
$j\,n'\,q$ and divide it into any number of equal parts (here six)
in the points l', m', n', o', and p'. From these points draw lines
perpendicular to $j\,q$, cutting $j\,q$ in the points l, m, n, o and p
respectively (these lines are not shown in the figure), and
through l, m, n, o, and p draw L′ L, M′ M, N′ N, O′ O and P′ P,
parallel each to K′ K.

Next draw (Fig. 42) an indefinite line $k\,q$, and set off on it
distances $k\,l$, $l\,m$, $m\,n$, $n\,o$, $o\,p$ and $p\,q$, each equal to $j\,l'$
(Fig. 40), that is, to one of the divisions of the semicircle
$j\,n'\,q$, and through k draw K′ K at the same angle to $k\,q$ that
K′ K (Fig. 40) is to $k\,q$. Now through l, m, n, o, p and q
(Fig. 42) draw L L′, M M′, N N′, O O′, P P′ and Q Q′ parallel
each to K K′. Make k K, l L, m M, n N, o O, p P and q Q
equal respectively to k K, l L, m M, n N, o O, p P and q Q
(Fig. 42). Also make k K′, l L′, m M′, n N′, o O′, p P′ and q Q
equal respectively to k K′, l L′, m M′, n N′, o O′, p P′ and q Q′
(Fig. 42). Joining K, L, M, N, O, P and Q, and K′, L′, M′,
N′, O′, P′ and Q′ by unbroken curved lines gives in K Q Q′ K′
the pattern required.

In practice it is found desirable to leave the ends H′ H
(Fig. 41) and Q′ Q (Fig. 42) of the pattern full long. It is

easier to cut a piece off if a pattern is too long than to add
a piece if too short.

The pattern for the segment K B' H' Q (Fig. 40) need not

FIG. 41.   FIG. 42.

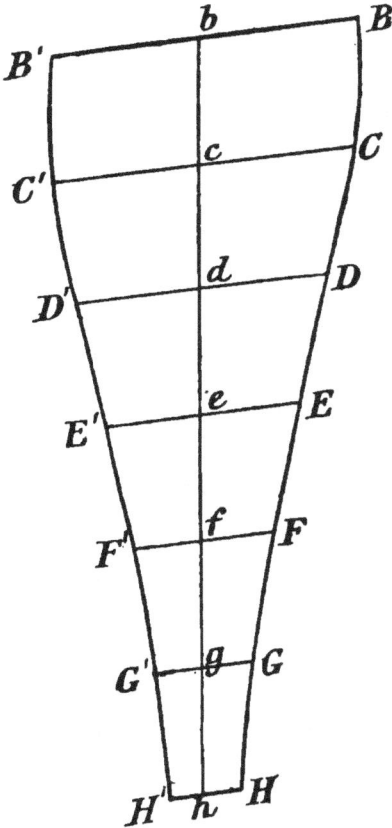   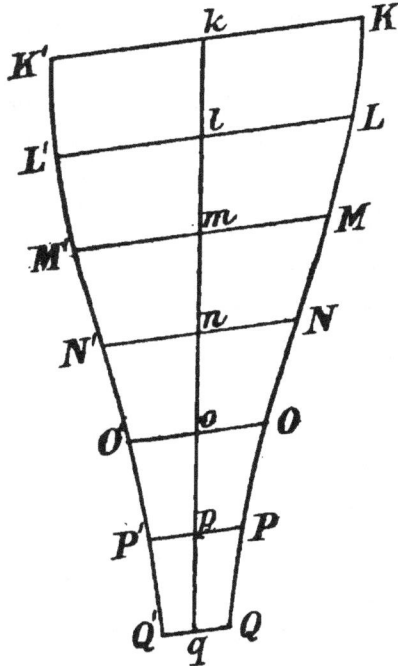

be drawn here. To add the construction for it would make
a confusion of lines in the Fig. The pattern can be obtained
in similar manner exactly to that by which the patterns are
obtained for the other two segments.

The patterns drawn are of course halves only of the full

2 D

circular segments ; and the remark that concludes Case V. of Problem I., p. 381, applies also here. The dimensions of the bend decide whether its segments shall be in one or two pieces each. The present Problem corresponds with that of Case V. just referred to.

CASE II.—The bend to be made in four pieces, namely, two part-sides (alike), one back-piece, and one throat-piece. The seams will follow the curve of the bend, two on each side of it.

Let D A E H (Fig. 43) be a side elevation of the bend. Divide the arc D A into any convenient number of equal parts, here three, in the points C and B; and the arc H E into the same number, in the points G and F ; and join C G, B F. On each of the lines D H, C G, B F, A E describe a semicircle, and divide each semicircle into any convenient number of equal parts, here four, in the points a', b', c' for the semicircle on D H, the points d', e', f' for the semicircle on C G, the points g', h', j' for the semicircle on B F, and the points k', l', m' for the semicircle on A E. From the points a', b' and c' draw lines perpendicular to D H, intersecting it in the points a, b and c. Similarly obtain the points d, e, f on C G, g, h, j on B F, and k, l, m on A E. From a through d and g to k draw an unbroken curved line ; also from b through e and h to l, and from c through f and j to m. In shaping up a part-side piece the line $b e h l$ will be a neutral line. The seams of the bend will correspond with the lines $a d g k$ and $c f j m$.

To draw the pattern for a part-side piece.

Draw (Fig. 44) two lines O a', O k' at right angles to each other. Also draw a line O d' at the same angle to O a' that O C is to O D (Fig. 43), and a line O g' at the same angle to O k' that O B is to O A (Fig. 43). On O a' set off a distance O b equal to O b (Fig. 43); on O d' a distance O e equal to O e (Fig. 43); on O g' a distance O h equal to O h (Fig. 43); and on O k' a distance O l equal to O l (Fig. 43). From the point b through e and h to l draw an unbroken curved line ;

this line is a reproduction of the curved line *b e h l* (Fig. 43).
From *b* set off *b c′, b a′* equal to *b′ c′, b′ a′* (Fig. 43); from *e* set

FIG. 43.

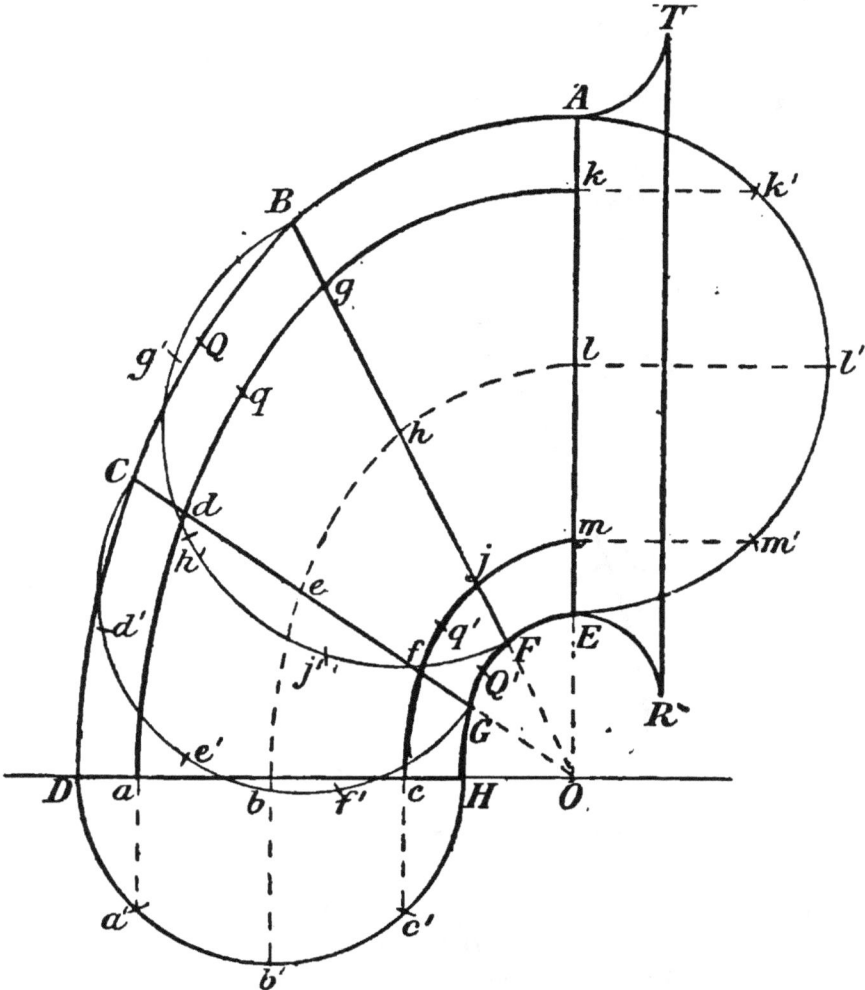

off *e f′, e d′* equal to *e′ f′, e′ d′* (Fig. 43); from *h* set off *h j′, h g′*
equal to *h′ j′, h′ g′* ; and from *l* set off *l m′, l k′* equal to *l′ m′, l′ k′*

2 D 2

(Fig. 43). From a' through d' and g' to k' draw an unbroken curved line, and from c' through f' and j' to m' draw an unbroken curved line; the figure $a' d' g' k' m' j' f' c'$ will be the pattern required.

To draw the pattern for the back-piece.

Bisect the curve B C (Fig. 43) in Q, and the curve g d in q. Now draw (Fig. 45) an indefinite straight line A D, and

FIG. 44.

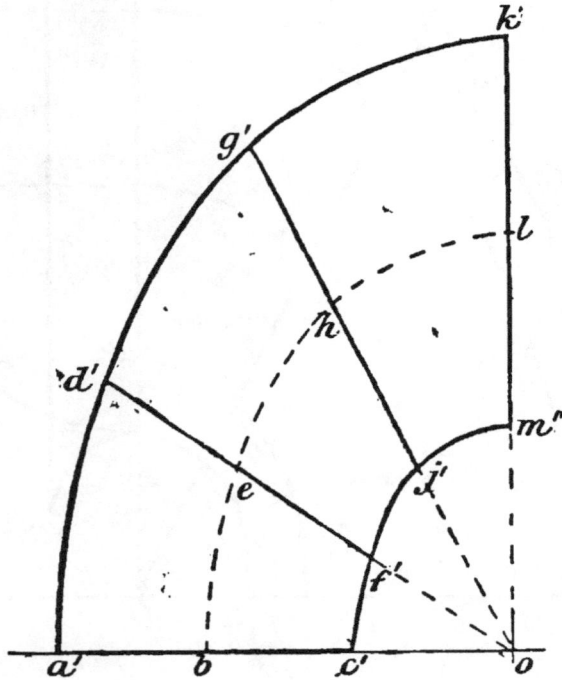

about midway in it take a point Q. From Q on Q A set off a distance Q B equal to Q B on the curve C B (Fig. 43), and from B on same line set off B A equal to B A (Fig. 43). Also from Q on Q D set off a distance Q C equal to Q C (Fig. 43) and from C on same line a distance C D equal to C D (Fig. 43). Next from Q set off Q g equal to q g (Fig. 43), and

from *g* set off *g k* equal to *g k* (Fig. 43) ; and again from Q a
distance Q *d* equal to *q d* (Fig. 43), and from *d* a distance *d a*

FIG. 45.

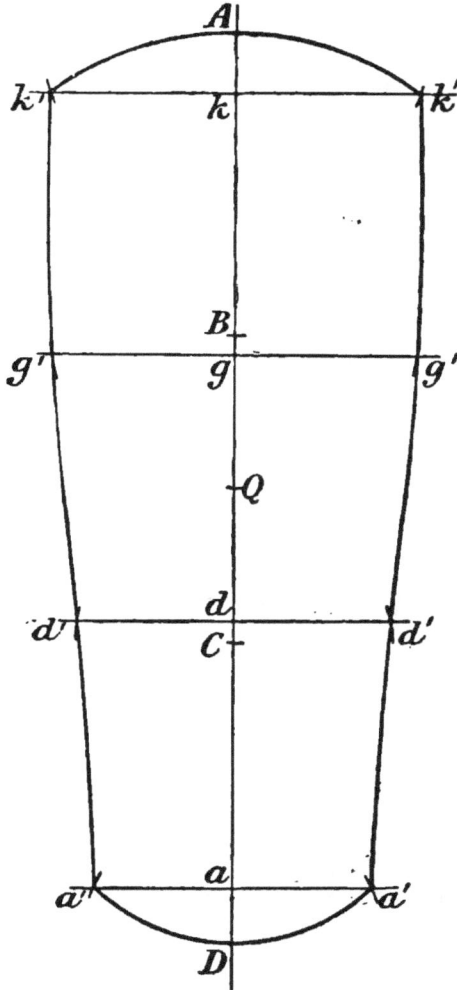

equal to *d a* (Fig. 43). Through the points *k*, *g*, *d* and *a*
draw indefinite lines perpendicular to A D. With A as

centre and radius A k' (Fig. 43) describe arcs intersecting the
line through k in the points k', k'; with B as centre and
radius B g' (Fig. 43) describe arcs intersecting the line
through g in the points g', g'; with C as centre and radius C d'
(Fig. 43) describe arcs intersecting the line through d in the
points d', d'; and with D as centre and radius D a' (Fig. 43)
describe arcs intersecting the line through a in the points a',
a'. From k' through g' and d' to a', on each side of the line

FIG. 46.

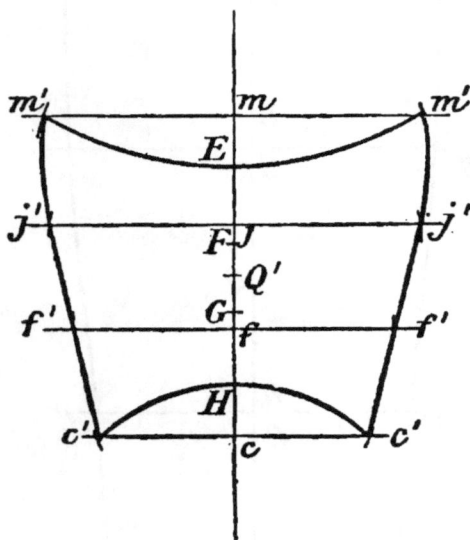

A D, draw an unbroken line; also from k' on the left of A D
an unbroken line through A to k' on the right of A D; and
similarly from the point a' an unbroken line through D to a'.
The figure D a' k' A' k' a' will be the figure required.

To draw the pattern for the throat-piece.

Bisect (Fig. 43) the curve F G in the point Q' and the
curve fj in q'. Now draw (Fig. 46) an indefinite straight line
$m\,c$, and about midway in it take a point Q'. From Q' on
Q' m set off a distance Q' F equal to Q' F (Fig. 43), and from

F a distance F E equal to F E (Fig. 43); also from Q' set off
Q' G equal to Q' G (Fig. 43), and from G set off G H equal to
G H (Fig. 43). Next from Q' set off Q' j equal to $q'j$ (Fig. 43),
and j m equal to j m (Fig. 43); and again from Q' a distance
Q' f equal to $q'f$ (Fig. 43), and from f a distance fc equal to
fc (Fig. 43). Through the points m, j, f and c draw lines
perpendicular to mc. With E as centre and radius E m'
(Fig. 43) describe arcs intersecting the line through m in
the points m, m'; with F as centre and radius F j' (Fig. 43)
describe arcs intersecting the line through j in the points j',
j'; with G as centre and radius G f' (Fig. 43) describe arcs
intersecting the line through f in the points f', f'; and with H
as centre and H c' (Fig. 43) as radius describe arcs intersecting the line through c in the points $c' c'$. From the point m
through j' and f' to c', on each side of mc, draw an unbroken
line; also from m' on the left of mc an unbroken line through
E to m' on the right of mc; and similarly from the point c' an
unbroken line through H to c'. The figure H c' m' E m' c' will
be the figure required.

PROBLEM VII.

To draw the pattern for a circular tapering ventilator.

Let (Fig. 43) D A T R E H be a side elevation of the
ventilator. The preceding Problem deals with the D A E H
portion of it, which portion is a circular tapering bend. The
A T R E mouth-piece may be stretched out of flat sheet-
metal or be made as a cone-frustum, and then stretched as
required.

BOOK V.

CHAPTER I.

METALS AND THEIR PROPERTIES; ALLOYS; SOLDERS; SOLDERING FLUXES.

IN the pages that follow, we deal almost exclusively with the metals that are used largely in plate or sheet; our business being the setting-out of patterns on sheets or surfaces.

CHARACTERISTIC FEATURES.

(92.) Metals are natural elementary substances, as far as is known. They are *opaque* (not transparent), reflect light from their polished surfaces, and have a characteristic lustre, known as the *metallic lustre*. With the exception of mercury, they are all solid at the ordinary temperature of the atmosphere.

(93.) Silver, tin, lead, mercury, antimony, zinc, cadmium, and bismuth, have a whitish or greyish colour. Gold stands alone as a metal having a yellow colour; copper is the only red metal.

(94.) Metals differ widely in their behaviour under the influence of heat; some, as tin and lead, are fusible below red heat; others, as copper, gold, and silver, fuse readily in ordinary furnaces; nickel, iron, and manganese fuse with great difficulty; platinum is practically infusible. Arsenic, cadmium, zinc, and mercury are *volatile*, that is, vaporise easily. An interesting example of volatility is that of zinc, which when at a bright red heat takes fire, burns with a greenish flame, and *oxidises* (unites with the oxygen gas of the atmosphere), being thereby converted into a dense white flocculent substance, called formerly 'philosopher's wool.'

(95.) The fracture of metals is often characteristic; we get crystalline, granular, fibrous, silky, and other fractures. We have fractures in iron, depending upon its kind, of the first three forms. In copper we find the silky fracture.

Properties; Specific Gravity; Melting-points.

Metals have various properties. Some remarks on these, and other particulars respecting metals, now follow.

(96.) *Malleability.*—A property which is possessed by metals in varying degree is that of malleability, that is, of permitting extension of surface without rupture, by, for example, hammering, pressure, or rolling. Gold, which is capable of being hammered into leaves of extreme thinness, is the most malleable of all metals. Other metals, though malleable to a considerable degree, require to be *annealed* (heated red and allowed to cool down slowly) once or even several times during the operation of rolling out, or extending by the hammer as in raising and stretching. Copper is an example; though, curiously enough, copper is equally malleable whether, after heating, it is allowed to cool slowly or is cooled suddenly by dipping while at red heat in cold water. Zinc is in its most malleable condition at a temperature a little above the boiling-point of water: when less than half as hot again as this, it is brittle and unworkable.

(97.) Of the theory of annealing nothing definite appears to be known; but it is supposed that on rolling out or hammering a piece of metal the particles or molecules of which the metal is composed, become strained and disarranged ('the grain closed'), and the metal is hardened; and that on heating, the metal expands, and the strain being removed, the molecules rearrange themselves. This, however, does not explain many matters connected with

annealing; for example, why one heated metal is hardened
by being suddenly dipped in cold water, and another metal
softened when treated in the same manner.

The following, which illustrates the effect of ' hammer
hardening' on iron, may be of interest :—

(98.) In 1854, at the meeting of the British Association in
Liverpool, a paper on the crystallisation of iron under certain
circumstances was read by Mr. Clay (Mersey Iron and Steel
Works), who stated that he selected a piece of good, tough,
fibrous bar-iron, which he heated to a full red heat, and then
hammered by light, rapid, tapping blows, until it was what
is called ' black-cold.' After complete cooling he broke it,
and found that the structure of the iron was entirely changed,
and that, instead of bending nearly double without fracture,
as it should have done, and breaking with a fine silky fibre
when fracture did occur, an entire alteration had taken place,
and the bar was rigid, brittle, and sonorous, incapable of
bending in the slightest degree, and breaking with a glassy,
crystallised appearance. By simply heating the bar to full
red-heat again, the fibre was restored exactly as before.

(99.) *Tenacity.*—The property in metals of resisting being
torn asunder by a tensile or stretching force, is called
tenacity.

The tenacity of metals varies with their purity and molecular condition, as due to modes of treatment or preparation :
for example, the tenacity of steel is much influenced by its
' temper,' and that of cast iron made by the *cold-blast* process
is greater than when the process is that of the *hot-blast.*

(100.) *Ductility.* — The property of being permanently
lengthened by a tensile or stretching force, as in wire
drawing, is called *ductility.* All the malleable metals are
more or less ductile, though the most malleable metals are
not necessarily the most ductile ; ductility being influenced
more by tenacity than by malleability.

The table shows how some of the metals, starting from
gold (see ' Malleability,' above, § 96), rank under the headings
given.

Malleability.	Ductility.	Tenacity.
Gold	Gold	Steel
Silver	Silver	Iron
Copper	Platinum	Copper
Tin	Iron	Platinum
Platinum	Copper	Silver
Lead	Aluminium	Gold
Zinc	Zinc	Zinc
Iron	Tin	Tin
Nickel	Lead	Lead

(101.) *Conductivity.*—Of all solids, metals are the best conductors of heat. The order of conductivity for a few important metals, beginning with the best conductors, is—silver, copper, gold, tin, iron, lead, platinum, and bismuth.

(102.) *Welding.*—An important property of some of the metals is that pieces can be ' welded ' together, that is, incorporated with each other. Iron at a white heat is in a pasty condition and can be ' welded '; that is, if two white-hot and clean surfaces of iron be brought into contact and pressed or hammered together, they thus are ' welded,' that is, become part of one and the same mass. If lead and gold in a fine state of division be strongly pressed together at the ordinary temperature of the atmosphere, they will form one mass.

(103.) *Hardness.*—The comparative hardness of metals is usually estimated by the force required to draw the metals into wires of equal diameter. In order of hardness we have —steel, iron, copper, silver, tin, antimony, and lead.

(104.) *Specific Gravity.*—Some substances are, in their nature, more weighty bulk for bulk than others. Thus, a cubic inch of lead is heavier than a cubic inch of iron; and a cubic inch of iron than a cubic inch of water.

By their *specific gravity* the weights, relatively to each other, of substances, are known. The standard of comparison is an equal bulk of pure distilled water, and if the specific gravity of a body is, say, 2, this means that it is twice as heavy as the same bulk of water.

The specific gravity of platinum is 21 ; platinum therefore bulk for bulk is 21 times heavier than water.

The specific gravity of antimony is 6·7, and a cubic foot of pure distilled water weighs very nearly 1000 ounces. Therefore a cubic foot of antimony weighs 6·7 times 1000 ounces, that is, 6700 ounces.

Knowing the specific gravity of a substance, we can find the weight of any volume of it, by multiplying the given volume, in cubic inches, by its specific gravity, and by 62·4 the weight in lbs. of a cubic foot (1728 cubic inches) of water, and dividing by 1728. Thus the weight of 48 cubic inches of cast copper, the specific gravity of which is 8·6, is

$$\frac{48 \times 8 \cdot 6 \times 62 \cdot 4}{1728}$$, is, that is to say, 14·9 lbs.

As the relative weights of equal volumes of metals have often to be taken into consideration in using metals for constructive purposes, for example, in the covering of roofs, where weight is sometimes a matter of importance, a table of specific gravities follows.

TABLE OF SPECIFIC GRAVITIES AND MELTING-POINTS.

Metals.	Specific Gravities.	Melting-points. (Centigrade.)	Authority for Melting-points.
Antimony	6·7	432	Buillet.
Bismuth	9·8	268·3	Reimsdyk.
Copper (cast)	8·6	} 1054	Violle.
„ (wrought)	8·8		
Steel	7·8	{ 1300 to 1400 }	Buillet.
Cadmium	8·6	320·7	Person.
Iron (wrought)	7·8	{ 1500 to 1600 }	Buillet.
Lead	11·36	326·2	Person.
Tin	7·29	232·7	Person.
Zinc	7·19	433·3	Person.
Aluminium	2·67	1045	Violle.
Nickel	8·30	1450	Pictel.
Platinum	21·25	1775	Violle.

The melting-points of the metals named are added to the table, as it is often useful to be able to refer to these. The degrees of heat are according to the Centigrade thermometer. This portion of the table is taken from 'Melting and Boiling Point Tables,' by Thos. Carnelley, 1885.

We now proceed to notice more particularly the metals iron (including cast iron and steel), tin, zinc, and copper.

IRON AND STEEL.

(105.) Iron in a state of purity is comparatively little known; the ores of it are various and abundant. In its commercial forms, as plate or sheet, bar, and cast iron, it is well known. As sheet it can be cut into patterns and bent into desired forms; as bar it can be made hot and ' wrought,' that is, shaped by means of the hammer; and when molten it can be run or cast into all sorts of shapes. Cast iron is brittle, crystalline in fracture (§ 95), and not workable by the hammer. In sheet and bar form, iron is malleable, mostly fibrous in fracture, and capable of being welded (§ 102). The presence of impurities in bar iron, that is, the presence of substances not wanted in it at the time being, seriously affects its malleability. Thus the presence of phosphorus, or tin, renders it brittle when cold ('cold-short'), and the presence of sulphur makes it unworkable when hot ('hot-short'). Iron quickly *rusts* (oxidises, § 94) if exposed to damp air, as in the case of iron exposed to all weathers; or to air and water, as with vessels in which barely sufficient water is left to cover the bottoms, the rusting (oxidation) being then much more rapid than when the vessels are kept full. Heated to redness and above, 'scale' (oxide of iron) rapidly forms and interferes greatly with welding. It is impossible to enter here into any consideration of the processes by which iron is prepared from its ores. To two modern processes, however, we shall presently have par-ticularly to refer.

(106.) The effects of the presence of several foreign sub-

stances in iron as impurities has just been alluded to, but
the presence in it of 'carbon' we have not spoken of. This
is a substance which in its crystalline form is known as the
diamond, and in its uncrystalline form as charcoal. The
presence of carbon in iron destroys its malleabili'y, but at
the same time gives to it properties various, so remarkable
and useful to mankind, that to say, as a defect, of a piece of
iron with carbon in it, that it is not malleable, is simply
equivalent to saying when we have a piece of brass, that it
is not a piece of copper. Quite the reverse of being 'matter
in the wrong place,' carbon in iron furnishes a compound so
valuable on its own account, so entirely of its own kinds (in
the plural because its kinds are several), that, if there were
other substances not metals, the compounding of which with
a metal gave products at all resembling those of iron and
carbon, then all such compounds would form a class of their
own. The iron and carbon compound, however, stands incon-
veniently alone. There we shall not leave it, but as aiding
the full comprehension of it, notwithstanding that we define
alloys (§ 115) as compounds of *metals*, shall consider it not
as outside but as within this class of substances, as well as
also shall speak of iron as being *alloyed* with it. We shall
deal with it, however, under the present heading, treating
compounds of actual metals later on.

(107.) Iron is alloyed with carbon in proportions varying
from say $\frac{1}{2}$ to 5 per cent. When in the proportion of from
2 per cent. upwards, the compound is cast iron, that is, iron
suitable for casting purposes; in other proportions it is
known as steel. In cast iron the metallic appearance is
somewhat modified; in steel it is maintained. Originally
steel was made by the addition of carbon to manufactured
iron, and the word had then a fairly definite signification;
meaning a material of a high tensile strength; that by being
heated dull red and suddenly cooled could be made so hard
that a file would not 'touch' it, that is, would slide over it
without marking it; and that could have that hardness
modified or '*tempered*' by further application of heat. But

with the introduction of the Bessemer process of steel making, and of the Siemens' process of making steel direct from the ores, processes by which any desired percentage of carbon can be given, the signification of the word has become enlarged, and now includes all alloys of iron and carbon between malleable iron and cast iron; except that the term ' mild ' steel is sometimes applied to those alloys that approach in qualities to malleable iron. Steel plates are now produced equal in toughness, and it is said even excelling the best ' charcoal' plates, and as they are much cheaper, the old charcoal-plate-making process is very generally giving way to the direct process. In practice, however, these plates are found to be more *springy* than good charcoal plates, and not so soft and easy to work.

(108.) As iron is very liable to rust, surface protection is given to it by a coating of tin, or of an alloy of lead and tin, (lead predominating), or of zinc. Plates coated with tin are termed ' *tin* ' plates ; with lead and tin have the name of '*terne*' plates, and if coated with zinc are said to be 'galvanised.' Terne plates are used for lining packing-cases, also for work to be japanned. Usual sizes of tin and terne plates are $14'' \times 10''$, $20'' \times 14''$, $20'' \times 28''$, and they are made up to $40'' \times 28''$.

(109.) Large iron sheets of various gauges coated with tin and having the same appearance as a 'tin' plate are called *Manchester plates*, and sometimes *tinned iron*. But the latter term is more generally applied to sheets of iron which are coated with lead and tin, and are dull like terne plates.

(110.) Iron coated with zinc is not so easily worked as when ungalvanised. In galvanising, the zinc 'alloys' with the surface of the iron, and this has a tendency to make the iron brittle. Galvanised iron is useful for water tanks and for roofing purposes, as the zinc coating prevents rust better than a tin coating. For roofing, however, 'terne' plates are largely used in America, and, kept well painted, are found to be very durable. Owing to the ease with which zinc is attacked by acids, galvanised iron is not suitable for vessels exposed to acids or acid vapours.

COPPER.

(111.) This, the only red metal (§ 93), is malleable, tenacious, soft, ductile, sonorous, and an excellent conductor of heat. For this reason, and because of its durability, it is largely made use of for cooking utensils. It is found in numerous states of combination with other constituents, as well as 'native' (uncombined). Its most important ore is *copper pyrites*. Copper melts at a dull white heat and becomes then covered with a black crust (oxide). It burns when at a bright white heat with a greenish flame. No attempt at explanation of its manufacture will here be made, as any description not lengthy would be simply a bewilderment. For the production of sheet copper it is first cast in the forms of slabs, which are rolled, and then annealed and re-rolled, this annealing and re-rolling being repeated until the copper sheet is brought down to the desired thickness. In working ordinary sheet copper, it is hammered to stiffen it, and 'close the grain.' *Hard-rolled* copper is, however, nowadays produced that does not require hammering.

(112.) In the course of the manufacture of copper it undergoes a process termed '*poling*' to get rid of impurities. We mention this because we shall find (§ 129) a similar process gone through in preparing solders. The poling of copper consists in plunging the end of a pole of green wood, preferably birch, beneath the surface of the molten metal, and stirring the mass with it. Violent ebullition takes place, large quantities of gases are liberated, and the copper is thoroughly agitated. It is doubtful if this poling process is fully understood, for, though it is quite obvious that there may be insufficient poling ('*underpoling*'), it is not easy to explain '*overpoling.*' But overpoling, as a fact, is fully recognised in the manufacture of copper, and the metal is brittle both if the poling is too long continued or not long enough. If duly poled, the cast slab when set displays a comparatively level surface; if underpoled a longitudinal furrow forms on the surface of the slab as it cools; if over-poled, instead of a furrow, the surface exhibits a longitudinal ridge. Copper, duly poled, is known as ' best selected,' and as ' tough cake ' copper.

ZINC.

(113.) Of this metal, known also very commonly as ' spelter,' *calamine* is a very abundant ore ; another abundant ore is *blende*. The metal is extracted from its ores by a process of distillation, the metal volatilising (§ 94) at a bright red heat, and the vapour, passing into tubes, condenses, and is collected from the tubes in powder and in solid condition. If required pure, further process is necessary. This metal does not appear to have been known till the sixteenth century. Henkel, in 1741, was the first, at least in Europe, who succeeded in obtaining zinc from calamine. Zinc is hardened by rolling, and requires to be annealed at a low temperature to restore its malleability. Until the discovery of the malleability of zinc when a little hotter than boiling water, it was only used to alloy copper with, and sheet zinc was unknown. Zinc expands $\frac{1}{340}$th by heating from the freezing to the boiling point of water. The zinc of commerce dissolves readily in hydrochloric and in sulphuric acid ; pure zinc only slowly. If zinc is exposed to the air, a film of dull grey oxide forms on the surface; it suffers afterwards little further change. Zinc alloys with copper and tin, but not with lead ; it also alloys with iron, for which it is largely used as a coating ; iron so coated being known as ' galvanised ' iron (§ 108).

LEAD.

(114.) Another metal that is prepared in sheet is lead. This metal was known in the earliest ages of the world; it is soft, flexible, and has but little tenacity. One of its principal ores is *galena*. Being a soft metal, it is worked ('*bossed*') by the plumber into various shapes by means of special tools, which often saves the making of joints. As it is compara-

tively indestructible under ordinary conditions, it is largely
used for roofing purposes and for water cisterns. It is also
used for the lining of cisterns for strong acids, in which case
the joints are not soldered in the ordinary way with
plumber's solder, but made by a process termed '*autogenous
soldering*' or '*lead burning.*' Lead prepared in sheet by
casting is known as *cast* lead, but when prepared by the
more modern method of casting a small slab of the metal
and then rolling it to any desired thickness is called *milled*
lead.

ALLOYS.

(115.) An alloy is a compound of two or more metals.
Alloys retain the metallic appearance, and whilst closely
approximating in properties to the metals compounded, often
possess in addition valuable properties which do not exist in
either of the constituent metals forming the alloy. An alloy
of copper and zinc has a metallic appearance and working
properties somewhat similar to those of the individual metals
it is made up of, and so with an alloy of gold, or silver, and
a small percentage of copper. But the latter alloys have
the further property of hardness, making them suitable for
coinage, for which gold, or silver, unalloyed, is too soft.
Like to this addition of copper to gold or silver is the
addition of antimony to lead and to tin, by which alloys are
obtained harder though more brittle than either lead or tin by
itself. The alloy of lead and antimony is used for printer's
type, for which lead alone is too soft.

(116.) Alloys are often more fusible than the individual
metals of which they are composed. Thus while lead melts at
326° C., tin at 233° C., bismuth at 268° C., and cadmium at
321° C., an alloy of 8 parts bismuth, 4 tin, and 4 lead, forms
what is technically known as a ' fusible ' alloy, meaning an
alloy very readily fusible, the particular alloy stated melting
indeed at 95° C., that is, below the boiling-point of water. The
addition of a little cadmium to the above forms a still more
' fusible ' alloy, called Wood's alloy, which melts at about 65° C

(117.) We give here the names of some of the more important alloys, with those of the metals of which they are made up.

Bronze .. .. ⎫
Bell-metal .. .. ⎬ Copper and tin.
Gun-metal .. .. ⎪
Speculum-metal .. .. ⎭

Brass .. .. ⎫
Dutch-metal .. .. .. ⎬ Copper and zinc.
Muntz' metal .. .. .. ⎪
Hard solders (see p. 287).. ⎭

German silver .. .. .. Copper, zinc, and nickel.

Britannia metal .. .. { Tin, antimony, copper, bismuth, and zinc.

Pewter .. .. ⎫ Lead and tin.
Soft solders (see p. 287) .. ⎭

Type-metal .. .. .. Lead and antimony.

ALLOYS OF COPPER AND TIN.

(118.) Alloys of copper and tin are known as bronze, gun-metal, speculum-metal, &c. Some of these alloys possess the property of becoming soft and malleable when cooled suddenly while red-hot, by dipping into cold water. but of being hard and brittle when cooled slowly.

Name of Alloy.	Composition per cent.	
	Copper.	Tin.
Bell-metal	78·0	22·0
Gun-metal	90·0	10·0
Speculum-metal	66·6	33·0
Bronze coinage	95·0	4·0 and 1 zinc.

(119.) Speculum-metal is a very hard, brittle, steel-grey alloy, capable of receiving a very smooth and highly-polished surface.

2 E 2

ALLOYS OF COPPER AND ZINC.

(120.) *Brass.*—Brass is the general name given to alloys of
copper and zinc (ordinary brass consists of two of copper to
one of zinc); by some writers also to alloys of copper and
tin, now better known as *bronze.* Brass was known to the
ancients, who prepared it from copper and calamine (§ 113),
as they were unacquainted with the metal contained in
calamine. We are said to have learned the fusing of copper
with calamine from the Germans, and until a comparatively
recent date brass, called calamine brass, was thus prepared
in this country. Between 1780 and 1800, various patents
were taken out for improving the manufacture of brass, by
fusing copper and zinc direct instead of employing calamine.
The calamine method, however, did not at once die out, as
it was thought by some that calamine brass was better than
that made direct.

In the manufacture of brass, the copper is first melted,
because of its high melting-point, and the zinc, warmed, is
then let down by tongs into the crucible containing the
molten copper, plunged under its surface, and held there till
melted. The mass is then stirred with a hot brass or iron
rod, so as to mix the metals, great care being taken not
to introduce any cold or damp matter. A little sulphate
of sodium 'salt-cake,' or carbonate of sodium 'soda-ash,'
thrown into the crucible at the moment of pouring, assists in
raising any impurities to the surface, which can then be
skimmed off as the mass is poured. With proper management,
the loss of zinc is not so great as might be expected, considering the comparatively low temperature at which it
volatilises, and the relatively high temperature necessary to
melt the copper.

(121.) Brass is harder than copper, and therefore stands
wear better; it is very malleable and ductile, may be rolled
into thin sheets, shaped into vessels by 'spinning' (see § 124),
stamping, or by the hammer, and may be drawn into fine
wire. It is well adapted for casting, as it melts easily at a

lower temperature than copper and is capable of receiving very delicate impressions from the mould. It is said to resist atmospheric influences better than copper, but when its surface is unprotected by lacquer, it rapidly tarnishes and becomes black. It has a pleasing colour, takes a high polish, and is cheaper than copper.

(122.) The malleability of brass varies with its composition, and the heat at which it is worked. The malleability is also affected in a very decided degree by the presence of various foreign metals in its composition, even though these are present in but minute quantity. Brass intended for door-plate engraving is improved by the presence of a little tin; or by the presence of a little lead if to be used in the lathe or for casting. Brass for wire-drawing, however, must not contain lead; nor must brass intended for rolling contain antimony, which renders it brittle. Some kinds of brass are only malleable while cold, others only while hot, others are not malleable at all. A good example of the remarkable malleability of certain kinds is furnished by *Dutch Metal*, which contains a large proportion of copper, and which can be hammered into leaves of less than $\frac{1}{57000}$th of an inch in thickness.

Though extremely tenacious, brass loses its tenacity in course of time by molecular change, especially if subject to vibration or continued tension. It is therefore unfitted for chains or for the suspension of weights. Chandelier chains have been known to lose their tenacity, become brittle, and break; and fine brass wire, which is of course brass in a state of tension, will, in time, become quite brittle, merely hanging in a coil.

(123.) *Muntz' Metal* is a variety of brass consisting of above three parts of copper to two of zinc, with about one per cent. of lead. The alloy is yellow, and admits of being rolled at a red-heat. It is extensively applied for the sheathing of ships, as it is said to keep a cleaner surface than copper sheathing.

Hard Solders.—These are treated further on.

The number of alloys of copper and zinc is considerable, and there is great confusion in respect of their names and composition. The table on opposite page, showing the proportions of some alloys of copper and zinc, is an extract from a table by Dr. Percy.

(124.) *Britannia Metal.*—We follow on with this alloy although not an alloy of copper and zinc alone. Britannia metal is highly malleable and one of the best of the substitutes for silver. It is composed of tin, antimony, copper, bismuth, and zinc, in various proportions, according to the purpose for which it is required. If the alloy is to be 'spun,' that is, worked into shape by specially formed tools whilst revolving in a lathe, a greater proportion of tin is used than when the alloy is only to be rolled. If it is to be cast, the proportion of tin is much less.

ALLOYS OF TIN AND LEAD.

(125.) Alloys of tin and lead furnish us with our soft solders, and are therefore of great practical value. Under the heading 'solders' further particulars of these alloys will be found. The melting points of the soft solders in the Table of Solders (p. 287) can be compared with the melting points (p. 274) of the individual metals of which the solders are composed, as illustrating what has already been stated about the fusing points of alloys often being below those of the metals forming them.

(126.) *Pewter.*—The composition of pewter varies considerably. Common pewter consists of tin and lead alone; the best contains also small percentages of antimony, copper, and bismuth; varying indeed but little from Britannia metal.

SOLDERS.

(127.) A solder is a metallic composition, by the fusion of which metals are united. The requirements of a good solder are twofold. (1) Its melting point must be below that of

No.	Composition per cent. Copper.	Composition per cent. Zinc.	Specific Gravity.	Colour.	Qualities, &c.
1	86·3	12·7	8·705	Reddish yellow	Not so good for rolling, wire-drawing, or hammering as common brass.
2	85·4	14·6	8·591	Yellowish red.	Behaves irreproachably under hammer, rolls, and in wire-drawing.
3	83·02	16·98	8·415	Yellowish red	
4	74·58	25·42	8·397	Pale yellow	Working qualities the same as those of common brass. It is said this alloy was puffed up in England, under the name of Mosaic Gold.
5	66·18	33·82	8·299	Full yellow	Works excellently under the hammer, rolls, and in wire-drawing.
6	52·0	48·0	8·229	Nearly gold yellow ..	Much to be recommended on account of its golden colour. Well adapted for cast ware; after casting very flexible while still hot; hard to cut when cold. Is unsuitable for wire-drawing, hammering, or rolling.
7	32·85	67·15	8·283	Deep yellow	A tough alloy.
8	31·52	68·48	..	Silver white	Too hard to file or turn.
9	30·3	69·7	..	Silver white	
10	29·17	70·83	..	Silver white to silver grey.	Lustre nearly equal to speculum metal (see § 119).

whatever metal is to be joined; (2) it must run easily when melted. It is comparatively easy (§ 116) to fulfil the first condition of a good solder. To fulfil the second requirement, one of the constituents of the solder must be either the same metal as that to be soldered, or a metal which will readily alloy with it, or which will readily coat its surface.

(128.) Solders are 'soft' or 'hard,' according to the temperature at which they melt. Hard solders fuse at a red heat. Soft solders are those which can be applied with a 'soldering iron,' that is to say, a 'copper bit,' or 'plumber's iron,' or with a mouth blowpipe; these solders melt below say 300° C. Hard solders have a much higher fusing point, and require either a forge or a blast blowpipe to apply them. Soldering with hard solders is termed 'brazing.'

(129.) An important particular in the preparation of solders is that they should be well stirred before pouring, preferably with a piece of green wood (§ 112), and the surface of the molten metal exposed as little as possible to the air, so that 'dross' (oxide) shall not form on the surface. A few knobs of charcoal on the molten metal will to a very great extent prevent the formation of dross.

(130.) Examining the soft solders, we see that plumber's solder melts at 227° C., that is to say, at a lower melting point than the metal (lead pipe), for soldering which it is used. Further, it is largely composed of lead. It thus fulfils both the requirements of a good solder. Tinman's solder melts at 160° C. It is used for soldering tin-plate, which, remember, is iron coated with tin. Tin melts at 233° C., a higher temperature than that of its solder, and tin is a constituent of the solder. Again the conditions of a good solder are fulfilled. Tinman's solder is also used for soft-soldering copper, because an alloy of lead and tin will readily coat copper, as also readily alloy with it.

TABLE OF SOLDERS.

SOFT SOLDERS.

No.	Name.	Lead.	Tin.	Melting-point. (Centigrade.)	Uses.
1	Plumber's solder	2	1	227°	For joining lead-pipe.
2	Coarse ,,	3	1	250°	Blowpipe or gas-fitter's solder; sometimes used by tinmen.
3	Fine ,,	1	2	171°	
4	,, ,,	1	1½	168°	First class tinman's solder; used also for soldering zinc, and the 'metal' wheels, floats, &c., of wet gas-meters.
5	Very fine ,,	1	3	180°	Ordinary tinman's solder.
6	Very fusible ,,	1	1	160°	Pewterer's solder.
7	,, ,, soft	1	1 and 2 bismuth	95°	

HARD SOLDERS.

No.	Name.	Copper.	Zinc.	Silver.	Melting-point. (Centigrade.)	Uses.
1	Spelter, hardest	2	1	0		For ironwork, gun-metal, &c.
2	,, hard	1½	1	0		For copper and iron.
3	,, soft	1	1	0		For ordinary brasswork.
4	,, finer	2	2	¼		For finer kinds of brasswork.
5	Silver-solder, hardest	1	0	4		Hardest, but makes very neat joints.
6	,, ,, hard	1	0	1		Makes a sound joint, and will not burn.
7	,, ,, soft	1	0	2		For general use.

SOLDERING FLUXES.

(131.) Substances that 'flux' or aid the flow of metals when melting or melted are termed '*fluxes.*' The general subject of fluxes is outside our province; we are, however, specially interested in what we have designated '*soldering fluxes,*' namely, those fluxes that facilitate the flow of the solders and of the metals of which they are composed.

(132.) Essentially this 'fluxing' consists in the prevention of the formation of oxide (§ 94) to which metals are very prone when highly heated or molten. The black scale (§ 111) that forms on the surface of copper, for instance on copper 'bits,' when highly heated, is an oxide; also the scale that falls off red-hot iron when hammered (§ 105); and also the 'dross' that forms on the surface of molten lead or molten solder (§ 129).

(133.) The employment of charcoal (carbon) for the purpose of preventing the formation of dross we have already alluded to in speaking of the preparation of solders. Sometimes a layer of it is spread over the surface of the molten metal to keep it from contact with the air; sometimes a layer of grease.

In their character of aiding the flow of metals, fluxes are further applied to the surface of the metals to be soldered, which they clean, as well as aiding the flow of the molten solder when that is applied.

(134.) '*Spirits of salts*' (hydrochloric or muriatic acid) when 'killed' is a most useful flux for soft solders. The 'killing' is done by dissolving zinc in the acid till gas is no longer given off. As the gas is most offensive, the dissolution of the zinc should be effected in the open air. This flux is not one to be used where rust would be serious; though there is very little danger of this, if, after soldering, the joint is wiped with a clean damp rag, and further cleaned with whiting.

(135.) *Resin, or resin and oil* is a good flux for almost any kind of soft soldering. The surface to be soldered must, however, be well cleaned before applying the flux.

(136.) '*Killed spirits*' (chemically, *chloride of zinc*) is specially useful for tin-plate soldering, because it assists in cleaning the edges to be joined; whereas if resin, or resin and oil, is used, the edge must, as stated, be cleaned previously.

(137.) Spirits of salts not killed is used for soldering zinc because it cleans the surface of the zinc; it acts indeed as chloride of zinc, for this is what it becomes on the application to the zinc, in fact the cleaning is the result of this action. The killed spirits, however, answers equally well as the strong acid if the zinc is bright and clean, so far as the experience of the writer has gone. The 'raw' (unkilled) spirits of salts is improved, as a flux for soldering zinc, by adding a small piece of soda to it.

(138.) Powdered resin, or resin and oil, as a flux, possesses the great advantage over chloride of zinc, that there is no risk of rust afterwards. For this reason resin, or resin and oil, is much used in the manufacture of gas-meters. It is also used, or should be, for the bottoms and seams of oil bottles. The resin and oil flux can easily be wiped off joints immediately after soldering; it is for this reason better than dry resin which has to be scraped off. Even this trouble, however, can be got over if the hot copper bit is dipped in oil before application to the joint to be soldered.

(139.) In 'tinning' a copper bit, that is, coating its point with solder before using it in soldering (a piece of manipulation of much importance as regards the easy working of the bit), the best thing to use is a lump of *sal-ammoniac*. In a small hollow made in the sal-ammoniac, the point of the bit, after having been filed smooth and bright, should be well rubbed, while hot, along with some solder; the point of the bit will then become coated with solder ('tinned'). For 'tinning' copper utensils, that is, coating them with tin, sal-ammoniac both in powder and lump is largely used. Sal-ammoniac water is also used for cleaning copper bits;

the hot bits being dipped into it prior to being used for soldering. Killed spirits, however, acts better. Sal-ammoniac and resin, mixed, is used as a flux for soldering 'sights' on gun-barrels.

(140.) As a flux for lead soldering, plumbers use tallow ('touch'). For pewter, Gallipoli oil is the ordinary flux.

(141.) For hard soldering, the flux is *borax*. This flux is also made use of in steel welding.

TINNING AND RE-TINNING.*

The remarks that follow have mainly in view the needs of the ordinary workshop. The object of covering the surface of a piece of sheet metal with tin is to protect the sheet from chemical action. In the case of iron, tinning prevents oxidation (rusting); in the case of copper cooking vessels, it protects the copper from the action of acids, which might prove injurious to health.

To ensure the proper union or *alloying* of the tin with the iron or other surface to be covered, this surface must be made perfectly clean, and to assist the flow of the molten tin over it a suitable flux must be employed. According to the purpose for which the sheet metal under treatment is to be used, it has to be tinned on one or on both sides. If the sheet is to be tinned on one side only, the tin is placed, either in a molten or a solid state, on the cleaned surface, and heat, from a forge fire or a gas blowpipe, is applied to the under surface, so as to heat the sheet of metal to a temperature sufficient to keep the tin melted if already molten or to melt the tin if this is applied solid. Greater care is needed with the flux when the tinning is on one side only than when on both sides, because there is a tendency for fluxes such as chloride of zinc (killed spirits) to dry and form a skin on the surface of the sheet before the tin has time to cover and unite

---

* The writer originally prepared this article for the *Ironmonger Diary*, in which it appeared.

with it. Hence fluxes such as sal-ammoniac, resin and tallow
are the most suitable fluxes. Where a sheet has to be tinned
on both sides the like care is not required, as it can be dipped
bodily into molten tin. Care *is* required, however, that the
sheet metal to be dipped is not wet or damp or even very cold,
for if it is the tin will fly out, perhaps even to the serious in-
jury of the operator. The sheet should be dipped gradually
so that any dampness is steamed off and the sheet warmed
before it is finally plunged and submerged in the molten tin.

RE-TINNING COPPER.

In the re-tinning of copper saucepans and other vessels
that have been used for cooking, they should first be warmed
and all grease, especially near rivets, carefully wiped off.
They can then be cleaned by scouring thoroughly until
bright with either wet forge scale, or better still, silver sand
which has been moistened with lemon juice, and which may
be applied either with a cork to secure friction, or with a
piece of rough moleskin, or with tow or similar material.
It is sometimes necessary, when articles are very dirty and
where there are rivets, to first clean them with warm killed
spirits before scouring thoroughly bright. In the case of
copper moulds, it is often difficult and inconvenient to
thoroughly scour the inside crevices. It will be found very
helpful in such cases to half fill the moulds with a solution
of equal parts lemon juice and spirits of salts, and after a
few minutes to rub with a strip of cork. This greatly assists
in securing a perfectly bright surface. Better results are
obtained with the mixture of lemon juice and raw spirits
of salts than with either of these liquids separately. When
the vessel to be re-tinned has been thoroughly cleansed, it
should be well washed for about a minute with chloride of
zinc, and then if it is to be tinned on both sides, it should be
gradually and carefully dipped into a bath of molten tin.
For re-tinning a vessel on one side only it is advisable to
prepare the surface that is to go next the fire by coating it

with a mixture of wet whiting and salt, as the untinned part can then be cleaned more readily afterwards. The vessel being placed on the fire, the tin is applied molten or solid. If molten or after melting, a little fresh mutton fat or a little powdered resin or powdered sal-ammoniac may be dusted over the surface to prevent oxidation. Then the vessel should be moved about gently so that the molten tin passes over the whole of the surface that is to be covered. If there are any spots to which the tin does not readily adhere, it may be wiped over them by means of a piece of tow or by a piece of rod wire that has been previously prepared by being coiled up at one end and tinned. It is useful to have at hand a lump of sal-ammoniac to make similar use of, wiping as it were with it the tin over the spots that do not readily cover. A large cork is also handy to rub the tin over the surface that does not willingly take it. The tin invariably clings about the edges of the vessel, and forms a list which must be shaken off. If it will not shake off, it must be wiped off gently and quickly with a piece of moleskin cloth, this is better than tow. When tinned either on one or both sides, the vessels should at once be washed with hot water, dried with bran, and polished bright, and it is important that this should be done immediately after the tinning if a bright surface is to be ensured. Scouring and pickling liquids may be drained off and saved, and used over and over again, as well as the sand, which will settle at the bottom.

RE-TINNING WROUGHT IRON.

Wrought iron is more difficult to tin than copper, and the surface requires longer time to prepare it to take a coating of tin, and the preparation is more troublesome. The article to be tinned should be heated to redness and afterwards placed in dilute sulphuric acid for about twelve hours and then immersed in chloride of zinc (killed spirits) for about six hours, or, if the article cannot be annealed then it should be "pickled" in raw spirits of salts for twenty-four hours (it is better to do this in a closed vessel) and afterwards immersed in chloride of zinc in which sal-ammoniac has been dissolved. In either case, if there are patches on the surface which are difficult to clean or tin, it will be advantageous to well rub them with a piece of wet pumice stone. A wrought-iron stewpan which has been previously tinned and is in fair condition, may only require warming to clean off the grease, and judgment must be used as to the length of time required for the pickling, the article being examined at intervals to see whether the surface is clean and bright. The wrought iron vessel thoroughly cleaned by either method, should now be gradually dipped in a bath of molten tin, and allowed to remain there for two or three minutes. If required to be tinned on one side only, the tinning may be done in the same way as already explained for copper.

RE-TINNING CAST IRON.

Cast iron is most difficult to tin, and for this reason one would never recommend an attempt being made to re-tin cast-iron cooking vessels as it pays better to substitute new ones. If cast iron is to have anything like a good appearance when tinned, it must not be spongy, and its surface should be rendered smooth either by grinding, filing or machining, whichever it is most convenient to do. If possible the casting should be made red hot, and then pickled for twenty-four hours in slightly diluted spirits of salts (hydrochloric acid), and immediately on being removed should again be completely immersed for about two hours in chloride of zinc. If the casting cannot be made red hot, it should be immersed for about ten minutes in dilute sulphuric acid and warmed to about 90° Fahrenheit, then pickled in raw spirits of salts for two days, and afterwards allowed to soak for two hours in a mixture of chloride of zinc and sal-ammoniac, about two ounces of the latter to a gallon of spirits.

The casting, by whichever method cleaned, is now ready to be dipped gradually into the tinning pot. Should it not be sufficiently tinned the first time, the surface should be well rubbed with a piece of cork directly the casting is withdrawn from the melted tin, and be again dipped in chloride of zinc, and then once more into the tinning pot. Sometimes it is required to tin a part of a casting and to use the copper bit. When such is the case it is essential that the surface be cleaned and smoothed by scraping or otherwise, then soaked with raw spirits of salts, and then with chloride of zinc in which sal-ammoniac has been dissolved. Then the tin should be applied by means of the copper bit, and a piece of cork used to rub in the tin where there is a difficulty to get it to adhere. Tinning in this way, though not so strong as tinning in the way just described, is sufficiently effective for most purposes.

CHAPTER II.

SEAMS OR JOINTS.

WE notice here and illustrate the more important seams or joints used in metal-plate work. The drawings are intended to aid in the intelligent comprehension of the formation of joints, and not as exact representations of them.

Lap Seam.—In No. 1 is shown how metal plates are arranged for a lap seam which is to be soldered.

Circular Lap Seam.—No. 11 shows how the edge of the bottom of a cylindrical article is bent up previous to soldering. It is evident that this seam is essentially No. 1 seam adapted to the fitting a bottom to a cylinder. Such bottom is called a ' snuffed on ' or ' slipped on ' bottom.

Countersunk Lap Seam.—This is represented in No. 2. It will be seen that the edge of one of the plates is bent down, so that the edge of the plate to be joined to it may lie in the part bent down, and that the two plates when joined may present an unbroken surface.

Rivetted Lap Seam.—This is shown in No. 8. The amount of lap should not be less than three times the diameter of the rivet.

Folded Seam.—No. 3 shows how the edges of plates are prepared for folded seam.

Circular Folded Seams.—With a circular article the folded seam is sometimes in the form of No. 12, which shows a ' paned down ' bottom to a cylinder. This seam is essentially No. 13.

Another form of circular folded seam is shown in No. 13. It is really No. 12 seam turned up, so as to lie close against the cylinder (see reference letter A in Nos. 12 and 13).

2 F

A bottom thus fitted is called a 'knocked up' bottom. Here again comparison with No. 3 should be made.

Double Folded Seam.—This is shown in No. 6, and needs no explanation. It is used with thick plates, where these when

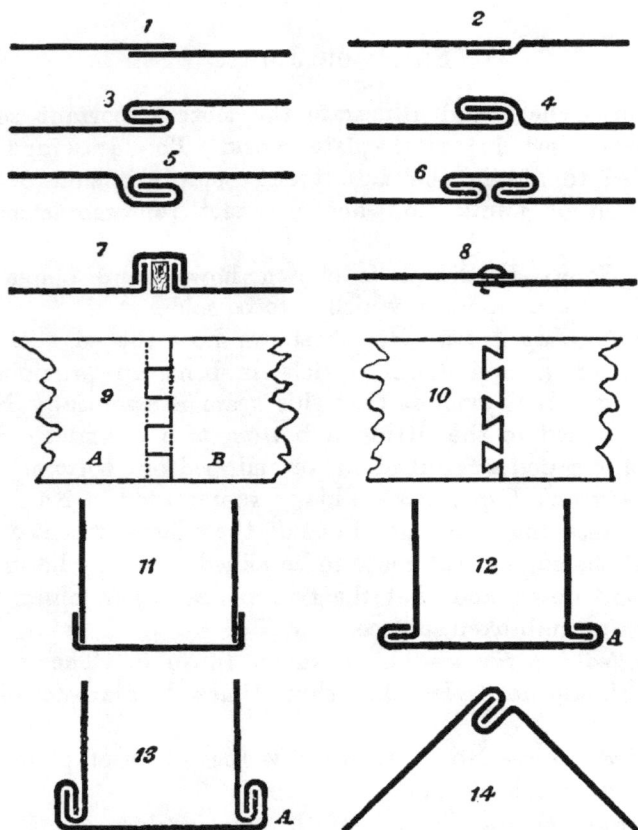

joined have to present to the eye an unbroken surface; as in the hot-plates of large steam-closets.

Grooved Seam.—This is represented in No. 4. It will be seen that the seam is the same as No. 3, but one plate is countersunk. In fact No. 3 shows the seam as prepared for countersinking ('grooving') with a 'groover.' Seam No. 6 is used where plates are too thick for grooving.

Countersunk Grooved Seam.—This seam (No. 5) is used when an unbroken surface is required on the outside of an article, for example, in toilet-cans, railway-carriage warmers. It is prepared as No. 3 and then countersunk the reverse way to No. 4.

Box Grooved Seam.—This seam, shown in No. 14, is used for joining plates in 'square work,' as for example where the ends and sides of a deed-box are joined together. It is essentially No. 3 seam.

Zinc - roofing Joint.—The arrangement for this joint is seen in No. 7. This joint admits of the expansion and contraction of the zinc sheets. The edges of two sheets, the wood 'roll,' and the 'roll cap' are shown. The zinc 'clip,' by which usually the sheets of zinc are held down, is not represented.

Brazing Joints.—A brazing joint for thin metal is shown in Fig. 9. The edge of plate A is cut to form laps as represented, and these laps are arranged alternately over and under the edge of plate B.

For thick metal the brazing joint is shown in Fig. 10. It is essentially the same thing as the 'dovetail' joint of the carpenter.

CHAPTER III.

Useful Rules in Mensuration.

To find the circumference of a circle, the diameter being given.

Multiply the diameter by $3\frac{1}{7}$. In other words, multiply the diameter by 22 and divide by 7.

Or, should closer accuracy be required, multiply the diameter by 3·1416.

Example I.—The diameter of a circle is 8 inches; to find the circumference.

$8 \times 22 = 176$, which divided by 7 gives $25\frac{1}{7}$ inches, the circumference required.

Or, $8 \times 3·1416 = 25·13$ inches.

To find the area of a circle, the diameter being given.

Multiply one-quarter of the diameter or, which is the same thing, half the radius, by the circumference.

Example II.—The diameter of a circle is 8 inches; to find its area.

One-quarter of the diameter is 2 inches; the circumference is $25\frac{1}{7}$ inches.

$2 \times 25\frac{1}{7} = 50\frac{2}{7}$ square inches, the area required.

To find the area of an ellipse, the axes being given.

Multiply the axes together, and multiply the result by ·7854.

Example III.—The major axis of an ellipse is 6 inches and the minor 4 inches; to find its area.

$6 \times 4 = 24$, which multiplied by ·7854 gives 18·85 square inches nearly.

To find the area of a rectangle.

Multiply the length by the breadth.

Example IV.—The length of a rectangle is 16 inches and the breadth 9 inches; to find its area.

$16 \times 9 = 144$ square inches, that is, one square foot, the area required.

To find the volume of a circular, elliptical, rectangular, or other tank, or vessel, of which the sides are perpendicular to the base.

Multiply the area of the base by the height.

If the answer is required in cubic inches, all the dimensions must be multiplied in inches. If in cubic feet, the dimensions must be in feet. (See Examples.)

Example Va.—The height of a circular tank is 6 feet, and the diameter of the base 8 feet; to find its volume.

By Example II. the area of the base is $50\frac{2}{7}$ square feet, which multiplied by 6 feet gives $301\frac{5}{7}$ cubic feet, the volume required.

Example Vb.—The height of an elliptical tank is 1 foot 6 inches, the base is 6 inches by 4 inches; to find its volume.

By Example III. the area of the base is 18·85 square inches, which multiplied by 18 inches (that is to say, by the height in inches, as the answer is to be in cubic inches) gives 339·3 cubic inches, the volume required.

Example Vc.—The height of a rectangular vessel is 2 feet 3 inches, the length 1 foot 4 inches, and the breadth 9 inches; to find its volume.

We will suppose the answer is required in cubic feet. This being so, it is in feet that the dimensions must be multiplied. Stated in feet, the height is $2\frac{1}{4}$ feet, the length $1\frac{1}{3}$ feet, and the breadth $\frac{3}{4}$ foot.

By Example IV. the area of the base is $1\frac{1}{3}$ feet multiplied by $\frac{3}{4}$ foot, that is, is $\frac{4}{3} \times \frac{3}{4} = \frac{12}{12} = 1$ square foot, which multiplied by $2\frac{1}{4}$ feet gives $2\frac{1}{4}$ cubic feet, the volume required.

To find the volume of a right cone.

Multiply the area of the base by the height and divide by 3.

Example VI.—The height of a cone is 6 inches, and the diameter of the base $3\frac{1}{2}$ inches; to find the volume.

The circumference of the base is $\dfrac{3\frac{1}{2} \times 22}{7}$ = 11 inches.

The area of the base is one-half of $1\frac{3}{4}$ inches (the radius) $\times$ 11 = $\frac{7}{8} \times 11 = \frac{77}{8} = 9\frac{5}{8}$ square inches.

And the area $9\frac{5}{8} \times 6$ inches (the height) = $57\frac{3}{4}$, which divided by 3 gives $19\frac{1}{4}$ cubic inches, the volume required.

To find the volume of a frustum of a right cone.

From the volume of the complete cone of which the frustum is a part subtract the volume of the cone out off.

For example, the volume of the frustum C A B D (Fig. 8a, p. 33) is equal to the volume of the complete cone O A B less the volume of O C D the cone cut off.

The height of the complete cone and that of the cone cut off from it to form the frustrum can be found by Problem V., p. 36.

To find the volume of a sphere, the diameter being given.

Multiply the diameter of the sphere by the area of a circle of same diameter, and take two-thirds of the product.

Example VII.—The diameter of a sphere is 8 inches. The area of a circle of same diameter is $50\frac{2}{7}$ (see Example II.); which multiplied by 8 gives $402\frac{2}{7}$ cubic inches, two-thirds of which ($268\frac{1}{3}$ about) is the volume required.

Or, Multiply the cube of the diameter by ·5236.

Given the volume of a vessel, any vessel, to find the number of gallons, quarts, or pints that it will hold.

If the volume is in cubic feet, as in Example V *a*, then, to bring it to gallons, multiply by $6\frac{1}{4}$, there being in a cubic foot of water $6\frac{1}{4}$ gallons about. If the volume is in cubic inches, divide by 277. The number of cubic inches in a gallon of water is $277\frac{1}{4}$ nearly; but in ordinary calculations, the quarter may be omitted.

If the volume is required in quarts, multiply it, if in cubic feet, by 25; if in cubic inches, divide it by 69. The number of cubic inches in a quart of water is about $69\frac{1}{4}$; in our examples here we have disregarded the fraction.

If the volume is required in pints, multiply it, if in cubic feet, by 50; if in cubic inches, divide it by 35. The number of cubic inches in a pint of water is rather more than $34\frac{1}{2}$; in our examples we have taken it as 35.

Example VIII.—To find the number of gallons, quarts, or pints, contained in the tank of Example V*a*.

Gallons.—$301\frac{5}{7} \times 6\frac{1}{4} = 1885\frac{5}{7}$ gallons.

Quarts.—$301\frac{5}{7} \times 25 = 7542\frac{6}{7}$ quarts.

Pints.—$301\frac{5}{7} \times 50 = 15085\frac{5}{7}$ pints.

Example IX.—To find the number of gallons, quarts, or pints, contained in the tank of Example V*b*.

Gallons.—$339 \cdot 3 \div 277 = 1 \cdot 22$ gallons about.

Quarts.—$339 \cdot 3 \div 69 = 4 \cdot 92$ quarts about.

Pints.—$339 \cdot 3 \div 35 = 9 \cdot 7$ pints about.

Given the number of gallons, quarts, or pints, that a tank or other vessel contains, any vessel, of which the sides are perpendicular to the base, also the dimensions of the base, to find its height.

Divide the number of gallons by $6\frac{1}{4}$; this will give the volume of the required tank in cubic feet. If the quantity is given in quarts, then to ascertain the required volume, multiply by 69. If the quantity is given in pints, multiply by 35.

To find the required height for the tank, divide the volume found as just shown, by the area of the base.

COMPARATIVE WEIGHTS AND GAUGES OF SHEET IRON, COPPER, AND ZINC, AND OF TIN-PLATE.

Gauge.	Equivalent by Birmingham Wire-Gauge.	Tin-Plates of Equivalent Strength.	Approximate Weight per Square Foot.		Zinc. Approximate Weight.		
			Iron.	Copper.	Per Square Foot.	Per Sheet 7 ft by 3 ft.	Per Sheet 8 ft. by 3 ft.
			lb. oz.	lb. oz.	lb. oz.	lb. oz.	lb. oz.
4	33	..	..	..	0 4¾	6 4	
5	31	..	..	..	0 5¾	7 9	
6	30	1 C	0 8	0 9¼	0 6¾	9 0	10 5
7	29	..	0 9	0 10¼	0 7¾	10 6	11 10
8	28	{ 1 X and D C }	0 10¼	0 12	0 9	11 12	13 5
9	27	X X	0 11½	0 14	0 10¼	13 5	15 8
—	26	S D C	0 13	0 13½			
10	25	{ S D X / D X }	0 15	1 0	0 11½	15 1	17 3
11	24	{ 1 XXXXX / DXX / SDXXX }	1 0	1 2½	0 13¾	17 4	20 0
12	23	DXXX	1 2	1 5	0 15	19 12	22 11
13	22	DXXXX	1 4	1 8	1 1	22 4	25 7
14	21	DXXXXX	1 6½	1 10	1 2¾	24 12	28 2
15	20	DXXXXXX	1 9	1 12	1 5¾	28 11	32 10
16	19	..	1 12	2 0	1 8¾	32 10	37 2
17	18	..	1 14½	2 4	1 11¾	36 8	41 8
18	—	..	..	..	1 14¾	40 7	
19	17	..	2 3	2 8	2 3¼	45 0	
20	16	..	2 8	2 14½	2 4½	49 8	
21	15	..	2 13	3 4	2 8¾	55 0	
22	..	..	..	..	2 12¾	59 0	
23	..	..	..	..	3 1	64 8	

STRENGTH, SIZES, AND WEIGHTS OF TIN-PLATES.

Strength.	Sizes of Sheets in Inches.	Thickness in Birmingham Wire-Gauge.	Weights per Box and Number of Sheets.			
				cwt.	qrs.	lb.
1 C	14 × 10		225 sheets weigh	1	0	0
	14 × 20	30	112 ,, ,,	1	0	0
	28 × 10					
	28 × 20		112 ,, ,,	2	0	0
1 X	14 × 10		225 ,, ,,	1	1	0
	14 × 20		112 ,, ,,	1	1	0
	28 × 10					
D C	28 × 20	28	112 ,, ,,	2	2	0
	17 × 12½		100 ,, ,,	0	3	14
	34 × 12½		50 ,, ,,	8	3	14
	25 × 17					
1 X X	14 × 10		225 ,, ,,	1	1	21
	14 × 20		112 ,, ,,	1	1	21
	28 × 10	26 easy				
S D C	28 × 20		112 ,, ,,	2	3	14
	15 × 11		200 ,, ,,	1	1	27
	22 × 15		100 ,, ,,	1	1	27
1 X X X	14 × 10		225 ,, ,,	1	2	14
	14 × 20		112 ,, ,,	1	2	14
	28 × 10					
D X	28 × 20	25 easy	112 ,, ,,	3	1	0
	17 × 12½		100 ,, ,,	1	0	14
	34 × 12½		50 ,, ,,	1	0	14
S D X	25 × 17					
	15 × 11		200 ,, ,,	1	2	20
	22 × 15		100 ,, ,,	1	2	20
1 X X X X	14 × 10		225 ,, ,,	1	3	7
	14 × 20		112 ,, ,,	1	3	7
	28 × 10	25 full				
S D X X	28 × 20		112 ,, ,,	3	2	14
	15 × 11		200 ,, ,,	1	3	13
	22 × 15		100 ,, ,,	1	3	13
1 X X X X X	14 × 20		112 ,, ,,	2	0	0
	28 × 10					
	28 × 20		112 ,, ,,	4	0	0
D X X	17 × 12½	24	100 ,, ,,	1	1	7
	34 × 12½		50 ,, ,,	1	1	7
	25 × 17					
S D X X X	15 × 11		200 ,, ,,	2	0	6
	22 × 15		100 ,, ,,	2	0	6
S D X X X X	15 × 11	24 full	200 ,, ,,	2	0	27
	22 × 15		100 ,, ,,	2	0	27

STRENGTH, SIZES, AND WEIGHTS OF TIN-PLATES—*continued.*

Strength.	Sizes of Sheets in Inches.	Thickness in Birmingham Wire-Gauge.	Weights per Box and Number of Sheets.				
					cwt.	qrs.	lb.
1 X X X X X X	14 × 20 28 × 10 28 × 20	23 easy	112 sheets weigh	2	0	21	
			112 ,, ,,		4	1	14
D X X X	17 × 12½ 34 × 12½ 25 × 17	23	100 ,, ,,		1	2	0
			50 ,, ,,		1	2	0
D X X X X	17 × 12½ 34 × 12½ 25 × 17	22	100 ,, ,,		1	2	21
			50 ,, ,,		1	2	21
D X X X X X	34 × 12½ 25 × 17	21	50 ,, ,,		1	3	14
D X X X X X X	34 × 12½ 25 × 17	20	50 ,, ,,		2	0	7

INDEX.

References are to pages, except those in parentheses, which are to paragraphs.

A.

ALLOYS, defined, 418; fusible, 418; important, 419; of iron and carbon (steel), 414; copper and tin, 419; copper and zinc, 420; tin and lead, 422

Angle, defined, 3; to draw, equal to a given angle, 6; to bisect, 10.

Angles, measurement of, 20.

Annealing, 409.

Apex, of cone, 24, 105 (43); of pyramid, 66 (29).

Aquarium base pattern, of one moulding, 310; of more than one, 314.

Arc, defined, 5; to complete circle from, 9; to find if given curve is, 10.

Arcs, proportionate and similar, 124.

Area, of circle, 436; ellipse, 436; rectangle, 437.

Articles, equal tapering and other, see Bodies.

Athenian hip-bath, plan, 137.

Axis, of cone, 24 (6), 105 (43); of ellipse, 16; of pyramid, 66 (29).

B.

BAKING-PAN pattern, 77.

Bath, Athenian hip, plan, 137.

——, 'equal-end,' equal taper, pattern, 84–90; four pieces, 84; two pieces, 86; one piece, 87; short-radius method, 89.

——, ——, unequal taper, plan, 130; pattern, 157–68; four pieces, 158; two pieces, 161; one piece, 162; when ends are cylindrical, 164; short-radius method, 166.

——, oblong taper, plan, 140; representation (plate), 237; plate explained, 231; pattern, 230–6; short-radius method, 235.

——, Oxford hip, see that heading.

——, oval, plan, 131; representation (plate), 181; plate explained, 169; pattern, 168–83; four pieces, 171; two pieces, 175; one piece, 177; short-radius method, 177.

Bath, sitz, plan, 137.

Bends, patterns for, 359–407 ; equal circular section, made up of two like pieces, 360 ; of throat and back piece, 364 ; four pieces (two patterns) 369 ; four pieces (three patterns), 375 ; circular segments, 379 ; lobster back cowl, 382 ; of square section, 389–98 ; circular tapering, of circular segments, 398 ; four pieces (three patterns), 402 ; see also Elbow Patterns ; Round-neck T.

Bevel (angle), 6.

Bisect a line, to, 8 ; an angle, 10 ; practically, 310 (107A).

Bodies, of equal taper, see Equal-tapering bodies ; of unequal taper, see Unequal-tapering bodies ; of rectangular base and circular top, 293.

Brass, 420.

Brazing, 424 ; joints for, 435.

Britannia metal, 422.

Bronze, 419, 420.

C.

CANISTER-TOP, oval, plan, 132 ; representation (plate), 203 ; plate explained, 193 ; pattern, 192–205 ; four pieces, 193 ; two pieces, 197 ; one piece, 198 ; short-radius method, 200.

Cap, conical, and concentric square pipe, pattern, 286 ; and concentric rectangular pipe, 290.

Carbon, 414 ; and iron, 413, 414

Cast iron, 413, 414 ; re-tinning, 431.

Centre, of circle, 5 ; of ellipse, 16.

Chord, defined, 5.

Chords, scale of, to draw, 21 ; how to use, 22.

Circle, defined, 5 ; sector of, 28 (9) ; to find centre, 9 ; to describe, that shall pass through three points, 9 ; to complete from arc, 10 ; to find if given curve is arc of, 10 ; to inscribe polygon in, 10 ; to find length of circumference geometrically, 12 ; to find same arithmetically, 436 ; to find area, 436.

Circular pipe and cone, pattern of elbow formed by, 284.

——, pipes, meeting at any angle, pattern, 239 ; one being an oblique cylinder, T formed by, pattern, 267.

—— ——, inclined, extreme cases of oblique cone frusta, 112 (61) ; pattern for, 121, 255 ; pattern for, in Y-piece, 261.

—— ——, unequal, bend to connect, 398.

——, tapering bend, 398.

—— ——, ventilator, 407.

——, vessel, volume, 437.

Circumference of circle, defined, 5 ; to find length of, geometrically, 12 ; arithmetically, 436.

Classification of patterns, 2.

Coffee-pots, 34 ; hexagonal, 70 (35).

Colour of metals, 408 (93).

Conductivity of metals, 411.

Cone, defined, 105 (43); axis, radius, apex, base, 24 (6), 105 (43); elevations of generating lines, 108 (55).

——, right, defined, 24 (6), 105 (44); basis of patterns for articles of equal taper, 24 (5); compared with oblique cone, 106 (from 47); representation (plate), 227 (and see p. 217); development of, by paint, 28; generating lines, 106 (47); corresponding points of generating lines, see Corresponding points; to find height, 26; to find slant, 27; to find dimensions of, from frustum, 36; volume, 438.

Cone, right, pattern, 29–31; one piece, 29; more than one piece, 30.

——, ——, frustum (round equal-tapering body), defined, 33; representation, 50; representations of round equal-tapering articles, 34; relations of, with complete cone, 34 (from 13); development, 34 (15); volume, 438.

——, ——, ——, plan, 50 (23); characteristic features, 55 (c, d); to draw, see equal-tapering bodies (plans, to draw).

——, ——, ——, pattern, 37–43; ends and height given, 37; ends and slant given, 39; pattern for parts of, 39; short-radius method, 41.

——, oblique, defined, 106 (45); basis of patterns for articles of unequal taper, 105 (42); compared with right cone, 106 (from 47); obliquity, how measured, 107 (52); representations (plates), 181, 203, 213, 227, 237; development of, 108 (56); generating lines, 106 (46–7); longest and shortest generating lines, 107 (from 51); lines of greatest and least inclination, 107 (52); height of elevations of generating lines, 108 (55); true lengths of elevations of generating lines, 107 (54).

——, ——, plan of axis, 125–6 (a); of generating lines, 126 (d, e); of longest and shortest generating lines, 126 (b); of apex, 126 (from c).

——, ——, pattern, 108.

——, ——, frustum (round unequal-tapering body), defined, 111 (58–9); representation, 125 (plate), 259; generating lines, 126 (b); generating lines of, when circumscribing oblique pyramid frustum, 150 (78); corresponding points of generating lines, see Corresponding points; oblique cylinder an extreme case of, 112 (61).

——, ——, ——, plan, 125 (69), 128 (from 73); of axis, 126 (a); of lines of greatest and least inclination (longest and shortest generating lines), 126 (b); characteristic features, 127 (71–73).

——, ——, ——, pattern, 113–23; ends, height, and inclination of longest generating line given, 113; height and plan given, 116; short-radius method, apex accessible or inaccessible, 118; extreme case of (oblique cylinder), 121.

—— cap, with pipe attached, see Conical Cap.

Cone ellipse, size of circular pipe to fit, 284.

—— penetratings and penetrations, see Penetrations.

—— sections, and patterns of sectioned cones, 271–82.

—— ——, elliptic, 271 (93A); to measure and draw the ellipse, 274 pattern of the cut cone, 276; case of elliptic section, 284.

—— ——, parabolic, 272 (94A); to measure and draw the parabola, 278 pattern of the cut cone, 279.

—— ——, hyperbolic, 272 (95A); to measure and draw the hyperbola, 281; pattern of the cut cone, 282; cases of hyperbolic section, 286, 290.

—— and circular pipe, pattern of elbow of, 284.

Conical cap, and concentric square pipe, pattern, 286; and concentric rectangular pipe, 290.

Cooking vessels, copper and iron, re-tinning, 429–31.

Copper, 408, 409, 416; gauges and weights of, 440; and tin alloys, 419; and zinc alloys, 420; table of, 423; copper bits, tinning, 427; cooking vessels, re-tinning, 409.

Corresponding points, of right cone frustum, 35, 51; of equal-tapering body, 52 (25); of oblique cone frustum, 124 (68), 126 (b).

—— —— in plan, of right cone frustum, 51 (23, 24), 54; of equal-tapering body, 52 (25), 53 (26); of oblique cone frustum, 124 (68), 129 (g); of oblique pyramid frustum, 145 (76); of oval equal-tapering body, 65 (28); distance between, is equal in plans of equal-tapering bodies, 55 (from a); to find distance between, height and slant given, 56; height and inclination given, 57.

Cowl, lobster-back, pattern, 323.

Cubic dimensions, conversion into gallons, 439.

Curved elbow, see Elbow.

Cylinder, right, defined, 112 (61); extreme case of right cone frustum; 271 (93A).

——, ——, circular sections of, 269 (92A); elliptical sections of, 271 (93A); cut obliquely, to measure and draw the ellipse. 274; pattern of the cut cylinder, 274.

——, oblique, defined, 112 (61); an extreme case of oblique cone frustum, 112 (61); pattern for, 121, 255; pattern for, in Y-piece, 261; pattern for, in T-piece formed by oblique and right cylinders, 267.

——, ——, elliptical sections, 269 (92A).

——, penetrated by conical body, 328.

——, sections, and patterns of sectioned cylinders, 271–74.

Cylinders, equal, meeting at any angle, pattern, 239.

D.

DEFINITIONS, of straight line, angle, perpendicular, 3; parallel lines, triangle, hypotenuse, polygon, pentagon, hexagon, heptagon, octagon, quadrilateral, square, oblong, rectangle, 4; circle, circumference, arc quadrant, semicircle, radius, chord, diameter, 5; ellipse, 17; ellipse focus, axis, diameter, centre, 16; cone, 105 (43); cone axis, radius, apex, base, 24 (6), 105 (43); right cone, 24 (6) 105 (44); right cone frustum, 33; oblique cone, 106 (45); oblique cone frustum, 111 (58-9); cylinder, right and oblique, 112 (61); pyramid and axis, 66; pyramid frustum, 143 (74); proportional arcs, similar arcs, 124; corresponding points, see Corresponding Points; sections of cylinder and cone, 271 (93A).

Degree (angle) explained, 21.

Diameter of circle, 5.

—— of ellipse, 16.

Dripping-pan pattern, with well, 295.

Ductility of metals, 410; table of, 411.

Dutch metal, 421.

E.

EDGING, allowance for, 33.

Egg-shaped oval, to draw, 14.

Elbow-patterns, of cylindrical pipes, any angle, 239; rectangular pipes, any angle, 264; circular pipe and cone, 284; curved, to join two equal rectangular pipes, 266; to join two unequal square pipes, 389-97, (throat an angle of the square, 389; throat a side of the square, 394).

Elevation, explained, 48.

Ellipse, defined, 17; focus, axes, diameter, centre, 16; to describe mechanically, 15, 18; geometrically, 17; area, 436.

—— of cylinder cut obliquely, to measure and draw, 274.

—— of cone cut elliptically, to measure and draw, 274; size of circular pipe to fit, 284.

Elliptical vessel, volume, 437.

Equal end bath, see Bath.

Equal-tapering bodies, 24.

Round, 28-45; essentially right cone frusta, 34 (13); to find slant or height of cone of which body is portion, 36; to find slant of body, ends and height given, 43; to find height, ends and slant given, 44; slant and inclination given, 44; see also Cone (right, frustum).

Of flat surface, 66-83; essentially right pyramid frusta, see Pyramid (right frustum).

Equal tapering bodies.

Of flat and curved surfaces, 84–96; curved surfaces, portions of right cone frusta, 55 (d), 84; see also Cone (right frustum).

——— ——— , plans,

Characteristic features of, of round bodies, 51; characteristic feature, body oblong with round corners, 52; features of plans summarised, 55; how to find from plan if article is of equal taper, 55 (b).

——— ——— ——— , ——— , to draw.

Either end, height, and slant given, 57; either end and distance between corresponding points ('out of flue') given, 59.

Oval bodies, 64

Of flat and curved surfaces. Oblong bodies, 59, 62.

See also Corresponding points.

——— ——— ——— , patterns.

Round bodies (frusta of right cones), 28–45; see also Cone (right, frustum, pattern).

Oval bodies. Patterns, 96–104; in four pieces, 96; in two pieces, 99; in one piece, 100; short radius method, 102.

——— ——— ——— . Bodies having flat surfaces.

Ends of body and height given, 71; short radius method, 73; body oblong or square (pan) pattern in one piece with bottom, 77, 80; same with bottom, sides, and ends in separate pieces, 82; baking-pan pattern, bottom, width of top, and slant given, 77; bottom, length of top, and slant given, 80; top, slant, and height given, 81; top, slant, and inclination of slant given, 81.

——— ——— ——— . Bodies of flat and curved surfaces combined.

Body having flat sides and semicircular ends, see Bath (equal-end and equal-taper); flat sides and ends, and round corners (oblong or square with round corners), see Oblong Pan.

F.

FLAT surface penetrated by conical body, patterns, 323

Flue, out of, 55, 61.

Fluxes, 426

Focus of ellipse, 16.

Frustum of cone, see Cone (right, frustum; and oblique, frustum); o pyramid, see Pyramid (right, frustum; and oblique, frustum).

G.

GALVANISED IRON, 415.

Gauges, tables, 440.

Generation of cone, 24

Gravy strainers, frusta of right cones, 34 (13).

Grooved seam, 32, 434.

H.

HARDNESS OF METALS 411; table, 412.

Heptagon, 4.

Hexagon, 4.

Hexagonal pyramid, 66 (29); pattern of right, see Pyramid (right, frustum, pattern); oblique, see Pyramid (oblique, frustum, pattern); coffee-pots, 70 (35).

Hip baths, see Athenian Hip Bath, Oxford Hip Bath.

Hoods, their relation to truncated pyramids, 70, 143 ; pattern, 154.

Hoppers, 143; see also Hoods.

Hyperbola, defined, 273 (95A); to draw, length, major axis, and width given, 346

——, cone cut in, to draw the cut, 281 ; pattern of cone so cut, 282 ; case of section in hyperbola, 286.

Hypotenuse, 4.

I.

INCLINATION OF SLANT, defined, 24 (4), 55 (b), 61 (Case II.); articles of equal, see Equal-tapering Bodies; of unequal, see Unequal-tapering Bodies.

INTRODUCTORY PROBLEMS, 3.

Iron, 408, 413; fusion, 408 ; fracture, 409, 410; hammer hardening, 410; impurities in, 413; oxide, 413; alloys with carbon, 414; galvanised, 415; tinned, 415; tinning, 428, 430–1.

——, sheet gauges and weights of, 440.

J.

JOINTS, 433.

K.

KILLING SPIRITS OF SALTS, 426.

L.

LAP, allowance for, 32.

——, seam, 433.

Lead, 408, 417.

Line, to divide in equal parts, 7 ; into two parts, 8.

Lines, straight, defined, 3; points joined by, 7; parallel defined, 4; true lengths of, 48.

Lobster-back cowl pattern, 382.

Lustre of metals, 408.

M.

MALLEABILITY OF METALS, 409 ; table, 411.

Manchester plates, 415.

Melting points, 412.

Mensuration, rules in, 436.

Metals and their properties, 408.

Miscellaneous patterns, 239.

Mitre of mouldings, 302 (97A)

Mouldings, 302–23 ; jointing and mitreing, 302 (97A) ; shape, how represented, depth and span, 302, (98A), 305 (101A) ; patterns, shape
and length given, 303 ; shape or section given, to draw section or
shape, and pattern of the moulding, 305 ; division points in curves
of, 308 (105A) ; pattern for joining two pieces at any angle, shape
being given, 310 ; aquarium base pattern, of one moulding, 310 ;
same, of more than one moulding, 314 ; raking moulding pattern,
318.

Muntz's metal, 421

O.

OBLIQUE CONE, see Cone (oblique) ; cylinder, see Cylinder.

Oblong, defined, 4 ; with round corners, to draw, 19 ; with semicircular
ends, to draw, 20 ; oblong equal-tapering body, see Equal-tapering
Bodies.

—— (or square) pan with round corners, pattern, 90 ; in four pieces, 90 ;
in two pieces, 92 ; in one piece, 93 ; short-radius method, 94.

—— taper bath, see Bath (oblong taper).

Octagon, 4.

Opaqueness of metals, 408.

Oval, to draw, 13, 344 ; egg-shaped, to draw, 14.

—— bath, see Bath

—— canister top, representation (plate), 203 ; plate explained, 193 ; plan,
132 ; pattern, 192 ; in four pieces, 193 ; in two pieces, 197 ; in one
piece, 198.

—— equal-tapering bodies, see Equal-tapering bodies (plans, to draw ;
and patterns) ; Corresponding points (in plan).

—— unequal-tapering bodies, see Unequal-tapering Bodies ; also Unequal-
tapering Bodies (plans ; patterns).

Oxford hip-bath, representation (plate), 227 ; plate explained, 217 ; plan,
134 ; pattern, 216 ; short-radius method, 223.

Oxide, oxidising, 408, 413, 426.

P.

PAILS, frusta of right cones, 34 (13).

Pan, baking, see Baking-pan; dripping, see Dripping-pan; oblong with round corners, see Oblong Pan.

Parabola, defined, 273 (94A); to draw, axis and width given, 345.

——, cone cut in, to draw the cut, 281; pattern of cone so cut, 282.

Parallel lines, 4.

Patterns, setting out, is development of surfaces, 2 (3), 34 (14).

Penetration, hole of; where cone penetrates flat surface, 327; penetrates cylinder, 330; penetrates cone, 337; short radius method of cone penetrating cone-frustum, 340.

Penetrations, problems of, 323-343; conical body penetrating flat surface, 323; penetrating cylinder, 328; penetrating cone, 333; and see Conical Cap, Elbow Patterns, Tall-boy, T-pieces, Y-pieces.

Pentagon, 4.

Perpendicular, 3.

Pewter, 422.

Pipe, circular, to fit cone ellipse, 284.

——, ——, and cone, pattern of elbow of, 284.

—— , square, fitted concentrically to conical cap, pattern, 286; rectangular, so fitted, 290.

—— bending, 347; iron, 347; copper and softer metals, 348; lead, 357; removing dents, 350; bending stout tubes, 351; coils, 351; sharp bends, 352; musical instrument tubes and bends, 352; see also Bends.

—— bends, patterns for, 359-407.

Pipes, circular, meeting at any angle, 239; joining equal pipes not in line, pattern, 255; joining unequal pipes, representation (plate), 259; plate explained, 255; pattern, 255; tapering, are frusta of oblique cones, 112 (59); pattern of T formed by, one pipe being an oblique cylinder, 267; see also Circular Pipes.

Pipes, rectangular, meeting at any angle, pattern, 264; pattern of curved elbow, to join, 266.

Plan, explained, 46; of equal-tapering bodies, see Equal-tapering Bodies (plans); of unequal-tapering bodies, see Unequal-tapering Bodies (plans).

Plates, 181, 203, 213, 227, 237, 259; descriptions respectively, 169, 193, 206, 217, 231, 255.

Points, angular, 3; always joined by straight lines, 7; corresponding, see Corresponding Points.

Poling, 416.

Polygon, defined, 4; to inscribe in circle, 10; regular, to describe, 11.

Projection, explained, 47.

Properties of metals, 409.

Proportional arcs, 124.

Pyramid,

    Defined, 66 (29); base, apex, axis, 66 (29); triangular, square, hex-
agonal, 66 (29).

——, right,

    Defined, 66 (29); can be inscribed in right cone, 66 (from 30); model
of, 73 (37).

——, ——, pattern,

    Hexagonal, 68; any number of faces, 69.

——, ——, frustum,

    Defined, 60 (33), 70 (34); representation, 70; can be inscribed in
right cone frustum, 70 (36); the basis of articles of equal taper having
flat surfaces, 70 (35); model, 73 (37); slant of face of, 151 (Case II.).

——, ——, ——, pattern,

    Hexagonal, 71; any number of faces, 73; short radius method, 73.

——, oblique,

    Defined, 143 (74); can be inscribed in oblique cone, 144 (75).

——, ——, pattern,

    Plan and height given, 145; plan and axis given, 148.

——, ——, frustum (unequal-tapering body),

    Defined, 143 (74); the basis of numerous articles of unequal taper having
flat surfaces, 143; can be inscribed in oblique cone frustum, 150 (78);
slant of face of, 151; model, 73 (37).

——, ——, ——, plan,

    What it consists of, 144 (76); corresponding points in, 145 (76); how to
determine from plan of unequal-tapering body if body is oblique
pyramid frustum, 145 (77).

——, ——, ——, pattern,

    Plan and height given (frustum, hexagonal), 148; ends and slant and
inclination of one face given, 151; short radius method, 152.

Q.

QUADRANT, defined, 5; to divide, 11.

Quadrilaterals, 4.

R.

RADIUS, defined, 5; of cone, 24.

Rake (angle), 6.

Raking moulding pattern, 318.

Rectangle, defined, 4; area, 437,

Rectangular base and circular top, pattern for article of, 293.

—— vessel, volume, 437.

Rectangular pipe on conical cap, pattern, 290

—— pipes, meeting at any angle, pattern, 264; curved elbow to join, pattern, 266.

Ridge of roof, pattern of tall-boy base on, 301.

Right angle, 3.

——, cone, see Cone (right); pyramid, see Pyramid (right).

Roof, pattern of tall-boy base on slant of, 295; on ridge of, 301.

Round articles of equal taper, see Cone (right, frustum); of unequal taper, see Cone (oblique, frustum).

——, pipes, various junctions of, see Circular Pipes.

Round neck T to connect equal circular pipes, 385.

Rust, 413; see also Oxide.

S.

SEAMS, 32, 433.

Sections of cylinder and cone, and patterns of sectioned cones, 271-84.

Sector, 28.

Semicircle, 5; to divide, 11.

Short radius methods—for round equal-tapering body (frustum of right cone), 41; equal-tapering body of flat surfaces (frustum of right pyramid), 73, 75; oblong body of equal taper and semicircular ends, 89; oblong (or square) body with round corners, 94; oval equal-tapering body, 102; round unequal-tapering body (frustum of oblique cone), 118; frustum of oblique pyramid, 152; oblong body of unequal taper and semicircular ends, 166; oval unequal-tapering body, 177; body of oblong bottom with semicircular ends and circular top, 189; body of oval bottom and circular top, 200; body oblong bottom with round corners and circular top, 212; Oxford hip-bath, 223; oblong taper-bath, 235; hole of penetration where cone penetrates cone, 340.

Similar arcs, 124.

Sitz-bath, plan, 137.

Slant, a length not an angle, 24 (4); further defined, 52 (25); of face of frustum of pyramid, 151 (Case II.); inclination of, see Inclination of Slant.

Slant of roof, pattern of tall-boy base on, 295.

Solders, 422; table, 425; fluxes, 426.

Specific gravity, 411.

Speculum metal, 419.

Sphere, volume, 438.

Square, defined, 4; pyramid, 66 (29).

—— bodies, of equal taper, essentially right pyramid frusta, see Pyramid (right, frustum); also Equal-tapering bodies (patterns); and Oblong pan.

Square pipe fitted concentrically to conical cap, pattern, 286.
—— pipes, unequal curved elbow to join, 389.
—— section ventilator, 397.
Steel, 414; mild, 415.
Subtend explained, 20.

T.

T-PIECE PATTERNS; T formed by two pipes at right angles, 242; by two
   pipes at any angle, 246; by funnel-shape pipe joined square to circular
   pipe, 251; by two circular pipes, one of which is oblique cylinder,
   267; round neck, to connect equal circular pipes, 385
Table of properties of metals, 411; specific gravities and melting points,
   412; alloys, 419, 423; solders, 425; gauges, weights, sizes, 440, 441.
Tall-boy base pattern, 293; on slant of roof, 295; on ridge of roof, 301.
Tank, volume, 437.
——, dimensions, to hold given quantity, 439.
Taper, an angle, 24 (4).
—— bath, oblong, see Bath (oblong taper).
Tapering articles; equal, see Equal-tapering Articles; unequal, see
   Unequal-tapering Articles.
—— square bend, 389; square ventilator, 397; tapering circular bend,
   398; circular ventilator, 407.
Tea-bottle top pattern, 183–91; in four pieces, 183; in two pieces, 187; in
   one piece, 188; short-radius method, 189.
Tenacity of metals, 410; table, 411.
Terne plates, 414.
Tin, 408.
——, plates, 415; gauges and weights, 440, 441.
——, and lead alloys, 419.
Tinned iron, 415.
Tinning and re-tinning, 428.
—— copper bits, 427; copper cooking vessels, 429; wrought iron, 430;
   cast iron, 431.
Triangle, 4.
Triangular pyramid, 66 (29).
True lengths, need of care to ascertain, 277 (96A).
Truncated pyramid, 143 (74).
Tube bending, see Pipe Bending.

U.

UNEQUAL-TAPERING BODIES.
   Round, essentially oblique cone frusta, 105 (42); representation (plate),
   259; plate explained, 255; see also Cone (oblique, frustum).

Unequal tapering bodies.

Of curved surfaces. Made up mostly of portions of oblique cone frusta, as oval bath, 169; body of oval bottom and circular top, 193; Oxford hip-bath, 217.

Of flat surfaces. Essentially oblique pyramid frusta, see Pyramid (oblique, frustum).

Of flat and curved surfaces, 157; the curved surface portions of frusta of oblique cones, 157; in "equal-end" bath, 157; in body of oblong bottom, with round corners and circular top, 206; in body of oblong bottom, with semicircular ends and circular top, 183; in oblong taper bath, 230; see also Cone (oblique, frustum).

————— ——— ———, plans.

How to find from plan if article is of unequal taper, 129.

Round, see Cone (oblique, frustum, plans).

Of curved surfaces, body oval, top and bottom, see Bath (oval); bottom oval, top circular, see Oval Canister-top; and see also Bath (Oxford-hip; Athenian; and Sitz).

Of flat surfaces. See Pyramid (oblique, frustum, plan).

Of flat and curved surfaces. Body oblong, with semicircular ends, see Bath (equal-end and unequal-taper); oblong bottom, with semicircular ends and circular top, see Tea-bottle Top; body oblong (or square), with round corners and circular top, 133; see also Oblong Taper Bath.

———, ———, ———, patterns.

Round bodies, see Cone (oblique, frustum, pattern).

Of curved surfaces. Body oval, top and bottom, see Bath (oval): bottom oval, top circular, 200 (see also Oval Canister Top); Oxford hip-bath see Oxford Hip-bath.

Of flat surfaces, see Pyramid (oblique, frustum, pattern).

Of flat and curved surfaces. Body oblong, with semicircular ends, see Bath (equal-end and unequal-taper); oblong bottom, with semicirclar ends and circular top, 189 (see also Tea-bottle Top); body oblong (or square), with round corners and circular top, 205; in four pieces, 206; in two pieces, 209; in one piece, 210; short-radius method, 212; oblong taper-bath, see Bath (oblong taper).

V.

VENTILATOR, of square section, pattern, 397; circular tapering, pattern, 407.

Vessel. dimensions, to hold given quantity, 439.

Volatility of metals, 408.

Volume of vessels, to find, 428; of right cone, 438; right cone frustum, 438; sphere, 438; to convert into gallons, 438.

W.

WEIGHTS, table of, 440.
Welding, 411.
Wiring, allowance for, 32.

Y.

Y-PIECE PATTERNS; Y formed by tapering pipes uniting circular pipes,
    representation (plate), 259; plate explained, 255; pattern, 255;
    formed by circular pipes uniting circular pipes, pattern, 261.

Z.

ZINC, 408, 409, 417; gauges and weights of, 440.

LONDON: PRINTED BY WILLIAM CLOWES AND SONS, LIMITED,
STAMFORD STREET AND CHARING CROSS.

www.ingramcontent.com/pod-product-compliance
Lightning Source LLC
Chambersburg PA
CBHW020904210326
41598CB00018B/1764